Probability and Statistics

Volume I

Didier Dacunha-Castelle
Marie Duflo

Probability and Statistics

Volume I

Translated by David McHale

With 22 illustrations

Springer-Verlag
New York Berlin Heidelberg Tokyo

Didier Dacunha-Castelle
Université de Paris-Sud
Equipe de Recherche Associée
au C.N.R.S. 532
Statistique Appliqué Mathématique
91405 Orsay Cedex
France

Marie Duflo
Université de Paris-Nord
93430 Villetaneuse
France

David McHale *(Translator)*
Linslade, Leighton Buzzard
Bedfordshire LU77XW
United Kingdom

AMS Classification 60-01

Library of Congress Cataloging-in-Publication Data
Dacunha-Castelle, Didier.
Probability and statistics.
Translation of: Probabilités et statistiques.
Bibliography: v. 1, p.
Includes index.
1. Probabilities. 2. Mathematical statistics.
I. Duflo, Marie. II. Title.
QA273.D23 1986 519.2 85-25094

Printed and bound by R.R. Donnelley and Sons, Harrisonburg, Virginia.
Printed in the United States of America.

9 8 7 6 5 4 3 2 1

ISBN 0-387-96067-8 Springer-Verlag New York Berlin Heidelberg Tokyo
ISBN 3-540-96067-8 Springer-Verlag Berlin Heidelberg New York Tokyo

CONTENTS

INTRODUCTION

Who is a probabilist? Someone who knows the odds of drawing an ace of diamonds, or the waiting time at a ticket office? A mathematician who uses a special vocabulary? And who is a statistician? Someone who is capable of determining whether tobacco encourages cancer, or the number of votes which a certain candidate will poll at the next elections?

Each one perceives the link: chance. The probabilist, guided by his intuition of poker or queues, constructs an abstract model; having fixed the mathematical framework, he calmly follows through his logical reasoning, which sometimes takes him very far from his starting point. The statistician works on solid ground. When a doctor asks him if it is worth using a new drug, he uses the tools of a probabilist. However he must reach a decision, the least harmful option, based on analyzing the doctor's observations, while taking into account the various risks involved. In short, a probabilist keeps his hands clean while dreaming of models, while a statistician must dirty his hands while working with concrete facts. Relations between the two have often been difficult; but the barriers to their dialogue are broken down by the interest in the concrete to supplement theoretical dreams or in complicated models to describe various phenomena.

However, few students have a chance to overcome these barriers. All too often the future mathematician follows a course in which so much time is spent constructing a probabilistic model that his studies end with barely a sprinkling of statistics. He would perhaps have liked this new mathematical vocabulary, but will remain convinced that his relation with the concrete is limited to red or black balls drawn out of an urn. Even for a mathematician, an axiomatic framework is only beautiful if it highlights and clarifies various problems.

The first two chapters of this book tackle the description of a census and the game of heads or tails. With elementary mathematical tools we can already study various questions: the game of poker, of course, but also the laws of heredity or the control of production of a machine. This part can be used on its own by first and second year undergraduate students or by readers who do not wish to follow all the developments but want to give themselves a fairly general idea of the subject.

The remaining chapters are intended for mathematics students. They assume a knowledge of certain classical subjects of analysis; the probabilistic model is founded on measure theory, the main methods of which we recall briefly in chapter 3. The foundations of probability are introduced at the same time as the most common problems of classical statistics, in chapters 3 to 6. Chapters 7 and 8 bring out the mathematical framework and the most common principles of statistics, thus generalizing the ideas of the preceding chapters. A future statistician will find in this volume a wide enough range of methods which can be of use to him. In line with tradition this volume also contains an introduction to probability which requires the usual calculations of the distributions of variables.

The exercises are very important in order to understand the aims of the text. Certain ones are routine, but most are openings, either to subjects treated further on, or to theoretical or applied subjects -- the theory of games, reliability, genetics, agronomics... .

Volume I, despite its title, treats (in the text or exercises) a certain number of dynamic problems, but the essentials of the problems of random evolution in time will be tackled in Volume II.

Those who have influenced the writing of this book are too numerous to mention. However we would like to thank J. P. Raoult and R. Tomassone who re-read part of the book and helped us by their criticisms.

Chapter 1
CENSUSES

Objectives

To begin a book on chance with a chapter in which chance does not appear may seem to be a little surprising. What concerns us here, in fact, is the response of a population to a questionnaire: which numerical parameters, which graphical representation help to give a better view? In this chapter we shall limit ourselves to touching briefly on **descriptive statistics.**

However, assuming that each individual of the population is equally important and has the same probability of being selected, chance is introduced. Descriptive statistics will be a first approach to probabilistic notions dealing with a finite number of observations. These ideas allow us to tackle, in exercises, various interesting probability problems, using only combinatorial analysis. Theories, more interesting than those of classical urns, can be reached in this way.

Sections 1.2.4 and 1.4 can be omitted on first reading.

1.1. Census of Two Qualitative Characteristics

In a population of n individuals, a *census of two qualitative characteristics* X, Y is carried out: e.g., the party for which they voted in a given election and their socio-economic class. Other qualitative characteristics could have been taken, such as type of holiday; cause of death... ; or characteristics having a more quantitative aspect, when the measured values are regrouped into a finite number of classes: number of children, level of income... .

The census asks each individual ω of the population Ω two questions: he answers $X(\omega)$, $Y(\omega)$. Let the political parties be numbered consecutively 1, ..., k and the social class by 1, ..., ℓ. The census of the two chracteristics is presented in the form of a **Contingency Table.**

If n_{ij} is the number of individuals voting for party i and in social class j, set $F_{ij} = n_{ij}/n$, the frequency of "case" (i,j) or of the "response" (i,j). In such a table each individual of the population Ω is treated as equally important, and the probability of individual ω is $P(\omega) = 1/n$. The probability that an individual responds $(X = i, Y = j)$ is

$$F_{ij} = P(X = i, Y = j),$$

the frequency of (i,j).

The frequency of the response $(X = i)$, or its probability, is denoted by $F_{i\cdot}$ or $P(X = i)$:

$$\begin{aligned} P(X = i) = F_{i\cdot} &= \frac{n_{i1} + \cdots + n_{i\ell}}{n} = \frac{n_{i\cdot}}{n} \\ &= F_{i1} + \cdots + F_{i\ell} \\ &= P(X = i, Y = 1) + \cdots + P(X = i, Y = \ell). \end{aligned}$$

If we are interested only in characteristic X, only the **marginal frequencies** $\{F_{i\cdot}\}_{1 \leq i \leq k}$ matter (in the table double

X \ Y	1	j	ℓ	Marginal total of X
1	F_{11}	F_{1j}	$F_{1\ell}$	$F_{1\cdot}$
i	F_{i1}	F_{ij}	$F_{i\ell}$	$F_{i\cdot}$
k	F_{k1}	F_{kj}	$F_{k\ell}$	$F_{n\cdot}$
Marginal total of Y	$F_{\cdot 1}$	$F_{\cdot j}$	$F_{\cdot \ell}$	1

bars indicate that a summation is being made). This sequence of k numbers (positive, summing to 1) is the **distribution** or **profile** of X. If we want to study the social composition of the voters for party i, only the n individuals $\{\omega;\ X(\omega) = i\}$ are required, each being equally important, while the others are ignored. Thus define on Ω the **probability conditional on** $X = i$:

$$P(\omega \mid X = i) = \frac{1}{n_{i\cdot}} \quad \text{for } X(\omega) = i$$
$$= 0, \quad \text{for } X(\omega) \neq i.$$

The relative frequency of social class j amongst the voters for party i is $F_{ij}/F_{i\cdot} = n_{ij}/n_{i\cdot}$ It is said that the probability of $(Y = j)$ conditional on $(X = i)$ is:

$$P(Y = j|X = i) = \frac{P(X = i,\ Y = j)}{P(X = i)} = \frac{n_{ij}}{n_i}$$
$$= \sum_{\{\omega;\ Y(\omega)=j\}} P(\omega|X = i).$$

The distribution of Y conditional on $(X = i)$ is the set of relative frequencies

$$\left\{\frac{F_{ij}}{F_{i\cdot}} = P(Y = j|X = i);\ j = 1, \ldots, \ell\right\};$$

this is also known as the profile of characteristic Y in party i.

Diagrams such as those that follow, called **histograms**, represent by descriptive statistics the distribution of X, or the profiles.

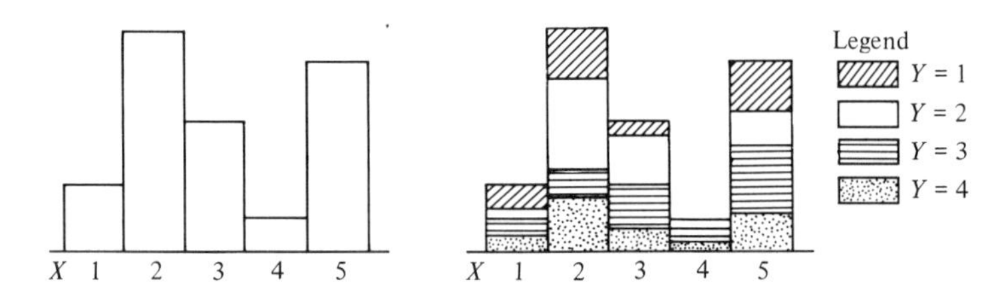

Scale: One cm^2 represents a frequency of 1/4

Under what conditions do the voters of the various parties have the same social composition? That is to say, for all j, $P[Y = j|X = i]$ is independent of i (and equals, for example, $P[Y = j|X = 1]$):

$$
\begin{aligned}
P(Y = j) &= \sum_{i=1}^{k} P(X = i, Y = j) \\
&= \sum_{i=1}^{k} P(Y = j|X = i)P(X = i) \\
&= P(y = j|X = 1) \sum_{i=1}^{k} P(X = i) \\
&= P(Y = j|X = 1) = P(Y = j|X = i) \\
&= \frac{P(X = i, Y = j)}{P(X = i)}
\end{aligned}
$$

Thus for all i and j:

$$P(X = i, Y = j) = P(X = i)P(Y = j).$$

It can be shown that *this property is equivalent to the social class profiles in each of the parties being the same, and coinciding with the social class profile in the entire population. It is also equivalent to all the political profiles of all the social classes being the same, identical to the political profile of the entire population. X and Y are then said to be "**independent**".*

Exercises 1.1

E.1. Consider the data of Table 1 below where the individuals here are the years. Let X be the characteristic equal to n for $2900 + (n\text{-}1)100 \leqslant X_1 < 2900 + n100$; let Q be the quality.

(a) Write down the contingency table of X and Q.
(b) Draw the histograms of X and Q.
(c) Give the profile of X in each quality and draw these profiles on the histogram of Q.

E.2. Consider Table 2 given below (where the individuals are thousands of days). Calculate and draw these profiles of the years in the various sectors of industry.

Table 1. A Study of Bordeaux Wines Classified by Quality: We Have 34 Climatic Observations from the Months of April to September Inclusive, for the Years 1924-1957. X_1 = the Sum of Average Daily Temperatures in Degrees (C°); X_2 = Hours of Sunshine; X_3 = the Number of Very Hot Days; X_4 = Level of Rainfall in *mm*. We Have, Moreover, Classified the Bordeaux Wines into 7 Categories According to Their Quality (going from +++, coded 1, to ---, coded 7)

Years	X_1	X_2	X_3	X_4	Quality
47	3478	1317	42	259	1
49	3458	1508	43	286	1
45	3381	1444	25	253	1 + + +
29	3267	1386	35	225	1
43	3308	1368	24	282	2
34	3317	1362	25	326	2
52	3270	1399	24	306	2
28	3245	1258	36	294	2 + +
55	3247	1277	19	375	2
37	3221	1424	21	382	2
53	3198	1299	20	367	2
44	3212	1289	17	302	3
50	3252	1361	26	346	3
33	3318	1310	29	427	3
42	3227	1331	21	414	3 +
48	3126	1248	11	315	3
26	3155	1133	19	393	3
24	3064	1201	10	361	3

Years	X_1	X_2	X_3	X_4	Quality
46	3061	1175	12	261	4
40	3094	1329	11	339	4
38	3019	1239	16	275	4
39	3022	1285	9	303	4
35	3182	1711	28	326	5
57	3043	1208	14	371	5 -
25	3000	1053	11	338	5
30	3080	966	13	417	6
36	2998	1102	9	349	6
56	3083	1195	5	441	6
27	3085	970	4	467	6 - -
51	3052	1186	14	443	6
54	2904	1164	6	311	6
31	2974	1185	12	488	6
41	3009	1210	15	536	6
32	3038	1103	14	677	7 - - -

Table 2. Number of Working Days (Thousands) Lost Due to Strikes per Year (1954-1974, 1968 Being Excluded) for 11 Sectors of Industry

Years	Energy	Mining	Metallurgy	Engineering	Building
74	102	8	110	879	144
73	909	71	126	508	211
72	151	158	108	938	205
71	148	1	187	1081	225
70	42	7	63	711	126
69	344	5	59	584	64
67	381	209	693	871	200
66	376	57	114	71	169
65	96	25	31	415	21
64	396	11	165	467	372
63	3925	190	97	548	143
62	549	9	75	290	50
61	295	37	35	576	46
60	108	111	32	410	39
59	54	60	48	440	151
58	291	44	16	208	88
57	493	56	186	1214	339
56	58	69	82	444	210
55	62	135	173	1409	458
54	119	30	53	528	198

Chemical	Textile	Paper	Transport	Commerce	Banking
91	130	25	304	39	755
78	105	74	330	0	1
180	101	49	436	20	110
159	130	75	1235	58	39
157	41	20	247	36	27
89	62	47	496	11	18
185	339	34	655	8	30
93	33	20	547	10	7
50	20	10	209	4	1
90	33	32	666	12	7
64	62	24	691	7	19
71	28	10	381	0	2
84	36	18	1332	2	24
30	79	10	199	11	0
84	767	8	57	1	21
45	34	12	302	4	0
103	212	22	705	16	418
58	44	17	118	10	1
261	61	27	193	4	9
67	66	35	187	4	14

1.2. A Census of Quantitative Characteristics

1.2.1. Quantitative Characteristics: Definitions

The census which has just been carried out on the population Ω of n equally probable individuals also included the following questions: "how old are you?"; "what is your annual income?" Individual ω's answers to these two questions will be denoted by $X(\omega)$, $Y(\omega)$, respectively. For each question there is a finite number of possible answers and a contingency table can be formed. But here the numerical value of the answers is important. The histogram of X is often replaced by a **line chart**: the various values of X are on the abscissa, and above each X a line is drawn, the length of which is proportional to $P(X = x)$.

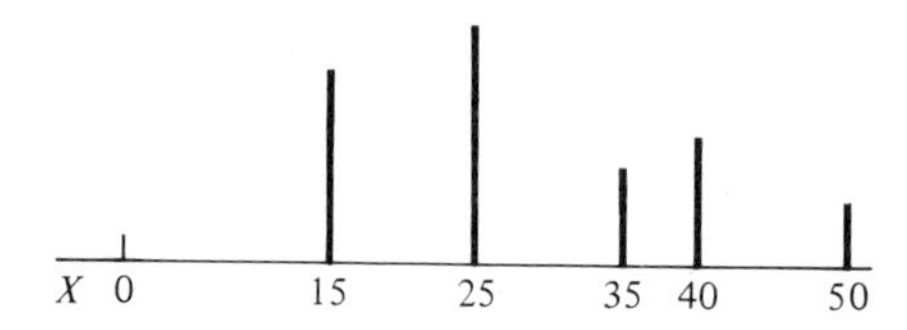

A line chart of X. (A line of 1 *cm* represents 1/5.)

A scatter diagram of n points can also be drawn in the Cartesian plane (**Ox**,**Oy**) (certain points may be coincident). To each individual ω corresponds a point $M(\omega)$ with abscissa $X(\omega)$ and ordinate $Y(\omega)$. The various values of $X(\omega)$ are also indicated on **Ox** and those of $Y(\omega)$ on **Oy**.

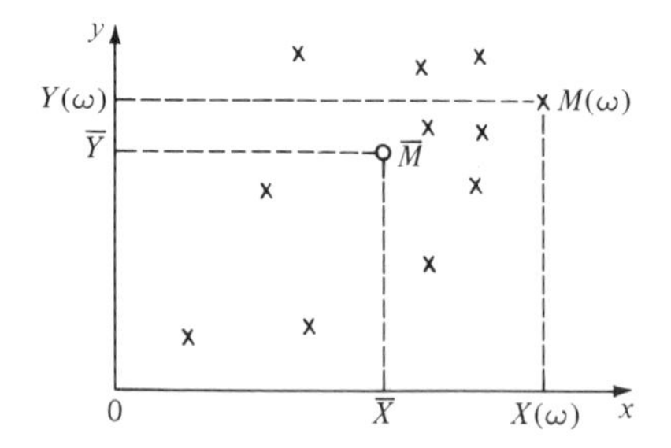

New questions can now be asked: what is the average age of the population? The answer is certainly $\overline{X} = (1/n)\Sigma_{\omega\in\Omega} X(\omega)$, and the mean annual income is $\overline{Y} = (1/n)\Sigma_{\omega\in\Omega}Y(\omega)$; $\overline{X}$ and $\overline{Y}$ are the means of the variables X and Y.

Demographers ask: is the age distribution very large or not? They also wonder if the economic inequalities are very great. This leads to a study of the dispersion of the values of X and Y around their means $\overline{X}$ or $\overline{Y}$. This dispersion is usually measured by the variances

$$\sigma^2(X) = \frac{1}{n} \sum_{\omega\in\Omega}(X(\omega) - \overline{X})^2, \ \sigma^2(Y) = \frac{1}{n} \sum_{\omega\in\Omega}(Y(\omega) - \overline{Y})^2.$$

The positive square root of $\sigma^2(X)$, $\sigma(X)$ is the **standard deviation** of X. A characteristic has zero variance if it is constant and equal to its mean. Finally, we would like to see the mutual influence of the characteristics X and Y. The point $\overline{M} = (\overline{X},\overline{Y})$ appears as the center of the cloud of data (it is the barycenter of the cloud of points, each given equal weight). If the characteristics X and Y have a tendency to change in the same direction $(X(\omega) - \overline{X})$ and $(Y(\omega) - \overline{Y})$ tend to have the same sign. To analyze this tendency the **covariance** $\Gamma(X,Y) = (1/n)\Sigma_{\omega\in\Omega}(X(\omega) - \overline{X})(Y(\omega) - \overline{Y})$ is introduced. A positive covariance signifies that X and Y have a tendency to change in the same direction; a negative covariance signifies that they have a tendency to change in the opposite direction. Below are two examples where the covariance is zero.

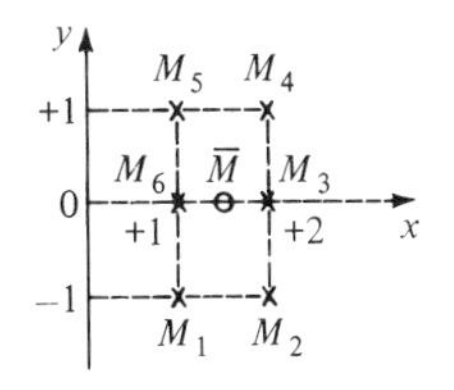

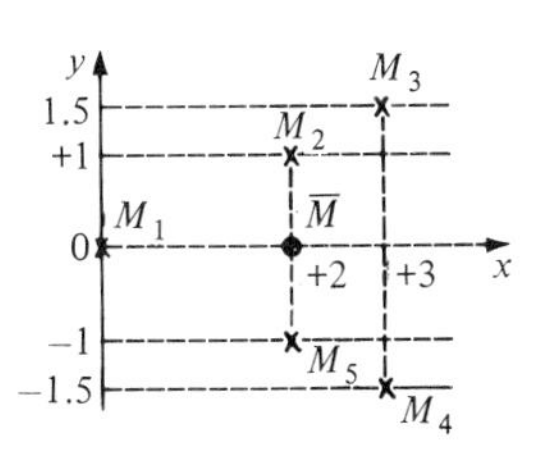

A change in the unit of measurement for one or other of the variables should not change the manner in which this link is measured. With this aim in mind, if neither of them is constant, $\rho(X,Y) = \Gamma(X,Y)/\sigma(X)\sigma(Y)$ is used, the **correlation** of X and Y, the sign being that of $\Gamma(X,Y)$.

Note. Roughly, a positive covariance signifies that the two variables have the tendency (on average) to grow simultaneously, but this is not always the case, which can be seen in the example below.

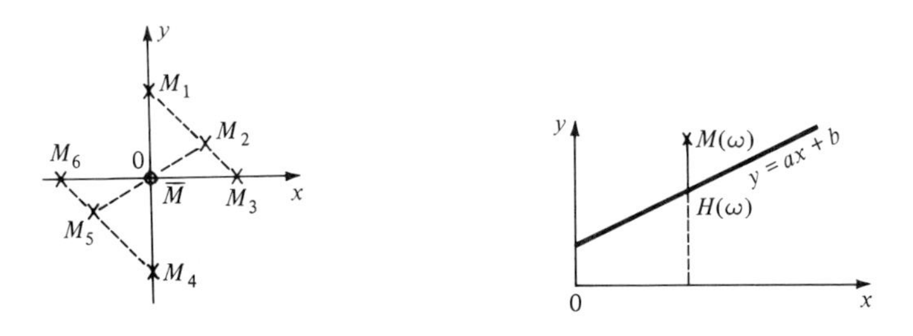

1.2.2. The Regression Line of Y on X

Let Y be the individual's annual income and take for X the annual income which the individual's parents had at the same age. A priori the covariance is thought to be positive. In trying to make the relationship precise, we ask in what way we are mistaken in supposing that there is a linear dependence $Y = aX + b$: this error may be measured by

$$\delta(a,b) = \frac{1}{n} \sum_{\omega\in\Omega} (M(\omega)H(\omega))^2 = \frac{1}{n} \sum_{\omega\in\Omega}(Y(\omega)-aX(\omega)-b)^2,$$

with $H(\omega) = (X(\omega), aX(\omega)+b)$. Consider then $\delta(a,b)$ as a distance from Y to the characteristic $aX+b$. We are trying to determine the regression line of Y on X, i.e., the line $y = ax+b$ which minimizes the expression $\delta(a,b)$. The gradient of δ vanishes for

$$\begin{cases} \dfrac{\partial}{\partial b}\,\delta(a,b) = -\dfrac{2}{n}\displaystyle\sum_{\omega\in\Omega}(Y(\omega) - aX(\omega) - b) = 0 \\[2ex] \dfrac{\partial}{\partial a}\,\delta(a,b) = -\dfrac{2}{n}\displaystyle\sum_{\omega\in\Omega} X(\omega)(Y(\omega) - aX(\omega) - b) = 0. \end{cases}$$

This system of equations is equivalent to

$$b = \overline{Y} - a\overline{X}$$

and

$$\sum_{\omega\in\Omega} X(\omega)(Y(\omega) - \overline{Y}) = a \sum_{\omega\in\Omega} X(\omega)(X(\omega) - \overline{X}).$$

Now

$$\overline{X} \sum_{\omega\in\Omega} (Y(\omega) - \overline{Y}) = \overline{X} \sum_{\omega\in\Omega} (X(\omega) - \overline{X}).$$

If X is nonconstant then we obtain

$$a = \frac{\Gamma(X,Y)}{\sigma^2(X)}, \quad b = \overline{Y} - a\overline{X}.$$

The line $y - \overline{Y} = [\Gamma(X,Y)/\sigma^2(X)](x - \overline{X})$ is, if X is nonconstant, the **regression line** of Y on X. The sign of its gradient is that of $\Gamma(X,Y)$, which is consistent with the interpretation already given.

What is the distance of Y from the characteristic $\overline{Y} + [\Gamma(X,Y)/\sigma^2(X)](X - \overline{X})$? It is

$$\frac{1}{n}\sum_{\omega\in\Omega}\left[Y(\omega) - \overline{Y} - \frac{\Gamma(X,Y)}{\sigma^2(X)}(X(\omega) - \overline{X})\right]^2$$

$$= \frac{1}{n}\sum_{\omega\in\Omega}(Y(\omega) - \overline{Y})^2$$

$$- 2\frac{\Gamma(X,Y)}{\sigma^2(X)}\frac{1}{n}\sum_{\omega\in\Omega}(Y(\omega)-\overline{Y})(X(\omega)-\overline{X})$$

$$+ \frac{\Gamma^2(X,Y)}{\sigma^4(X)}\frac{1}{n}\sum_{\omega\in\Omega}(X(\omega) - \overline{X})^2$$

$$= \sigma^2(Y) - \frac{\Gamma^2(X,Y)}{\sigma^2(X)} = \sigma^2(Y)[1 - \rho^2(X,Y)].$$

This distance is all the larger when the variance of Y is large and when the correlation between X and Y is small (it is then said that the variables are weakly correlated; if their covariance is zero the variables are said to be uncorrelated). If the absolute value of the correlation is 1, Y is an affine function of X. In all cases we obtain:

$$|\rho(X,Y)| \leqslant 1 \qquad (\text{or } |\Gamma(X,Y)| \leqslant \sigma(X)\sigma(Y)).$$

1.2.3. Principal Axes of a Scatter Diagram

The dispersion of the points in a scatter diagram about a line can be measured by the sum of squares of the distances from the points to the line. Let $\Delta_{a,b,\theta}$ be a line passing thrugh (a,b), orthogonal to the vector $(\cos\theta, \sin\theta)$. The distance of the point (x,y) to $\Delta_{a,b,\theta}$ is $|(x-a)\cos\theta + (y-b)\sin\theta|$. Consider therefore

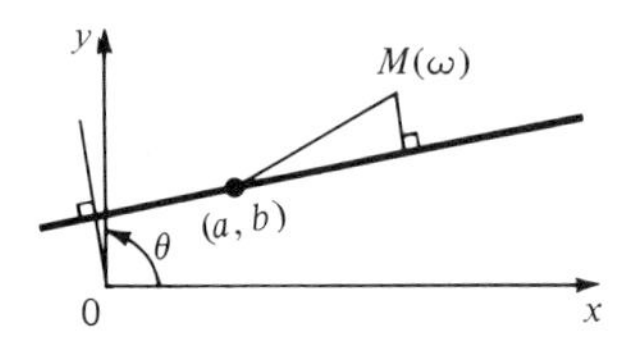

$$\phi(a,b,\theta) = \frac{1}{n} \sum_{\omega\in\Omega} [(X(\omega)-a)\cos\theta + (Y(\omega)-b)\sin\theta]^2.$$

The angle θ being fixed, the function $(a,b) \longmapsto \phi(a,b,\theta)$ is a minimum when $a = \overline{X}$ and $b = \overline{Y}$. This can be easily seen as above. Only the lines passing through the center of gravity $(\overline{X},\overline{Y})$ need be considered, denoted by $\Delta_\theta = \Delta_{\overline{X},\overline{Y},\theta}$ and $\phi(\theta) = \phi(\overline{X},\overline{Y},\theta)$:

$$\phi'(\theta) = \frac{2}{n} \sum_{\omega\in\Omega} [(X(\omega)-\overline{X})\cos\theta + (Y(\omega) - \overline{Y})\sin\theta] \cdot$$

$$\cdot [-(X(\omega)-\overline{X})\sin\theta + (Y(\omega)-\overline{Y})\cos\theta]$$

$$= 2\cos 2\theta\, \Gamma(X,Y) + \sin 2\theta[\sigma^2(Y) - \sigma^2(X)].$$

The function ϕ is constant if X and Y are uncorrelated and have the same variance: the scatter of points looks circular around its center. If not, ϕ' vanishes for two orthogonal lines: one corresponds to a maximum of ϕ, the other to a minimum. The one corresponding to the minimum is the orthogonal regression line of the pair (X,Y): it is around this line that the scatter of points is the greatest, it is also called the principal axis of the scatter diagram. The scatter is least around the line orthogonal to this, passing through $(\overline{X},\overline{Y})$.

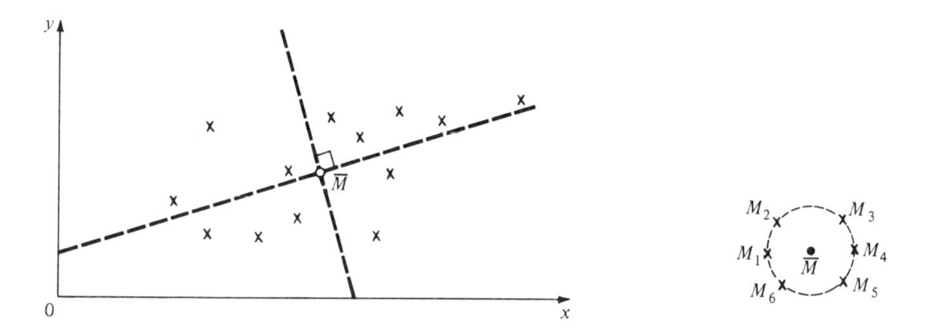

1.2.4. Description of a Census of p Quantitative Characteristics: Principal Components Analysis (PCA)

If the census asks p questions, the answers to which are quantitative $X_1 \cdots X_p$, the set of answers can again be visualized as a cloud of n points $X(\omega) = \{X_1(\omega), ..., X_p(\omega)\}$ in $\mathbb{R}^p$ for $\omega \in \Omega$. Unfortunately this is less easy to see when p is greater than 2 and we try to establish plots, i.e., orthogonal projections of the data on to a plane, the best possible representation of the data. As before, if you want to project the scatter diagram for two characteristics on to a line, the direction of the principal axis is the best choice, the orthogonal direction the worst (the image on the principal axis is good, the points being spread out; on the orthogonal axis the points are not as separated).

It is natural to think of reducing the number of characteristics. If, for example, a social survey includes the characteristic X_1 "annual income," X_2 "monthly income," X_3 "number of sons," X_4 "number of daughters,"... , we would tend to consider X_1 and X_2 as the same characteristic; X_3 and X_4 also, in so far as X_3 and X_4 play the same role on the economic level of the family. We could then come back to the two characteristics X_1 and $(X_3 + X_4)$.

Thus we study linear combinations of the characteristics $X_1, ..., X_p$: with $v = (v_1, ..., v_p) \in \mathbb{R}^p$ we associate the characteristic $v(X) = \Sigma_{i=1}^p v_i X_i$. The mean of $v(X)$ is

$$\overline{v(X)} = \frac{1}{n} \sum_{\omega \in \Omega} v(X(\omega)) = \sum_{i=1}^{p} v_i \overline{X}_i = v(\overline{X}),$$

where $\overline{X} = (\overline{X}_1, ..., \overline{X}_p)$ is the barycenter of the data.

Let v and w be in $\mathbb{R}^p$; the covariance of $v(X)$ and $w(X)$ is

$$\Gamma(v(X),w(X)) = \frac{1}{n} \sum_{\omega\in\Omega} (v(X(\omega))-v(\bar{X}))(w(X(\omega))-w(\bar{X}))$$

$$= \sum_{i=1}^{p} \sum_{j=1}^{p} v_i w_j \left[\frac{1}{n} \sum_{\omega\in\Omega} (X_i(\omega)-\bar{X})(X_j(\omega)-\bar{X})\right];$$

$$\Gamma(v(X),w(X)) = [v_1, \ldots, v_p]\Gamma \begin{bmatrix} w_1 \\ \vdots \\ w_p \end{bmatrix},$$

where Γ is the **covariance matrix** of $X = (X_1, \ldots, X_p)$, $\Gamma = \{\Gamma(X_i,X_j)\}_{1\leqslant i,j\leqslant p}$. It is a symmetric matrix, and the map $(v,w) \longmapsto \Gamma(v(X),w(X))$ is a symmetric bilinear form. The associated quadratic form is $v \longmapsto \Gamma(v(X),v(X)) = \sigma^2(v(X))$, the variance of $v(X)$. This quadratic form is positive and takes the value zero if $v(X)$ is a constant.

The characteristic $v(X)$ explains the dispersion of the individuals so much the better when its variance is large. v acts as a weighting of the components of X. In order to use a criterion which remains invariant if v is multiplied by a constant we try to maximize

$$\frac{\sigma^2(v(X))}{\sum_{i=1}^{p} v_i^2} = \frac{[v_1, \ldots, v_p]\Gamma \begin{bmatrix} v_1 \\ \vdots \\ v_p \end{bmatrix}}{\sum_{i=1}^{p} v_i^2}.$$

Since the space $\mathbb{R}^p$ has the scalar product $\langle v,w\rangle = \sum_{i=1}^{p} v_i w_i$ and the associated norm $\| \ \|$, there exists an orthonormal basis $(u_1, \ldots, u_p)$ of eigenvectors of Γ corresponding to the positive eigenvalues ranked in decreasing order $\lambda_1 \geqslant \lambda_1 \geqslant \ldots \lambda_p$. Interpreting this we have:

Proposition 1.2.1. *Let* $\lambda_1 \geqslant \ldots \geqslant \lambda_p$ *be the eigenvalues of* Γ *and* $u_1, \ldots, u_p$ *an orthonormal basis of the associated eigenvectors of* Γ. *Then*:

(a) $\sigma^2(u_i(X)) = \lambda_i$;

(b) $\Gamma(v(X),w(X)) = \sum^{p} \lambda_i \langle v,u_i\rangle\langle w,u_i\rangle$;

thus the characteristics $u_1(X)$, ..., $u_p(X)$ *are pairwise uncorrelated.*

(c) $$\sigma^2(u_1(X)) = \lambda_1 = \sup_{v \neq 0} \frac{\sigma^2(v(X))}{\sum_{i=1}^{p} v_i^2}$$

$$\sigma^2(u_i(X)) = \lambda_i = \sup\left\{\frac{\sigma^2(v(X))}{\sum_{i=1}^{p} v_i^2};\ v \neq 0,\ v(X) \textit{ uncorrelated with } u_1(X), ..., u_{i-1}(X)\right\}.$$

$u_i(X)$ is said to be an ith *principal factor (which is not determined uniquely, up to the sign, unless λ_i is a simple eigenvalue).*

Let V be a vector in $\mathbb{R}^p$. If the data are translated by adding V to each point (e.g., centering the data by adding $-\bar{X}$) the covariances do not change:

$$\Gamma[v(X + V), w(X + V)] = \Gamma[v(X), w(X)].$$

Hence the principal factors remain the same.

Dispersion of a Cloud of n Points $(X(\omega))_{\omega \in \Omega}$ Around its Barycenter $\bar{X}$. This dispersion (called the **inertia** of the cloud) can be measured by

$$I = \frac{1}{n} \sum_{\omega \in \Omega} \|X(\omega) - \bar{X}\|^2.$$

But $(u_1, ..., u_p)$ is an orthonormal basis, thus

$$I = \frac{1}{n} \sum_{\omega \in \Omega} \sum_{i=1}^{p} \langle X(\omega) - \bar{X}, u_i\rangle^2 = \sum_{i=1}^{p} \sigma^2(u_i(X)) = \sum_{i=1}^{p} \lambda_i .$$

The trace of Γ, therefore, measures the dispersion of the data about its center. Let $q < p$: suppose that we replace the points $(X(\omega) - \bar{X})_{\omega \in \Omega}$ by their projection on the subspace generated by $(u_1, ..., u_q)$. $X(\omega)$ is then replaced by

$$X^q(\omega) = \bar{X} + u_1(X)u_1 + ... + u_q(X)u_q .$$

The dispersion of the projected data $\{X^q(\omega)\}_{\omega \in \Omega}$ around its center is then $\lambda_1 + ... + \lambda_q$. If this sum is large with respect to

$\lambda_{q+1} + \ldots + \lambda_p$, the projection can be considered to be a good representation of the data.

If we call the ratio

$$\frac{\textit{dispersion of the projected data}}{\textit{dispersion of the original data}}$$

the **fidelity** of the projection, it can be seen that the space generated by $u_1, \ldots, u_q$ is the most faithful space of dimension q (unique if λ_q is a simple eigenvalue). If we are happy with $q = 2$, $X-\bar{X}$ can be replaced by $u_1(X)u_1 + u_2(X)u_2$: two factors alone can be used. This allows the data to be represented in the plane. With the origin as $\bar{X}$, two orthonormal axes associated with u_1 and u_2 can be taken, the individual ω being located by the point $(u_1(X(\omega)), u_2(X(\omega)))$. Each characteristic X_i appears as the point

$$\Gamma(X_i^2, u_1(X)), \ \Gamma(X_i, u_2(X)) = (\lambda_1 u_1(i), \lambda_2 u_2(i)),$$

where $u_1(i)$ and $u_2(i)$ are the ith components of u_1 and u_2. Thus two planar maps can be obtained: a *projection of the individual data points* and a *map of the characteristics* indicating their relationship with u_1 and u_2, and therefore their role in explaining the dispersion of the individuals.

Analysis in Normed Principal Components. In order to obtain an analysis independent of the units of measurement used for each of the factors, X_i is often replaced by $X_i/\sigma(X_i) = \hat{X}_i$ and X by $\hat{X} = (\hat{X}_1, \ldots, \hat{X}_p)$. This results in the matrix Γ being replaced by the correlation matrix $R = \{\rho(X_i,X_j)\}_{1 \leqslant i, j \leqslant p}$, where $\rho(X_i,X_j)$ is the correlation between X_i and X_j. The elements on the diagonal of R take the value 1 and the trace of R equals p: the total inertia of the cloud of points $\{\hat{X}(\omega)\}_{\omega \in \Omega}$ is p. For $\hat{\lambda}_1 \geqslant \ldots \geqslant \hat{\lambda}_p$ the eigenvalues of R, $\hat{u}_1, \ldots, \hat{u}_p$ an orthonormal basis of the associated eigenvectors and $1 < k < p$, we have, denoting $\hat{u}_k(X)$ by $\hat{u}_k$:

$$\Gamma(\hat{X}_i,\hat{u}_k) = \hat{\lambda}_k \hat{u}_k(i), \quad \sigma^2(\hat{u}_k) = \hat{\lambda}_k,$$

$$\sigma^2(\hat{X}_i) = \Sigma \ \hat{\lambda}_k \hat{u}_k^2(i) = 1.$$

Thus $\rho(\hat{X}_i,\hat{u}_k) = (\hat{\lambda}_k)^{1/2}\hat{u}_k(i)$. The ρ characteristics $\hat{X}_1, \ldots, \hat{X}_p$ can then be represented by their correlations with $\hat{u}_1$ and $\hat{u}_2$, thus by points with coordinates $((\hat{\lambda}_1)^{1/2}\hat{u}_1(i), (\hat{\lambda}_2)^{1/2}\hat{u}_2(i))$, $i = 1 \dots p$.

The points $((\hat{\lambda}_1)^{1/2}\hat{u}_1(i), \dots, (\hat{\lambda}_p)^{1/2}\hat{u}_p(i))$ lie on the unit sphere. Projection decreases the norm, and in this planar representation the characteristics are represented in a circle of unit radius centered on the origin. The correlations can be interpreted better than the covariances and the plot of the characteristics is clearer for the normed analysis than for the principal components analysis.

Example. Reconsider the data of Table 1 of Exercise 1.1 for the quantitative characteristics $X_1, \dots, X_4$. The table below gives the eigenvalues $\hat{\lambda}_1, \hat{\lambda}_2, \hat{\lambda}_3$ and the correlations of $\hat{X}_1, \dots, \hat{X}_4$ with the characteristics $\hat{u}_1, \hat{u}_2, \hat{u}_3$. The plot of the years (individual ω) is given by the points $(\hat{u}_1(\hat{X}(\omega)), \hat{u}_2(\hat{X}(\omega)))$. The data stretches out along $\hat{u}_1$, the eigenvector associated with the eigenvalue $\hat{\lambda}_1$, which is very large with respect to $\hat{\lambda}_2$. The qualities of the wines can be drawn out on this plot and the following interpretation put forward. The variability of the data is first of all largely explained by the first axis, which contrasts X_1, X_2, X_3 with variable X_4, the level of rainfall. The very good years are thus, above all, the years in which X_1, X_2, X_3 are very important (49, 47) as opposed to the bad years, when X_4 is very important. The second axis is again closely linked with the level of rainfall, but the hours of sunshine are not included and the other variables feature very little. This explains the catastrophic years such as 32.

i	$\hat{\lambda}_i$	$\rho(\hat{X}_1,\hat{u}_i)$	$\rho(\hat{X}_2,\hat{u}_i)$	$\rho(\hat{X}_3,\hat{u}_i)$	$\rho(\hat{X}_4,\hat{u}_i)$
1	2.707	0.91	0.87	0.87	-0.64
2	0.635	0.27	-0.02	0.28	0.75
3	0.364	-0.07	0.38	-0.17	-0.11

$$R = \begin{bmatrix} 1.000 & & & \\ 0.685 & 1.000 & & \\ 0.865 & 0.605 & 1.000 & \\ 0.409 & -0.481 & -0.401 & 1.000 \end{bmatrix}$$

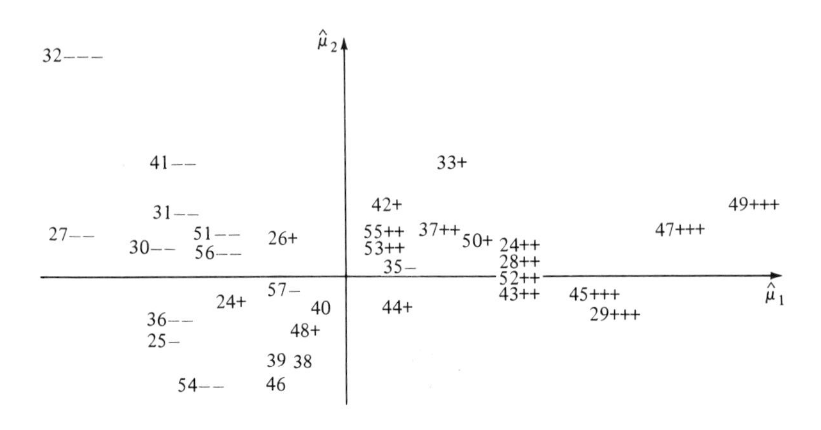

Exercises 1.2.

E.1. Consider the data of Table 1 of Exercise 1.1. Plot the data (X_1,X_4) and draw the line of regression of X_4 on X_1: find the principal axes of the scatter plot.

E.2. Two variables X and Y are observed each taking the value 0 or 1, and a contingency table is formed using the notation of para 1.1. Prove that their correlation is

$$\rho(X,Y) = \frac{F_{11}-F_{1.}F_{.1}}{(F_{0.}F_{.0}F_{1.}F_{.1})^{1/2}}$$

$$= \frac{F_{00}F_{11}-F_{01}F_{10}}{(F_{0.}F_{.0}F_{1.}F_{.1})^{1/2}}$$

E.3. A study is made of p measurements (height, chest size, etc.) $X_1, ..., X_p$ of n equally important individuals chosen from a large population. The correlation between any two of these characteristics $\rho(X_i,X_j)$ is expected to be positive. Why? In fact, experiment shows that for n large enough, values very close to one another can be obtained for each pair of characteristics.

The situation can be simplified by assuming that all the correlations $\rho(X_i,X_j)$, for $i \neq j$, are equal to $\rho > 0$. What are the eigenvalues of the correlation matrix? What is the principal factor in the normed PCA? This is called the size or shape factor.

E.4. *Reduction of data tables.* Consider the set $\mathcal{T}$ of $n \times p$ tables of real data, otherwise known as the set of real $n \times p$ matrices. For $T = (t_{ij})_{1 \leqslant i \leqslant n,\, 1 \leqslant j \leqslant p}$ and $S = (S_{ij})_{1 \leqslant i \leqslant n,\, 1 \leqslant j \leqslant p}$ in $\mathcal{T}$ let the scalar product be defined by $\langle T,S\rangle_2 = \Sigma_{i=1}^n \Sigma_{j=1}^p t_{ij} s_{ij}$; let $\|\cdot\|_2$ be the associated norm. Let H_q be the subset of $\mathcal{T}$ formed by linear combinations of q matrices of the form $u\,{}^t v$ with $u \in \mathbb{R}^n$ and $v \in \mathbb{R}^p$: a table of H_q depends on $q(n+q)$ data points, it is a **simple table of rank** at most q. Let $T \in \mathcal{T}$; if it exists we can study T_q, the best approximation of T by a simple table of rank q.

(a) Show that for $T \in \mathcal{T}$, ${}^tTT$ is a symmetric matrix; the associated quadratic form on $\mathbb{R}^p$ is given by $v \longmapsto {}^t v\,{}^tTTv = \|Tv\|^2$. Show that ${}^tTT$ and $T\,{}^tT$ have the same eigenvalues, that they are positive, and that at most $\inf(p,n)$ among them are strictly positive. Show that $\|T\|_2^2 = \mathrm{Tr}({}^tTT) = \mathrm{Tr}(T\,{}^tT)$.

(b) Show that T_1 equals $\lambda_1 u_1 {}^t v_1$, where λ_1 is the largest eigenvalue of ${}^tTT$, u_1 and v_1 the corresponding normed eigenvectors of $T\,{}^tT$ and ${}^tTT$, respectively, and that these vectors are related by

$$Tv_1 = \sqrt{\lambda_1}\, u_1, \quad {}^tTu_1 = \sqrt{\lambda_1}\, v_1.$$

(c) Assume that the nonzero eigenvalues $\lambda_1, \ldots, \lambda_r$ of ${}^tTT$ are simple. Let u_i be a normed eigenvector of $T\,{}^tT$ corresponding to λ_i $(1 \leqslant i \leqslant r)$, and $v_i = [1/\sqrt{\lambda_i}]\,{}^tTu_i$. Let $q = 2, \ldots, r$; prove that

$$T_q = \sum_{i=1}^{\inf(q,\,r)} \sqrt{\lambda_i}\, u_i {}^t v_i \quad \text{and} \quad T = \sum_{i=1}^{r} \sqrt{\lambda_i}\, u_i {}^t v_i.$$

For $1 \leqslant i \leqslant r$, set $u_i = (u_i^j)_{1 \leqslant j \leqslant n}$ and $v_i = (v_i^j)_{1 \leqslant j \leqslant p}$. Verify that the projection on v_i of the jth row vector of T is $\sqrt{\lambda_i}\, u_i^j v_i$ and the projection on u_i of the jth column vector of T is $\sqrt{\lambda_i}\, v_i^j u_i$.

(d) Show that the PCA of p characteristics $(X_1, \ldots, X_p)$ defined on $\Omega = \{1, \ldots, n\}$ is none other than the approximation of table T, of which the ith column vector is (with the notation of 1.2.4)

$$X_i - \bar{X}_i 1_n = \{X_i(1) - \bar{X}_i, \ldots, X_i(n) - \bar{X}_i\},$$

or $$\frac{X_i - \bar{X}_i 1_n}{\sigma(X_i)}$$ for the normed PCA.

E.5. *Spearman Correlation.* To compare the markings of two judges at the Olympic Games, the respective rankings R and S (without ties) of n competitors $\{1, ..., n\}$ in an event can be examined. The ith competitor is ranked $R_i = r_i$ by the first judge and $S_i = s_i$ by the second.

(a) Determine the mean value and the variance of the quantitative characteristic thus observed.
(b) Let ρ be the correlation between R and S. Set $D_i = r_i - s_i$ for $i = 1, ..., n$. Show that

$$\rho = 1 - \frac{6\sum_{i=1}^n D_i^2}{n(n^2-1)}.$$

1.3. First Definitions of Discrete Probabilities

1.3.1. Probabilities on a Countable Set

A certain phenomenon can be observed; let Ω be the set of possible observations, which we assume to be countable. To each $\omega \in \Omega$ a **probability** $P(\omega)$ can be associated. This number lies between 0 and 1 and we have $\Sigma_{\omega \in \Omega} P(\omega) = 1$. The choice of this probability is based on an intuitive or experimental idea of the phenomenon.

Thus, in the preceding census, taking for Ω a set of equally important individuals, each has been given the probability $1/n$, the **uniform probability** on Ω. If we play at heads or tails, Ω can be taken as {head, tail}. When the coin is regular, an idea of symmetry suggests $P(head) = P(tail) = 1/2$. Experiment shows that if heads or tails is played often, the frequency of heads is close to 1/2, which again suggests $P(head) = 1/2$.

Let us return to the general case. To a subset A of Ω, which we shall call an **event**, associate $P(A) = \Sigma_{\omega \in A} P(\omega)$, the probability of A. We identify $\{\omega\}$, the subset of Ω, and ω the element of Ω: $P(\{\omega\}) = P(\omega)$. If X is a function defined on Ω, then for each $x \in X(\Omega)$, we talk of the event $\{\omega; X(\omega) = x\}$, denoted by $(X = x)$ and of its probability $P(X = x)$. The family $\{P(X = x); x \in X(\Omega)\}$ is the **distribution** of X. Thus in the census of Section 1.1, the distribution of pairs of observations $(X,Y) = \{F_{ij};\ 1 \leqslant i \leqslant k,\ 1 \leqslant j \leqslant \ell\}$. If X is real valued, we call it a **random variable** (r.v.).

All this is closely linked with mechanics. If Ω is a set of point masses and if $P(\omega)$ is the mas of ω, the mass of A is $P(A)$. Place the point masses on an axis, putting ω at the point $X(\omega)$ on the abscissa (this may result in the same point on the axis receiving more than one point mass). Considering the axis as a bar of mass zero on which these masses are placed, it is important in mechanics to calculate the barycenter, i.e., the point at which the bar must be held in order to be in equilibrium. It is also important to know the amplitude of the oscillations about this equilibrium, and for this the inertia is used. The notion of barycenter corresponds in probability to that of the **mean** of X. If the barycenter exists, i.e., if $\Sigma_{\omega\in\Omega}|X(\omega)|P(\omega)$ converges, the variable X is said to be **integrable**. Then the **mean** or **expectation** of X is the sum $\Sigma_{\omega\in\Omega}X(\omega)P(\omega) = \Sigma_{x\in X(\Omega)}xP(X = x)$. This depends only on the distribution of X. This is denoted by $P(X)$ or $E_P(X)$, or if there is no doubt about the probabilities being used, $E(X)$. An integrable variable X for which the mean is zero is said to be centered. For X and Y integrable, a and b real, it is easily shown that:

$$E(aX + bY) = aE(X) + bE(Y).$$

For $A \subset \Omega$, the r.v. $\mathbf{1}_A$ can be introduced, the **indicator function** of A, equal to 1 on A and 0 on A^c. Then

$$\mathrm{E}(\mathbf{1}_A) = \sum_{\omega\in A} P(\omega) = P(A).$$

If X^p is integrable for $p > 0$, $E(X^p)$ is called the **moment of order** p of X.

1.3.2. Approximation of a Random Variable by a Constant

Suppose that X^2 is integrable. What error is made by replacing X by a constant a? Standard practice is to measure this error by $E(X-a)^2 = E(X^2) - 2aE(X) + a^2$, the average of $(X-a)^2$. This error is a minimum for $a = E(X)$. *The mean appears, in the* **least-squares** *sense, as the constant closest to X.* The distance from $E(X)$ to X is measured by the **variance** of

X, $\sigma^2(X) = E(X - E(X))^2 = E(X^2) - (E(X))^2$. This is the analogue of inertia. For a and b real, we have $\sigma^2(aX+b) = a^2\sigma^2(X)$. The standard **deviation** $\sigma(X)$ is the positive square root of $\sigma^2(X)$.

Let X be a positive r.v. and let $\varepsilon > 0$. We have

$$E(X) = \sum_{\omega\in\Omega} X(\omega)P(\omega) \geqslant \sum_{\omega\in\Omega} X(\omega)\mathbf{1}_{\{\omega;\ X(\omega)\geqslant\ \varepsilon\}}P(\omega).$$

We denote

$$(X \geqslant \varepsilon) = \{\omega;\ X(\omega) \geqslant \varepsilon\};$$

$$E(X) \geqslant \varepsilon \sum_{\omega\in\Omega} \mathbf{1}_{(X\geqslant\ \varepsilon)}(\omega)P(\omega) = \varepsilon P(X \geqslant \varepsilon).$$

Applied to $(X - E(X))^2$, the preceding inequality gives

$$P(|X - E(X)| \geqslant \varepsilon) \leqslant \frac{\sigma^2(X)}{\varepsilon^2}.$$

This is **Tchebychev's inequality.**

1.3.3. Approximation of One Random Variable by a Linear Function of Another

Let X and Y be two square integrable r.v.'s. The inequality $2|XY| \leqslant X^2 + Y^2$ shows that XY is integrable. We call the **covariance** of X and Y

$$\Gamma(X,Y) = E[(X-E(X))(Y-E(Y))] = E(XY) - E(X)E(Y)$$

and the **correlation**

$$\rho(X,Y) = \frac{\Gamma(X,Y)}{\sigma(X)\sigma(Y)}.$$

Their signs are positive if $(X - E(X))$ and $(Y - E(Y))$ tend to have the same sign, and they are negative in the opposite case. The function

$$\lambda \longmapsto \phi(\lambda) = E[(Y\text{-}E(Y)) - \lambda(X\text{-}E(X))]^2$$

$$= \sigma^2(Y) - 2\lambda\Gamma(X,Y) + \lambda^2\sigma^2(X)$$

is a minimum for $\lambda = [\Gamma(X,Y)/\sigma^2(X)](X-E(X))$, and this minimum equals $\sigma^2(Y)[1 - \rho^2(X,Y)]$.

Thus $\tilde{Y} = E(Y) + [\Gamma(X,Y)/\sigma^2(X)](X-E(X))$ is the affine function of X closest to Y in the least-squares sense. Notice th similarity to the calculations for the regression lines in Section 1.2. The error of approximation

$$E(Y-\tilde{Y})^2 = \sigma^2(Y)(1 - \rho^2(X,Y)$$

is **zero** for $\rho^2 = 1$ and maximum for ρ zero. (The r.v.'s are then said to be **uncorrelated**.) We always have $\rho^2 \leqslant 1$.

The covariance Γ is a symmetric bilinear form defined on the vector space of square integrable r.v.'s. Two r.v.'s, X and Y are **orthogonal** for Γ if they are uncorrelated. The r.v.'s orthogonal to constants are the centered r.v.'s. Moreover, $\tilde{Y}$ and $Y-\tilde{Y}$ are orthogonal in the above case.

The associated quadratic form of Γ is the variance, which is positive and takes the value zero for constants. If none of the points of Ω has zero probability, Γ vanishes only on constants and defines a scalar product on the space of centered r.v.'s. We have a version of Pythagonas' Theorem: if X and Y are orthogonal, $\sigma^2(X+Y) = \sigma^2(X) + \sigma^2(Y)$, since in general $\sigma^2(X+Y) = \sigma^2(X) + \sigma^2(Y) + 2\Gamma(X,Y)$.

1.3.4. Approximation of One r.v. by an Arbitrary Function of Another

Let B be an event with nonzero probability. If we consider only the observations $\omega \in B$ (observations which realize event B), ω is given its weight relative to the weight of B. If it is not in B, ω is given zero weight. We can thus define a new probability on Ω, the **probability conditional on the event** B, $P(\cdot|B)$:

$$P(\omega|B) = 0 \qquad \text{for } \omega \in B$$
$$= \frac{P(\omega)}{P(B)} \qquad \text{for } \omega \in B.$$

Let X be a random variable; if its expectation with respect to this new probability exists it is called the **conditional mean** (or **expectation**) on B. Denote this by $P(X|B)$ or $E_P(X|B)$ or, if there is no ambiguity about P, $E(X|B)$. If the variance with respect to the new probability exists, it is denoted by $\sigma^2(X|B)$:

it is the **variance of** X ***conditional on*** B.

Let X and Y be two square integrable random variables. Can we find a real function f such that Y equals $f(X)$? or the closest possible to $f(X)$ in a least-squares sense?

Let us consider the set Ω' of observations with nonzero probability. For a real-valued function f defined on $X(\Omega')$ we have

$$E([Y - f(X)]^2) = \sum_{\omega\in\Omega'} (Y(\omega) - f(X(\omega))^2 P(\omega)$$

$$= \sum_{x\in X(\Omega')} P(X=x) \sum_{\{\omega\,;\,X(\omega)=x\}} (Y(\omega) - f(x))^2 P(\omega|X = x).$$

It is necessary then for each x to minimize the function

$$\lambda \longmapsto \psi(\lambda) = \sum_{\omega\in\{X=x\}} (Y(\omega)-\lambda)^2 P(\omega|X = x).$$

This minimum is attained for $\lambda = f(x) = E(Y|X = x)$ and it equals $\sigma^2(Y|X = x)$.

The real function f defined on $X(\Omega')$, such that $f(X)$ is closest to Y in the least-squares sense, is therefore the function $x \longmapsto E(Y|X = x) = f(x)$. We call $f(X)$ the **expectation of** Y **conditional** on X and denote $f(X) = E(Y|X)$. *The mean of* $E(Y|X)$ *is* $E(Y)$; in fact,

$$E[E(Y|X)] = \sum_{x\in X(\Omega')} P(X = x)E(Y|X = x)$$

$$= \sum_{x\in X(\Omega')} E(Y1_{(X=x)}) = E(Y).$$

The error made in replacing Y by $E(Y|X)$ is

$$E[(Y-E(Y|X))^2] = \sum_{x\in X(\Omega')} P(X=x)\sigma^2(Y|X = x).$$

This can be denoted by $E[\sigma^2(Y|X)]$. It is definitely majorized by the error obtained in Section 1.3.3 for an approximation by an affine function. From the following proposition, *the error of approximation and best approximation are orthogonal variables.* This implies the relationship

$$\sigma^2(Y) = E[\sigma^2(Y|X)] + \sigma^2[E(Y|X)].$$

Proposition 1.4.3. *We have* $E[(Y-E(Y|X))\phi(X)] = 0$, *for all square integrable functions* $\phi(X)$.

Indeed

$$E[E(Y|X)\phi(X)] = \sum_{x \in X(\Omega')} \phi(x) \frac{\sum_{\{\omega;\ X(\omega)=x\}} Y(\omega)P(\omega)}{P(X = x)} P(X=x)$$

$$E[Y\phi(X)] = \sum_{\omega \in \Omega} Y(\omega)\phi(X(\omega))P(\omega)$$

$$= \sum_{x \in X(\Omega')} \phi(x) \sum_{\{\omega;\ X(\omega)=x\}} Y(\omega)P(\omega).$$

1.3.5. Independence

Let X and Y be two functions on Ω taking values in a given space and $\mathcal{X} = X(\Omega')$, $\mathcal{Y} = Y(\Omega')$. Knowledge of X does not carry any information on Y, when, for all $y \in \mathcal{Y}$,

$$P(Y = y|X = x)$$

does not depend on $x \in \mathcal{X}$. This common value is $P(Y = y)$:

$$P(Y = y) = \sum_{x \in \mathcal{X}} P(X = x, Y = y)$$

$$= \sum_{x \in \mathcal{X}} P(X = x)P(Y = y|X = x)$$

$$= P(Y = y|X = x).$$

This is equivalent to saying that for all $x \in \mathcal{X}$ and all $y \in \mathcal{Y}$,

$$P(X = x, Y = y) = P(X = x)P(Y = y).$$

Or again, that for all $x \in \mathcal{X}$, $P(X = x|Y = y)$ does not depend on y and $P(X = x)$. We then say that X and Y are **independent.**

Let f and g be functions defined on $\mathcal{X}$ and $\mathcal{Y}$, respectively. If X and Y are independent, it is easily shown that $f o X$ and $g o Y$ are also independent.

Let X and Y be two independent integrable random variables. The product XY is integrable; indeed,

$$\sum_{\omega\in\Omega} |X(\omega)||Y(\omega)|P(\omega)$$

$$= \sum_{x\in X(\Omega)} \sum_{y\in Y(\Omega)} |x||y|P(X = x, Y = y)$$

$$= \sum_{x\in X(\Omega)} |x|P(X=x) \sum_{y\in Y(\Omega)} |y|P(Y=y) = E(|X|)E(|Y|).$$

The same calculation shows $E(XY) = E(X)E(Y)$. Thus, *two independent square integrable random variables are uncorrelated.* The converse is false: in the two diagrams of uncorrelated random variables in Section 1.2.1, the first corresponds to independent characteristics, but not the second.

Moreover, $E(Y|X = x)$ is then, for all $x \in \mathcal{X}$, equal to $E(Y)$. The approximation by a function of X adds nothing more than the approximation by a constant. The conditional variance $\sigma^2(Y|X = x)$ is, for all $x \in \mathcal{X}$, equal to $\sigma^2(Y)$.

Exercises 1.3.

In this section we give some classical probability exercises, most of which are based on the idea of uniform probability on a set Ω of possible observations. Each ω is said to be a "random choice." What concerns us then is to specify Ω and to use combinatorial methods to find the number of elements of Ω and that of an event $A \subset \Omega$.

E.1. n guests seat themselves at random around a circular table. Let X be the number of guests that separate Peter and Paul (taken in the direction which makes X the smaller number). Calculate the distribution of X, $E(X)$, and $\sigma^2(X)$.

E.2. To play at French lotto, you pay one franc each time to tick off 6 squares on a grid of 49 squares. The number N of winners is the number of ticked squares which coincide with a reference grid. What is the distribution of N? The winnings announced for 1 franc on 24/9/1980 were

861,680 for $N = 6$; 6703 for $N = 5$;

114 for $N = 4$; 9 for $N = 3$.

What is the average win for 1 franc? and its variance?

E.3. On $\Omega = \{a,b,c\}$, define the probability P and the r.v.'s X and Y by

$$P(a) = \frac{1}{2}, \; P(b) = P(c) = \frac{1}{4},$$

$$X = \mathbf{1}_{\{a\}} - \mathbf{1}_{\{b,\, c\}}, \; Y = \mathbf{1}_{\{b\}} - \mathbf{1}_{\{c\}}.$$

Show that $\Gamma(X,Y)$ is zero and that X and Y are not independent.

E.4. *Bayes formula.* Let P be a probability on a countable set Ω and $(H_1, H_2, ..., H_n)$ a partition of Ω of n events of nonzero probability. Show, for $1 \leqslant i \leqslant n$ and A an event ofnonzero probability, that

$$P(H_i|A) = \frac{P(A|H_i)P(H_i)}{\sum_{1 \leqslant j \leqslant n} P(A|H_j)P(H_j)}.$$

E.5. *Intersection and union of events.* Let $A_1, ..., A_N$ be N events in the countable set Ω and let P be a probability on Ω.

(a) Show, if $A_1 \cap ... \cap A_{N-1}$ has a nonzero probability,

$$P[A_1 \cap ... \cap A_N] = P(A_1)P(A_2|A_1)...P(A_N|A_1 \cap ... \cap A_{N-1}).$$

(b) For $1 \leqslant n \leqslant N$, consider S_n, the sum of probabilities $P[A_{i_1} \cap A_{i_1} \cap ... A_{i_n}]$, for $(i_1, ..., i_n)$ a strictly increasing sequence of n integers between 1 and N. Show

$$P[A_1 \cup ... \cup A_N] = S_1 - S_2 + ... + (-1)^{k+1} S_k + ... + (-1)^{N+1} S_N.$$

E.6. *Mendel's laws in genetics: Mendel's first law.* According to Mendel, a genetic characteristic is often determined within each individual by two genes of two possible types (or *alleles*). Let us call them A and a. There are then three *genotypes* AA, Aa, aa. It is often the case that one of these types is *dominant*, i.e., AA and Aa correspond to the same physical aspect, giving two *phenotypes* (AA, Aa) and aa. Suppose that a child receives at random a gene from each of his parents and that the parents choose each other at random (or at least without taking into account the genotype under consideration). The populations are asumed to be large and we can speak of frequency or probability. There is no immigration or mutation, and we assumed that the generations are well separated.

(a) Let p_{AA}, p_{Aa}, p_{aa} be the probabilities in a first generation of three genotypes and let $p = 1\text{-}q$ be the probability $p_{AA}+ p_{Aa}/2$ that a gene is of type A. Show that the probabilities of the three genotypes in the following generation are p^2, $2pq$, q^2, as if the genes had been mixed at random. What are the probabilities of the various phenotypes in the second generation if A is dominant?
(b) The father of a child is not known. Give the probabilities of the various genotypes of the child conditional on the mother's genotype being AA (then Aa or aa). Give the probabilities of the various genotypes of the mother conditional on genotype AA (then Aa or aa) of the child.
(c) The genes which correspond to blood groups have three possible alleles, α, β, and ω. We can observe four phenotypes or blood groups, according to the individual's pair of genes: A for $\alpha\alpha$ or $\alpha\omega$, B for $\beta\beta$ or $\beta\omega$, O for $\omega\omega$. With the same conventions as above, assume that in one generation the probabilities of the genes α, β, and ω are p, q, and $r = 1\text{-}p\text{-}q$, respectively. Determine the probability of the various blood groups in the following generation.

E.7. *Mendel's Second Law.* We can observe two characteristics of beans. One, "colored or colorless" is determined by a gene having two alleles C or I; C is dominant, a colored bean can be seen for the genotypes CC and CI. The other, "smooth or wrinkled," is determined by two alleles of the same gene L or R; L is dominant, and we observe smooth for LL and LR. Mendel's second law states that the genes of two different characters are transmitted from the parents to the offspring independently. To test his hypothesis, Mendel crossed a large number of pure bred beans, one of the parents being (CC, LL), the other (II, RR). Which type are the beans in the first generation? Show, for the beans in the second generation, that if Mendel was correct,

$$P[\text{colored, smooth}] = \frac{9}{16},$$

$$P[\text{colored, wrinkled}] = P[\text{colorless, smooth}] = \frac{3}{16},$$

$$P[\text{colorless, wrinkled}] = \frac{1}{16}.$$

E.8. An absent-minded postman distributes letters at random in the n postboxes of a block of flats.

(a) There is a bag of rent bills, one for each tenant. The postman puts one in each box without looking at the name. Let A_i be the event "the letter addressed to tenant i is put in the correct box" ($i = 1, ..., n$). Calculate

$$P(A_1),\ P(A_1 \cap A_2 \cap ... \cap A_r),$$

$$P(A_1 \cup A_2 \cup ... \cup A_n)$$

(Exercise 1.3.6). Let N be the number of letters which arrive at the correct destination. Show that $P_n = P(N{=}0)$ is equivalent, for N large, to $1/e$. Show

$$P(N = r) = \frac{P_{n-r}}{r!},\quad r = 0, ..., n.$$

Calculate $P(A_i|N = r)$.

(b) p identical prospectuses are then distributed. The postman puts them at random one by one in the N boxes, forgetting as he goes along in which boxes he has already put one. Let B_i be the event "tenant i's box is empty." Calculate $P(B_1 \cup B_2 \cup ... \cup B_r)$, $1 \leqslant r \leqslant n$. Let M be the number of empty boxes. Calculate $P(M = 0)$, then $P(M = r)$ for $1 \leqslant r \leqslant n$.

E.9. *Wilcoxon's statistic and the hypergeometric law.*

(a) An individual is chosen at random from a population $\{1,2, ..., N\}$; let R_1 be his index (or rank). Calculate its mean and variance.

(b) *n distinct* individuals are chosen at random from the same population; let $R_1, ..., R_n$ be the observed ranks for the first individual chosen, ..., nth individual. What is the set of the sequence $(R_1, ..., R_n)$ (all of the same probability)? How many elements are there? Show that for $1 \leqslant i \leqslant n$, R_i has the same law as R_1. For two distinct indices i and j, show that (R_i,R_j) has the same law as (R_1,R_2).

(c) Let $W_n = R_1 + ... + R_n$ (Wilcoxon's statistic). Calculate its variance as a function of $\Gamma(R_1,R_2)$ and of n. What does W_N equal? What is its variance? Deduce from this $\Gamma(R_i,R_j)$, $1 \leqslant i, j \leqslant n$. Verify that

$$E(W_n) = \frac{n(N+1)}{2}, \quad \sigma^2(W_n) = \frac{n(N-n)(N+1)}{12}.$$

(d) n distinct objects are chosen at random from a sequence of N real numbers $(x_1,x_2, ..., x_N)$. This is equivalent to choosing the object $(x_1 = x_{R_1}, ..., x_n = x_{R_n})$ for the variables $R_1, ..., R_n$ defined in (b). Verify that, for $1 < i \leqslant n$, X_i has the same distribution as X_1 and that, for $i \neq j$, (X_i,X_j) has the same distribution as (X_1,X_2). Let $S_n = X_1 + ... + X_n$. Calculate the variance of S_n as a function of $\sigma^2(X_1)$, $\Gamma(X_1,X_2)$, and of n. What does it equal for $n = N$?

(e) *Hypergeometric distribution.* n different individuals are questioned at random in a population of N individuals, of whom N_1 smoke and $N_2 = N-N_1$ do not. Set $X_i = 1$ or $X_i = 0$, depending on whether the ith response is "smoker" or "nonsmoker." Let $S = X_1 + ... + X_n$ the number of smokers. Prove that, for $k = 0, ..., n$

$$P[S = k] = \frac{\binom{N_1}{k}\binom{N-N_1}{n-k}}{\binom{N}{k}} \quad \text{(hypergeometric distribution).}$$

Calculate the variance of X_i, and the covariance of (X_i,X_j) for $1 \leqslant i,j \leqslant n$, $i \neq j$. Verify

$$E(S) = n\frac{N_1}{N}, \quad \sigma^2(S) = \frac{N-n}{N-1}n\frac{N_1}{N}\left(1 - \frac{N_1}{N}\right).$$

(f) *Multinomial distribution.* In this case n different individuals are questioned from a population in which k different answers are possible: N_1 of type 1, N_2 of type 2, ..., N_k of type k. Denote by X_i the ith individual answer and Z_j the number of answers equal to $j(Z_j = \Sigma_{i=1}^n \mathbf{1}_{(X_i=j)})$.

Let $(j_1, ..., j_k)$ be a sequence of n integers summing to n; calculate $P[Z_1 = j_1, ..., Z_k = j_k]$. Calculate the covariance of Z_j and $Z_{j'}$ $(1 \leqslant j, j' \leqslant k, j \neq j')$.

E.10. *Sequences.*

(a) What is the number of possible configurations of n individual objects in r boxes arranged on a shelf. (Represent the objects by n crosses in a row, and the boxes by vertical strokes and count in how many ways you could place amongst the crosses the r-1 vertical strokes which separate the boxes.) What is the number of such configurations such that no box is empty?

(b) Two types of object are mixed in an urn, n_1 of type A, n_2 of type B $(n_1 \geqslant 1, n_2 \geqslant 1)$. These are then extracted one by one. A sequence of N objects such as $ABABAAAA$ is obtained. How many sequences are possible? Assume them all to be equiprobable.

Let R_1 be the number of consecutive sequences of A and R_2 the number of consecutive sequences of B (in the example $(R_1 = 3, R_2 = 2)$. Prove that, for $r_1 = 1, ..., n_1$ and $r_2 = 1, ..., n_2$

$$P[R_1 = r_1, R_2 = r_2] = \frac{c\binom{n_1-1}{r_1-1}\binom{n_2-1}{r_2-1}}{\binom{n_1+n_2}{n_1}},$$

with $c = 2$ for $r_1 = r_2$, $c = 1$ for $r_1 = r_2 \pm 1$ (and c zero otherwise). Calculate $P(R_1 = r_1)$ and $P(R_1 + R_2 = r)$, $2 \leqslant r \leqslant n$. For $n_1 = 4$, $n_2 = 3$, what are the probabilities of observing $AAAABBB$? $ABABABA$? $BBBAAAA$?

E.11. *Ranks and Spearman correlation.* Let $\mathfrak{G}$ be the set of $n!$ permutations of $\{1,2, ..., n\}$ given the uniform distribution. Observing $\sigma \in \mathfrak{G}$ is the same as observing n individuals in a certain order. We interpret the number associated with the individual as his rank in a decreasing order of beauty. To $\sigma = (\sigma(1), \ldots, \sigma(n))$ associate the **absolute rank** of the ith observed individual $R_i(\sigma) = \sigma(i)$ and his **relative rank** amongst those who have been observed before him, Y_i.

(a) What is the distribution of R_i?
(b) What is the distribution of Y_i and that of $(Y_1, \ldots, Y_i)$?
(c) Show that, for $i < n$, $1 = n_1$, $1 \leqslant n_2 \leqslant 2, \ldots, 1 \leqslant n_i \leqslant i$, $n_i \leqslant m \leqslant n$:

$$P(R_i = m | Y_1 = n_1, Y_2 = n_2, \ldots, Y_i = n_i) = \frac{\binom{m-1}{n_i-1}\binom{n-m}{i-n_i}}{\binom{n}{i}}.$$

Show that for $j > i$, R_j and Y_i are independent.
(d) To each permutation σ, associate $\Delta(\sigma) = \Sigma_{i=1}^{n}(\sigma(i) - i)^2$. Show that Δ is an even integer; calculate its mean and variance. For $n = 5$, calculate $P(\Delta \leqslant 4)$.
(e) In Exercise 1.2.5 assume that all the competitors are equally good. Assume that the two judges are independent of one another. Show that $\Sigma_{i=1}^{n} D_i^2$ then has the distribution of Δ studied in (d). What is the probability, for $n = 5$, of observing a Spearman coefficient $\rho \geqslant 0.8$.

E.12. *Jensen's inequality.* Let $(x_1, \ldots, x_n) \in \mathbb{R}^n$ and $(p_1, \ldots, p_n) \in \mathbb{R}_+^n$ such that $\Sigma_{i=1}^{n} p_i$ equals 1. If f is a convex function on an interval of $\mathbb{R}$ containing $x_1, \ldots, x_n$ in $\mathbb{R}$, show that

$$f(p_1 x_1 + \ldots + p_n x_n) \leqslant p_1 f(x_1) + \ldots + p_n f(x_n).$$

E.13. *Entropy.* Let Ω be a countable set and P a probability on Ω. Its entropy is defined by $H(P) = -\Sigma_{\omega \in \Omega} P(\omega) \log P(\omega)$. We need not specify the base of the logarithm and take by convention $u \log u$ equal to zero for $u = 0$ and $\frac{0}{0}$ equal to 0.

(a) Show that $H(P)$ is positive; when does it vanish?
(b) Let Q be another probability. Show that

$$-\sum_{\omega\in\Omega} P(\omega)\log[Q(\omega)/P(\omega)]$$

is positive and equals zero if and only if P and Q coincide. This number is the **Kullback information** of P on Q.

(c) Let $\Omega = \{1,2, ..., n\}$ and Q the uniform probability on Ω. Calculate $H(Q)$. Show that for every P different from Q, $H(P)$ is strictly majorized by $H(Q)$.

(d) Let X be a function defined on Ω, designate by $H(X)$ the entropy of its distribution. Show that $H(X)$ is majorized by $H(P)$.

(e) Let X and Y be two functions defined on Ω; $P(Y = \cdot|X = x)$ is the distribution of Y conditional on $\{X = x\}$, for $x \in X(\Omega')$, $\Omega' = \{\omega, P(\omega) > 0\}$. Prove

$$H(X,Y) = H(X) + \sum_{x\in X(\Omega')} P(X = x)H(P(Y = \cdot|X = x)).$$

The expression $H(X,Y) - H(X)$ is therefore an average of the entropies of the distribution of Y conditional on the various values of X. This is denoted by $H(Y|X)$ and is called the entropy of Y conditional on X. Show that $H(Y|X)$ vanishes if and only if Y is a function of X.

Show that $H(X,Y) \leqslant H(X) + H(Y)$, with equality if and only if X and Y are independent. The conditional entropy therefore lies between 0 and $H(Y)$; it measures the uncertainty remaining about Y when X is known (explain why).

(f) A player tries to find as quickly as possible a number Y, which his opponent has chosen at random between 0 and N-1. To do this he can ask any question he likes (even? equal to 47? greater than 50? ...) on condition that the answers are yes or no. After n questions he knows the value of a function X taking values in $\{\text{yes, no}\}^n$. Show that even if he is very astute he cannot be sure of determining Y before having asked more than $\log N/\log 2$ questions. (Compare the entropy of Y and the maximum entropy of X.)

(g) In a pile of N coins, all are identical except one counterfeit which is lighter. In order to find this coin, we use a set of scales with two balance pans and we can put as many coins on each balance as we like. Each experiment can have three possible answers: "the balance on the left is heavier"; the balance on the right is heavier"; or "they are equal." Show that the best approach will only guarantee success with a number of weighings larger than $\log N/\log 3$.

E.14.

(a) Let c and k be in $\mathbb{N}\backslash\{0\}$ and Ω the set of integers lying between 0 and c^k-1, given the uniform probability P. An integer ω of Ω can be written uniquely in the form

$$\omega = a_1(\omega) + ca_2(\omega) + \dots + c^{k-1}a_k(\omega)$$

with, for $1 \leqslant i \leqslant k$, $a_i(\omega) \in \{0,1, \dots, c-1\}$. Show that, for all $(i_1, \dots, i_k) \in \{0,1, \dots, c-1\}^k$, we have

$$P[a_1 = i_1, \dots, a_k = i_k] = \frac{1}{c^k}\,.$$

Show that, for $i \neq j$, a_i and a_j are independent and that they are uniformly distributed on $\{0,1, \dots, c-1\}$.

(b) For question (f) of Exercise 1.3.13, and $N = 2^k$, take $c = 2$ and ask these questions in succession "$a_1 = 0$?" "$a_2 = 0$?"... "$a^k = 0$?". Verify that this is the minimum number of questions to ask to be sure of determining Y. Generalize to any given N.

(c) For question (g) of Exercise 1.3.13 and $N = 3^k$, take $c = 3$, number the coins and put in the left-hand weighing pan those for which a_1 equals zero and in the right hand weighing pan those for which a_1 equals 1. The experiment determines the value of a_1 for the counterfeit coin. Follow this up and give a method for obtaining the counterfeit coin in a minimum number of weighings. Extend to arbitrary N.

E.15. *Codes.* We wish to transmit an alphabet of k letters A_1, ..., A_k with the aid of a binary alphabet $\{0,1\}$. We choose a code of k sequences of 0 and of 1, or words with respective lengths $\lambda_1, \dots, \lambda_k$. Only decipherable codes are used, where each word is not the beginning of another. Such a code can be visualized by a tree of the following kind, where a branch extends up to the right for a letter 0 and to the left for 1.

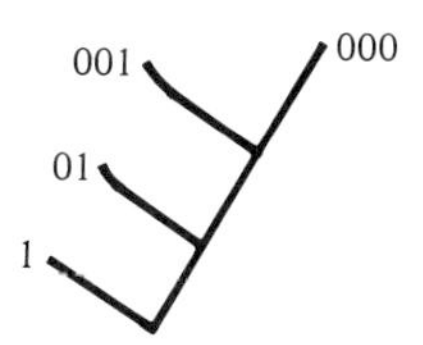

(a) Show that a necessary and sufficient condition such that a decipherable code exists with words of length $\lambda_1, ..., \lambda_k$ is $\Sigma_{i=1}^{k} 2^{-\lambda_i} \leqslant 1$ (notice that a word of length $\lambda' < \lambda$ can be prolonged to $2^{\lambda-\lambda'}$ words of length λ).

(b) Assume that the frequencies (or probabilities) of the letters $A_1, ..., A_k$ form the sequence $p = (p_1, ..., p_k)$. We wish to minimize the average length $\Sigma_{i=1}^{k} \lambda_i p_i$ of decipherable codes. Let $H_0(p) = H_0(p_1, ..., p_k)$ be this minimum. Show that, for $p_1 \geqslant p_2 \geqslant ... \geqslant p_k$ we are best to take $\lambda_1 \leqslant \lambda_2 \leqslant ... \leqslant \lambda_k$.

(c) Denoting by $\log_2$, logarithm to the base 2, show that there exists a decipherable code such that

$$-\log_2 p_i \leqslant \lambda_i < -\log_2 p_i + 1.$$

Show

$$H(p) = -\sum_{i=1}^{k} p_i \log_2 p_i \leqslant H_0(p) < H(p) + 1.$$

(d) Notice that to transmit the alphabet $A_1, ..., A_k$ we could transmit $A_1, ..., A_{k-2}$ and $(A_{k-1}$ or $A_k) = B_{k-1}$, then add in the case where B_{k-1} has been transmitted 0 or 1 to differentiate the elements. Show

$$H_0(p_1, ..., p_k) = H_0(p_1, ..., p_{k-2}, (p_{k-1} + p_k)) + (p_{k-1} + p_k)$$

if $p_1 \geqslant ... \geqslant p_k > 0$.

(e) We wish to transmit a word of n letters $X_1, ..., X_n$ of the initial alphabet. Assume these r.v.'s are independent and equal to A_i with probability p_i. Thus we code blocks of n letters of the initial alphabet. Let L_n be the minimum average length. Show $\lim_{n\to\infty} L_n/n = H(p)$ (Exercise 1.3.13(e)).

1.4. Pairs of R.V.'s and Correspondence Analysis

The contingency tables of Section 1.1 are analyzed here.

1.4.1. Functions of Y Well Explained by X

On $\Omega = \{1, ..., k\} \times \{1, ..., \ell\}$, let P be the probability defined by $P(i,j) = F_{ij}$. Let X: $(i,j) \longmapsto i$ *and* Y: $(i,j) \longmapsto j$ be functions of Ω into $\{1, ..., k\}$ and $\{1, ..., \ell\}$.

Assume $F_{i\cdot} > 0$ and $F_{\cdot j} > 0$, for all $(i,j) \in \Omega$, even if it does mean removing the unobservable states. For $f \in \mathbb{R}^k$, set $f(X) = \langle f,X\rangle$; and for $f \in \mathbb{R}^\ell$ set $f(Y) = \langle f,Y\rangle$. Since the best approximation and the error of approximation (Section 1.3.4) are uncorrelated, this implies

$$1 = \frac{E[(f(Y) - E(f(Y)|X))^2]}{\sigma^2(f(Y))} + \frac{\sigma^2[E(f(Y)|X)]}{\sigma^2(f(Y))}.$$

This relation is unchanged if f is multiplied by a constant. We are looking for the function $f(Y)$ (defined up to a factor) best explained by X. To do this we are led to maximize $\sigma^2[E(f(Y)|X)]$ when $\sigma^2(f(Y))$ equals 1.

We shall consider on $\mathbb{R}^\ell$ the bilinear form

$$(f,g) \longmapsto \Gamma[E(f(Y)|X), E(g(Y)|X)] = {}^t fLg,$$

where L is the matrix $L = \{\Gamma(E(1_{(Y=i)}|X), E(1_{(Y=j)}|X)\}_{1 \leq i, j \leq \ell}$.

The quadratic form associated with this is $f \longmapsto \sigma^2[E(f(Y)|X)]$, which is positive. Likewise, the bilinear form $(f,g) \longmapsto \Gamma(f(Y),g(Y))$ is associated with the positive quadratic form $f \longmapsto \sigma^2(f(Y))$. Let C_Y be the set of vectors f in $\mathbb{R}^\ell$ such that $f(Y)$ is centered. This is a complementary subspace of $1_\ell = (1, ..., 1)$ and, on C_Y, $(f,g) \longmapsto \Gamma(f(Y),g(Y))$ is a scalar product (Section 1.3.3). A base $(v_1, ..., v_{\ell-1})$ of C_Y can then be found, orthonormal for the scalar product $(f,g) \longmapsto \Gamma(f(Y),g(Y))$ and orthogonal for the bilinear form associated with L, formed from eigenvectors of the matrix L. Interpreting this we have:

Proposition 1.4.4. *There exist functions $v_1, ..., v_{\ell-1}$ from $\{1, ..., \ell\}$ into $\mathbb{R}$ and numbers $\lambda_1 \geq \lambda_2 ... > \lambda_{\ell-1} \geq 0$ such that*

(1) *for* $1 \leqslant j \leqslant \ell-1$: $E(v_j(Y)) = 0$, $\sigma^2(v_j(Y)) = 1$, $\sigma^2(E(v_j(Y)|X)) = \lambda_j$.

(2) *For* $1 \leqslant j, j' \leqslant \ell-1$ *and* $j \neq j'$: $\Gamma(v_j(Y), v_{j'}(Y)) = 0$ *and* $\Gamma(E(v_j(Y)|X), E(v_{j'}(Y)|X)) = 0$.

(3) *Each* $f \in \mathbb{R}^\ell$ *can be written uniquely as*

$$f = E[f(Y)] + \sum_{j=1}^{\ell-1} v_j \Gamma[f(Y), v_j(Y)].$$

For f *and* g *in* $\mathbb{R}^\ell$ *we have*

$$\Gamma[f(Y), g(Y)] = \sum_{j=1}^{\ell-1} \Gamma(v_j(Y), f(Y))\Gamma(v_j(Y), g(Y))$$

$$\Gamma[E(f(Y)|X),\ E(g(Y)|X)] = \sum_{j=1}^{\ell-1} \lambda_j \Gamma(v_j(Y), f(Y))\Gamma(v_j(Y), g(Y)).$$

We say that $v_j(Y)$ *is a* jth *principal factor for the analysis of* Y *by* X *(it is unique if* λ_j *is a simple eigenvalue).*

Note. While looking for the eigenvalues of L, we shall always find 1, the eigenvalue of the eigenvector $\mathbf{1}_\ell$, which is not in C_Y. Moreover, for each v_j $(1 \leqslant j \leqslant \ell-1)$,

$$\sigma^2[E(v_j(Y)|X)] = \lambda_j \leqslant \sigma^2(v_j(Y)) = 1.$$

Thus the eigenvalues of L are $1 \geqslant \lambda_1 \geqslant \dots \geqslant \lambda_{\ell-1} \geqslant 0$ and an associated basis of eigenvectors in $\mathbb{R}^\ell$ is $\{\mathbf{1}_\ell, v_1, \dots, v_{\ell-1}\}$.

It can happen, as for principal components analysis, that the eigenvalues are not all distinct. Assume that there exists a single eigenvector v_1 associated with λ_1 $(\lambda_1 > \lambda_2)$. Then $v_1(Y)$ is the function (of variance 1) of Y, of which the expectation conditional on X is the most dispersed. Assume, for example, $v_1(1) = v_1(2) > 0$ and $v_1(j) = 0$ for $j > 2$. These are the values of the social classes 1 and 2 in Section 1.1 which vary the most amongst electors in the same party and which explain the difficulty in finding a functional relationship between voting for a party and belonging to a social class. We are now going to develop the analysis of the relationships between the two characteristics.

1.4.2. Dispersion of Profiles

Let $x^i = (F_{ij}/F_{i\cdot})_{j=1,\ldots,\ell}$ be the profile of the social class Y in party i. If the weight F_i is given to x^i, the barycenter of the (x^i) is $\bar{x} = (F_{\cdot j})_{j=1,\ldots,\ell}$. It is thus the distribution of Y or mean profile. We study the **dispersion** or **inertia** of the profiles (x^i) allocated weights $(F_{i\cdot})$.

The profiles are probabilities on $\{1, \ldots, \ell\}$, thus points $\nu = (\nu(1), \ldots, \nu(\ell))$ of $\mathbb{R}^\ell$. These coordinates are positive and sum to 1. Instead of using Euclidean distance to measure the distance between two profiles we use the χ^2 **distance** defined by

$$\chi^2(\nu,\nu_o) = \sum_{j=1}^{\ell} \frac{(\nu(j)-\nu_0(j))^2}{\nu_0(j)}.$$

In spite of its name this is not a distance: it is not symmetric! The asymmetry hangs upon the fact that the problems posed will often be asymmetric: distance from a probability ν to a reference probability ν_o.

The change from Euclidean distance to χ^2 distance is a normalization intended primarily to bring out the differences between ν and ν_0 for values of j with a small weight $\nu_0(j)$, even if their absolute values are small. The choice of this distance allows the role of the smaller parties to be seen, and also allows us to interpret and analyze easily the profiles obtained when the parties are regrouped (e.g., into left, center, and right) (Exercise 1.4.2).

The reference measure here is the distribution $\bar{x}$ of Y. The distance of the profile x^i from $\bar{x}$ is $\chi^2(x^i,\bar{x})$ and the dispersion of the profiles, or the inertia, is given by

$$I = \sum_{i=1}^{k} F_{i\cdot}[\chi^2(x^i,\bar{x})] = \sum_{i=1}^{k} F_{i\cdot}\left[\sum_{j=1}^{\ell} \frac{1}{F_{\cdot j}}\left(\frac{F_{ij}}{F_{i\cdot}} - F_{\cdot j}\right)^2\right].$$

Note that if I is zero, we have for all (i,j), $F_{\cdot j} = F_{ij}/F_{i\cdot}$. The conditional distribution does not then depend on i, and the characteristics are independent.

Consider the function F^i defined by

$$F^i(j) = \frac{1}{F_{\cdot j}}\left[\frac{F_{ij}}{F_{i\cdot}} - F_{\cdot j}\right].$$

The r.v. $F^i(Y)$ is centered, the vector F^i is in C_Y and

$$F^i = \sum_{p=1}^{\ell-1} \nu_p \Gamma(F^i(Y), \nu_p(Y)),$$

$$\Gamma[F^i(Y), \nu_p(Y)] = \sum_{j=1}^{\ell} \left(\frac{F_{ij}}{F_{i\cdot}} - F_{\cdot j}\right) \nu_p(j) = E[\nu_p(Y)|X=i],$$

$$\sigma^2[F^i(Y)] = \sum_{j=1}^{\ell} F_{\cdot j}[F^i(j)]^2 = \sum_{p=1}^{\ell-1} (\Gamma[F^i(Y), \nu_p(Y)])^2.$$

Thus

$$\mathcal{I} = \sum_{i=1}^{k} F_{i\cdot}\, \sigma^2(F^i(Y)) = \sum_{i=1}^{k} F_{i\cdot} \sum_{p=1}^{\ell-1} (\Gamma[F^i(Y), \nu(Y)])^2$$

$$= \sum_{p=1}^{\ell-1} \sum_{i=1}^{k} F_{i\cdot} (E(\nu_p(Y)|X = i))^2$$

$$= \sum_{p=1}^{\ell-1} \sigma^2(E(\nu_p(Y)|X)) = \sum_{p=1}^{\ell-1} \lambda_p.$$

The dispersion of the profiles can therefore be measured by $\mathcal{I}$, the trace of the matrix L minus 1.

If, instead of studying the dispersion of the profiles of characteristic Y, we had studied the dispersion of characteristic $(\nu_1(Y), \ldots, n_r(Y))$ taking values in $\mathbb{R}^r$ $(r \leqslant \ell-1)$, we would have obtained $\lambda_1 + \ldots + \lambda_r$ (the principal factors are $\nu_1(Y), \ldots, \nu_r(Y)$). If we study the cloud of profiles obtained by replacing Y by the first r principal factors $(\nu_1(Y), \ldots, \nu_r(Y))$, the fidelity of this new cloud can be evaluated by the quotient of the inertias $(\lambda_1 + \ldots + \lambda_r)/(\lambda_1 + \ldots + \lambda_{\ell-1})$; the closer this quotient is to 1, the better.

Thus if we return to the situation studied in the example after Proposition 1.4.4, and if the social class profile is replaced by the approximated profiles of the individual social classes 1 and 2, we retain almost the same dispersion when $\lambda_1 + \lambda_2$ is much greater than $\lambda_3 + \ldots + \lambda_{\ell-1}$.

1.4.3. Duality Between Characteristics

The question of "correspondences" between X and Y ("How are they related?") is symmetric; but we have given only an asymmetric solution. Let us see what happens if the roles are exchanged. Recall the results of Proposition 1.4.4. For $1 \leq j, j' \leq \ell-1$ and $j \neq j'$, $E[v_j(Y)|X]$ and $E[v_{j'}(Y)|X]$ are uncorrelated and the variance of $E[v_j(Y)|X]$ is λ_j.

Let C_X be the subspace of $\mathbb{R}^k$ of vectors f such that $f(X)$ is centered. Let $r = \sup\{p;\ \lambda_p \neq 0\}$. For $p \leq r$, setting $w_p(\cdot) = [1/(\lambda_p)^{1/2}]E[v_p(Y)|X = \cdot]$, we obtain a system of r orthogonal vectors in C_X normed with respect to the scalar product $(f,g) \longmapsto \Gamma[f(X),g(X)]$. This system can be extended to obtain a basis $(w_p)_{1\leq p\leq k-1}$ of C_X equipped with this scalar product.

Let $f \in C_X$; then $E[f(X)|Y]$ is centered and for $g \in C_Y$ we can write

$$\begin{aligned}\Gamma[E(f(X)|Y),g(Y)] &= E[E(f(X)|Y)g(Y)]\\ &= E[f(X)g(Y)] = E[f(X)E(g(Y)|X)]\\ &= \Gamma[f(X),E(g(Y)|X)].\end{aligned}$$

Hence

$$\begin{aligned}E[f(X)|Y] &= \sum_{j=1}^{\ell-1} \Gamma[E(f(X)|Y),v_j(Y)]v_j(Y)\\ &= \sum_{j=1}^{r} \Gamma[f(X),w_j(X)]\sqrt{\lambda_j}\,v_j(Y),\end{aligned}$$

$$\sigma^2[E[f(X)|Y]] = \sum_{j=1}^{r}(\Gamma[f(X),w_j(X)])^2\lambda_j.$$

The vectors $(w_j)_{1\leq j\leq k-1}$ therefore form a basis orthonormal for the scalar product $(f,g) \longmapsto \Gamma(f(X),g(X))$ on C_X and orthogonal for the bilinear form $(f,g) \longmapsto \Gamma[E(f(X)|Y),E(g(X)|Y)]$. We thus have the following symmetric proposition.

Proposition 1.4.5. *There exist r numbers $\lambda_1 \geq \lambda_2 \geq \dots \geq \lambda_r > 0$ with $r \leq \inf(k-1, \ell-1)$, r vectors $v_1, \dots, v_r$ in $\mathbb{R}^\ell$ and r vectors $w_1, \dots, w_r$ in $\mathbb{R}^k$ related by, for $1 \leq i \leq r$,*

$$E[v_i(Y)|X] = \sqrt{\lambda_i}\,w_i(X)$$

and

$$E[w_i(X)|Y] = \sqrt{\lambda_i}\, v_i(Y)$$

and such that

(1) *For* $1 \leqslant i \leqslant r$, $v_i(Y)$ *and* $w_i(X)$ *are centered r.v.'s of variance* 1.

(2) *For* $1 \leqslant i, i' \leqslant r$, $i = i'$,

$$\Gamma[v_i(Y), v_i'(Y)] = \Gamma[w_i(X), w_{i'}(X)] = 0.$$

(3) *For* f *and* g *in* $\mathbb{R}^{\ell}$,

$$\Gamma[f(Y), g(Y)] = \sum_{i=1}^{r} \Gamma[f(Y), v_i(Y)]\Gamma[g(Y), v_i(Y)],$$

$$\Gamma[E(f(Y)|X), E(g(Y)|X)] = \sum_{i=1}^{r} \lambda_i \Gamma[f(Y), v_i(Y)]\Gamma[g(Y), v_i(Y)].$$

(4) *For* f *and* g *in* $\mathbb{R}^{k}$,

$$\Gamma[f(X), g(X)] = \sum_{i=1}^{r} \Gamma[f(X), w_i(X)]\Gamma[g(X), w_i(X)],$$

$$\Gamma[E(f(X)|Y), E(g(X)|Y)] = \sum_{i=1}^{r} \lambda_i \Gamma[f(X), w_i(X)]\Gamma[g(X), w_i(X)].$$

(5) *The inertia (for the* χ^2 *distance) of the cloud of profiles of* X *for* Y *is equal to the inertia of the cloud of profiles of* Y *for* X*: this communal value is* $\lambda_1 + \ldots + \lambda_r$.

For $q < r$, *the inertia of the profiles of* X *for* $\{v_1(Y), \ldots, v_q(Y)\}$ *is equal to the inertia of the profiles of* Y *for* $\{w_1(X), \ldots, w_q(X)\}$, *and this communal value is* $\lambda_1 + \ldots + \lambda_q$.

Note: The average variance of $v_q(Y)$ conditional on X is, for $1 \leqslant q \leqslant r$,

$$E[\sigma^2(v_q(Y)|X)] = \sum_{i=1}^{k} F_{i.}\, \sigma^2(v_q(Y)|X = i)$$

$$= \sum_{i=1}^{k} F_{i.}\, \lambda_q (w_q(i))^2.$$

The ratio

$$\frac{F_{i.}\,\lambda_q (w_q(i))^2}{\sum_{t=1}^{k} F_{t.}\,\lambda_q (w_q(t))^2}$$

is called the **relative contribution** of value i of characteristic X in the factor $v_q(Y)$. It measures the role of the observations $\{X = i\}$ in explaining $v_q(Y)$. Denote this by $\mathrm{RCT}_q^Y(i)$.

1.4.4. Graphical Representation

Let $q \leqslant r$. Consider for $j = 1, ..., \ell$, $M_j = (v_1(j), ..., v_q(j))$, and for $i = 1, ..., k$, $N_i = (w_1(i), ..., w_q(i))$. Let ϕ be the dilatation $(x_1, ..., x_q) \longmapsto (\sqrt{\lambda_1}x_1, ..., \sqrt{\lambda_q}x_q)$ (the transformation ϕ corresponds to a change of scale on the axes). For $q = r$, the point M_j is the image under ϕ of the barycenter of the points N_i allocated the weights $F_{ij}/F_{.j}$, with $1 \leqslant i \leqslant k$. The point M_j is close to $\phi(N_i)$ if $F_{ij}/F_{.j}$ is close to 1 (but this is not a necessary condition). The cloud $(M_j)_{1 \leqslant j \leqslant \ell}$ gives a representation of the profiles of the characteristic $(w_1(x), ..., w_q(x))$ in $Y = 1, ..., \ell$. Likewise N_i is the image under ϕ of the barycenter of the points M_j allocated the weights $F_{ij}/F_{i.}$, with $1 \leqslant j \leqslant \ell$; N_i is close to $\phi(M_j)$, when $F_{ij}/F_{i.}$ is close to 1; the cloud $(N_i)_{1 \leqslant i \leqslant k}$ gives a representation of the profiles of characteristic $(v_1(Y), ..., v_1(Y))$ in $X = 1, ..., k$.

A *simultaneous representation of the two characteristics* can be made with the help of the first q principal factors by considering $(M_1, ..., M_\ell)$ and $(\phi(N_1), ..., \phi(N_k)) = (\hat{N}_1, ..., \hat{N}_k)$. We can then (cautiously...) interpret the proximity of $\hat{N}_i$ and of M_j by the fact that the probability $P(X = i|Y = j)$ is large.

If $\lambda_1 + ... + \lambda_q$ is large with respect to $\lambda_1 + ... + \lambda_r$, these representations are the most faithful of the cloud of profiles of X in $Y = 1, ..., \ell$, or of Y in $X = 1, ..., k$. $q = 2$ is often taken as satisfactory; if not, the representation is made in $\mathbf{R}^q$, followed by various projections

$$(x_1, ..., x_q) \longmapsto (x_i, x_{i'}), \quad 1 \leqslant i < i' \leqslant q.$$

Example. Recall Table 2 of Exercise 1.1. We note first of all the importance of χ^2. The sectors with little industrial muscle but economically or sociologically significant (banks) are kept in the analysis.

With the above notations, the following tables are obtained (X is the year, Y the sector of industry). The fidelity is large for $q \geqslant 4$.

For $p = 1,2,3,4$, we read in the column $v_p(Y)$ (resp. $w_p(X)$) the value $v_p(j)$ of v_p in the sector $Y = j$ (respectively, the value $w_p(i)$ of w_p in $X = i$); COR is the square of the correlation of $\mathbf{1}_{(Y=j)}$ with v_p (resp. $\mathbf{1}_{(X=i)}$ with w_p), the sign being that of the correlations; finally RCT represents RCT_p^Y (sector) (resp. RCT_p^X (year)); a zero is understood to be in front of COR and RCT (i.e., read 0.240 instead of 240).

Simultaneous Representation of the Sectors and the Years by the First Two Principal Factors

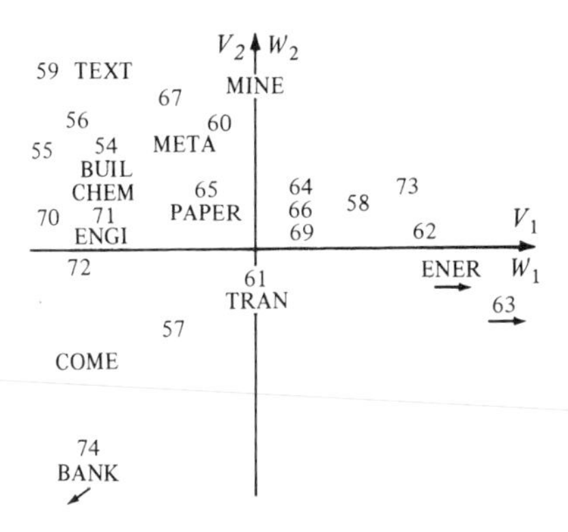

Eigenvalues Less Than 1		Fidelity
i	λ_i	$\dfrac{\lambda_1 + \dots + \lambda_i}{\lambda_1 + \dots + \lambda_{10}}$
1	0.29425	41.42
2	0.44390	61.68
3	0.13306	80.41
4	0.06682	89.88
5	0.03500	94.75
6	0.01788	97.26
7	0.01043	98.73
8	0.00369	99.25
9	0.00330	99.71
10	0.00200	100

Summarized Profile of the Years on the First Four Axes

Y	X	1 $w_1(X)$	COR	RCT	2 $w_2(X)$	COR	RCT
1	74	-727	240	103	-1169	621	543
2	73	461	753	38	97	33	3
3	72	-330	548	20	2	0	0
4	71	-283	250	20	14	1	0
5	70	-426	397	20	46	5	0
6	69	56	24	0	28	6	0
7	67	-193	76	10	323	211	58
8	66	26	9	0	56	41	1
9	65	-154	97	2	82	27	1
10	64	50	10	0	79	26	2
11	63	1194	941	617	-112	8	11
12	62	488	815	27	-11	1	0
13	61	14	0	0	-43	3	1
14	60	-167	73	2	264	180	11
15	59	-671	133	57	1019	306	269
16	58	258	501	7	72	29	1
17	57	-260	311	19	-332	506	64
18	56	-335	253	9	249	139	11
19	55	-436	310	40	200	71	19
20	54	-258	292	7	155	105	5

3			4		
$w_3(X)$	COR	RCT	$w_4(X)$	COR	RCT
-547	136	129	-10	0	01
-89	29	3	124	55	12
98	48	4	154	119	19
354	391	70	-301	284	102
272	162	13	187	76	17
204	344	13	-169	226	17
-51	5	2	38	3	2
260	874	24	0	0	0
298	355	13	6	0	0
254	265	23	45	8	1
-200	45	65	71	3	10
103	36	3	-138	66	10
513	362	108	-653	587	351
38	4	0	58	9	1
-1303	502	476	-374	42	70
141	113	3	-131	99	6
-156	112	15	-11	1	0
151	51	4	441	434	71
221	79	23	549	488	277
114	57	3	244	259	26

Summarized Profile of the Sectors on the First Four Axes

Sector	Y	1 $r_1(Y)$	COR	CTR	2 $r_2(Y)$	COR	CTR
1	ENER	1032	957	711	-91	8	12
2	MNE	15	0	0	353	183	25
3	META	-228	80	10	197	59	15
4	ENGI	-306	626	93	8	0	0
5	BUIL	-268	189	19	142	52	11
6	CHEM	-316	386	15	154	91	7
7	TEXT	-573	134	59	817	272	244
8	PAPER	-96	25	0	55	8	0
9	TRAN	-12	1	0	-34	4	2
10	COMM	-459	194	4	-360	120	5
11	BANK	-880	178	89	-1706	669	680

Sector	Y	3 $r_3(Y)$	COR	CTR	4 $r_4(Y)$	COR	CTR
1	ENER	-204	37	62	49	2	7
2	MINE	-85	11	2	357	188	54
3	META	55	5	1	169	43	23
4	ENGI	114	86	28	145	141	93
5	BUIL	75	15	3	359	335	147
6	CHEM	162	101	0	194	145	25
7	TEXT	-1135	527	510	-397	65	125
8	PAPER	231	143	5	24	1	0
9	TRAN	375	443	214	-414	543	523
10	COMM	213	42	2	-60	3	0
11	BANK	-808	140	163	-67	1	2

A possible interpretation: the first principle factor shows that the energy sector is essential for the study of the profiles of the years, it contrasts with nearly all the other sectors. The second shows that the second variability factor is the contrast between the tertiary sector and traditional industries (mining, textiles, ...). The proximity of 63 and energy or of 74 and banks is not surprising if the profiles Y = energy or Y = banks for the years is looked at.

Exercises 1.4.

E.1. For Table 2 of Exercise 1.1 and Z the number of days lost due to strikes each year:

(a) Calculate $E(Z|Y = \text{energy})$ and $\sigma^2(Z|Y = \text{energy})$. Repeat the same calculation leaving out the year 63. Notice that the mean is not a good description of the number of days lost due to strike in the energy branch when the year 63 is included.

(b) Calculate the χ^2 distance between the "energy" and "banking" sectors.

E.2. *Combination of profiles.* Let i and j be in $\{1, ..., \ell\}$. Two values i and j of a characteristic Y are said to be combined into the value $[i,j]$, if a characteristic $\tilde{Y}$ is defined equal to Y if Y is not equal to i or j and equal to $[i,j]$, if Y equals i or j. Under what conditions are the χ^2 distances for two profiles of X for the characteristic Y the same as for the combined characteristic $\tilde{Y}$?

E.3. *Correspondence analysis and approximation by simple tables.* Associate with $x = (x_i)_{1 \leqslant i \leqslant k}$ the point $\hat{x} = (x_i\sqrt{F_{i\cdot}})_{1 \leqslant i \leqslant k}$ and to $y = (y_j)_{1 \leqslant j \leqslant \ell}$ the point $\hat{y} = (y_j\sqrt{F_{\cdot j}})_{1 \leqslant j \leqslant \ell}$. Let

$$T = \left\{\frac{F_{ij} - F_{i\cdot}F_{\cdot j}}{F_{i\cdot}F_{\cdot j}}\right\}:$$

This is a $k \times \ell$ table (Exercise 1.2.4). Show that the best approximation to T by a simple table of order q is $\Sigma_{i=1}^{q} \sqrt{\lambda_i}\hat{w}_i{}^t\hat{v}_i$ and interpret the results obtained in Exercise 1.2.4 for correspondence analysis.

Bibliographic Notes

For a better understanding of this chapter, more time can be devoted to descriptive statistics. See Hodges, Krech, and Crutchfield (Stat. Lab.), for a clearly written introduction to the subject, based on a collection of data sets. See also Calot and, more difficult, Tukey.

Probabilities on a countable space can also be studied in greater depth, which leads us back to combinatorial analysis. Feller's book (vol. 1, chapters 1-5) contains many interesting developments. See also Hoel, Port and Stone (vol. 1, chapters 1, 2, and 4).

In order to study in greater depth more sophisticated methods of descriptive statistics (data analysis), which we touch on in this chapter by discussing principal components analysis and correspondence analysis, see Lebart and Fenelon, Caillez and Pages, and for more examples, Benzecri.

Data analysis covers these methods. It is only possible if fairly important means of calculations are available and it requires quite a critical approach. It leads quickly to the study of statistical programs and software (BMDP, SPSS, GENSTAT, etc.)

Lastly, certain exercises in Section 1.3 may make you wish to read:

(a) books on the theory of information or codes. Simple discussions of this may be found in Neveu [2], and in Renyi, Topsoe, and Yaglom; more difficult discussions in Ash and Wolfowicz.
(b) books on quantitative genetics: Ewens, Jacquard, and Falconer.

Other exercises in Section 1.3 tackle problems of rank. Bibliographic notes on this will be given in Chapter 4.

Chapter 2
HEADS OR TAILS
QUALITY CONTROL

Objectives

Heads or tails? Zero or one? Are goods produced by a machine defective? ... . Similar questions. By repeating identical and "independent" experiments a certain number of times, we will be able to have an idea of the "probability" of these events.

The game of heads or tails, or quality control, allows us to tackle nearly all probability and statistical questions in this book, without complicated mathematical tools (except the existence of the probability P defined in Section 2.5, which will be assumed for the moment). This chapter, like the previous one, can be understood by a less specialized reader. Moreover, it should clarify the table of contents. Section 2.5 can be left until later; it is an introduction to Volume 2.

2.1. Repetition of *n* Independent Experiments

2.1.1. Independent Experiments

n successive observations are made, taking values in a countable space E. This amounts to observing an element ω of $\Omega = E^n$. For $\omega = (\omega_p)_{1 \leqslant p \leqslant n}$ in E^n, set $X_p(\omega) = x_p$: X_p is the

pth coordinate function, $X_p(\omega)$ is the pth observation. It is assumed, for each observation X_p, that we know the probability $F_p(x)$ of observing x. F_p is a probability on E. Finally, assume that each experiment is independent of the previous ones. This leads to defining on Ω a probability P such that for $1 \leqslant p \leqslant n$ and $(x_1, ..., x_p) \in E^p$,

$$P(X_p = x_p) = F_p(x_p)$$

$$P(X_p = x_p | X_1 = x_1, ..., X_{p-1} = x_{p-1}) = F_p(x_p).$$

From which

$$\begin{aligned} P(X_1 = x_1, ..., X_p = x_p) &\\ &= P(X_p = x_p)P(X_1 = x_1, ..., X_{p-1} = x_{p-1}) \\ &= P(X_1 = x_1)P(X_2 = x_2) \ldots P(X_p = x_p). \end{aligned}$$

In other words, P is the probability on E^n which gives the weight $F_1(x_1) \ldots F_n(x_n)$ to the point $(x_1, ..., x_n)$.

For each p and each function f_p defined on E, integrable for the distribution F_p with mean $F_p(f_p)$, we have

$$P[f_p(X_p)] = \sum_\omega f_p(X_p(\omega))P(\omega) = \sum_{x \in E} f_p(x)F_p(x) = F_p(f_p)$$

$$\begin{aligned} P[f_1(X_1) \ldots f_n(X_n)] &\\ &= \sum_{(x_1,...,x_n) \in E^n} f_1(x_1) \ldots f_n(x_n)F_1(x_1) \ldots F_n(x_n) \\ &= F_1(f_1) \ldots F_n(f_n). \end{aligned}$$

When all the observations have the same distribution F, $(X_1, ..., X_n)$ is an n-sample of F, obtained by repeating n times, independently of the past, an experiment with distribution F.

2.1.2. Distributions on $\mathbb{Z}$

Consider a probability F on $\mathbb{Z}$ which we shall call a distribution. It is therefore a positive sequence $(F(n))_{n \in \mathbb{Z}}$ such that: $\Sigma_{n \in \mathbb{Z}} F(n) = 1$.

Let Ω be a countable set given a probability P and X a function from Ω into $\mathbb{Z}$, with distribution F. For f a function from Ω into $\mathbb{R}$, to say that $f(X)$ has a mean with respect to P signifies that the series $\Sigma_\omega |f(X(\omega))|P(\omega) = \Sigma_n |f(n)|F(n)$ converges. Then the mean of $f(X)$ with respect to P is the mean of f with respect to F,

$$P(f(X)) = \sum_\omega f(\omega)P(\omega) = \sum_n f(n)F(n) = F(f).$$

Thus if the distribution of X for P is known, the existence and the possible value of the mean of $f(X)$ with respect to P are known. In numerous problems, the probability P will be fixed and we shall talk at the same time about several functions taking values in $\mathbb{Z}$ with known distributions. In these cases, we often suppress P, and we talk about the **mean** of $f(X)$, also called the **expectation** of $f(X)$, without referring to P. We denote $P(f(X)) = E_p(f(X))$, or again by $E(f(X))$. We shall talk of the **mean** m and the **variance** σ^2 of the distribution F, these being the sums

$$m = \Sigma\, nF(n) = E(X),$$

$$\sigma^2 = \Sigma(n - m)^2F(n) = \Sigma\, n^2F(n) - m^2 = \sigma^2(X)$$

if these series converge absolutely. The **moment of order** p **of** F is $\Sigma\, n^pF(n)$, if this series converges absolutely.

Examples of Distributions on $\mathbb{Z}$. These are defined on their **support**, the set of points for which the weight is nonzero. They are represented by line charts. (Above each point on the abscissa a line is drawn the height of which measures the weight.)

(a) *The Dirac distribution on a, denoted δ_a:*

$$\delta_a(a) = 1: \quad m = a, \quad \sigma^2 = 0.$$

(b) *The Bernoulli distribution with parameter p, $p \in [0,1]$:*

$$b(\cdot;p) \quad \text{or} \quad b(p)$$

$$b(1;p) = p, \quad b(0;p) = 1 - p = q,$$

$$m = p, \quad \sigma^2 = p(1 - p).$$

If p equals 0 (resp. 1), this is the distribution δ_0 (resp. δ_1).

(c) *The Poisson distribution with parameter* λ, $\lambda > 0$, $p(\cdot,\lambda)$ or $p(\lambda)$:

$$p(n;\lambda) = e^{-\lambda}\frac{\lambda^n}{n!} \qquad (n \geq 0)$$

$$m = \lambda, \qquad \sigma^2 = \lambda.$$

(Only points with probabilities greater than 0.01 are drawn on the line chart.)

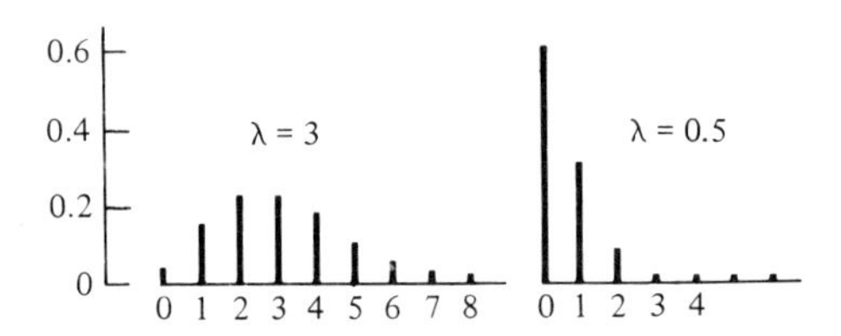

(d) *The geometric distribution with parameter* p, $p \in]0,1[$:

$$F(n) = (1 - p)p^n \quad \text{for} \quad n \geq 0.$$

(e) *The hypergeometric distribution.* Choose at random n objects from N, of which N_1 are of type T_1 and $N - N_1$ of type T_2. If all the $\binom{N}{n}$ possible draws are equally probable, the probability of obtaining k objects of type T_1 is

$$F(k) = \frac{\binom{N_1}{k}\binom{N - N_1}{n - k}}{\binom{N}{n}}.$$

F concentrated on $\{0,1, \dots, n\}$ is a hypergeometric distribution (see Exercise 1.3.9).

(f) *The Uniform Distribution on* A, *a finite subset of* $\mathbb{Z}$. This is the uniform probability on A.

We shall talk of Dirac, Bernoulli, Poisson, ... r.v.'s to indicate their distributions.

The study of distributions on $\mathbb{Z}$ will often be facilitated by the following tool. The function g_F defined on $[-1,+1]$ by $g_F(z) = \Sigma_{n\in N} z^n F(n) = E(z^X)$ is called the generating function of F (or of X, an r.v. with distribution F). When F is defined on $\mathbb{N}$, we define on $[0,\infty[$ **the Laplace transform ϕ_F of F** (or of X):

$$\phi_F(s) = \sum_{n\in N} e^{-sn}F(n) = E(e^{-sX}) = g_F(e^{-s}).$$

The function g_F characterizes the coefficients $(F(n))$ of the whole series of which it is the sum; *therefore g_f (or ϕ_F) characterizes the distribution of F.* Consider two independent r.v.'s X and Y with distributions F and G, taking values in $\mathbb{Z}$. What is the distribution, denoted by $F*G$, of $X+Y$? We have

$$\begin{aligned} P(X+Y=n) &= \sum_{p\in Z} P[(X=p),(Y=n-p)] \\ &= \sum_{p\in Z} F(p)G(n-p) = (F*G)(n). \end{aligned}$$

When F and G are defined on $\mathbb{N}$,

$$\begin{aligned} P(X+Y=n) &= \sum_{p=0}^{n} P[(X=p),(Y=n-p)] \\ &= \sum_{p=0}^{n} F(p)G(n-p) = F*G(n). \end{aligned}$$

This distribution $F*G$ is called the convolution of F and G. It is often easier to characterize $F*G$ by its generating function $g_{F*G} = g_F \cdot g_G$. In fact

$$g_{F*G}(z) = E(z^{X+Y}) = E(z^X z^Y) = g_F(z)\cdot g_G(z).$$

Example. For a Poisson distribution $p(\lambda)$, the generating function is

$$g_\lambda(z) = \sum_{n\geqslant 0} e^{-\lambda} z^n \frac{\lambda^n}{n!} = e^{\lambda(z-1)}.$$

Therefore, $g_\lambda \cdot g_\mu = g_{\lambda+\mu}$ and the sum of two independent Poisson r.v.'s with parameters λ and μ is a Poisson r.v. with parameter $\lambda+\mu$.

Of course, if the distributions F and G have mean m_F and m_G and variances σ_F^2 and σ_G^2 we obtain

$$E(X+Y) = m_{F*G} = m_F + m_G.$$

Since the r.v.'s X and Y are independent, $\Gamma(X,Y)$ is zero, and

$$\sigma^2(X + Y) = \sigma^2_{F*G} = \sigma^2_F + \sigma^2_G.$$

Note. *For r.v.'s X and Y taking values in* $\mathbb{Z}$, *the mean of X plus Y is always the sum of the means, but the variances only add together if the r.v.'s are uncorrelated!*

Recall the n independent experiments $X_1, \ldots, X_n$ of Section 2.1.1, taking values in $\mathbb{Z}$, with distribution F, mean m, and variance σ^2. We often associate with them the sum $S_n = X_1 + \ldots + X_n$, or their **sample mean** $\overline{X} = S_n/n$. What can we say about the distribution F_{S_n} of S_n? $F_{S_2} = F*F$, $F_{S_3} = F*F*F$, ... and $F_{S_n} = F^{*n}$. The mean of S_n is $E(S_n) = nm$ and its variance is $\sigma^2(S_n) = n\sigma^2$. Thus

$$E(\overline{X}) = m, \; \sigma^2(\overline{X}) = \frac{\sigma^2}{n} = \frac{E(S_n - nm)^2}{n^2}.$$

The generating function equals $g_{F*n} = (g_F)^n$.

Exercises 2.1.

E.1. *The Geometric Distribution.* Calculate the mean, variance, and generating function of a geometric distribution with parameter p. What is the entropy of such a distribution (Exercise 1.3.12)? Show that it is a maximum amongst the distributions on $\mathbb{N}$ with the same mean.

E.2. *Generating Function.* Let X be an r.v. taking values in $\mathbb{N}$ and let g be its generating function. Show that g is infinitely differentiable on $]-1,1[$. Show that X has a mean if and only if $\lim_{z\to 1} \uparrow g'(z)$ is finite; in this case $E(X)$ equals this limit. Show that X is square integrable if and only if $\lim_{z\to 1} g''(z)$ exists and is finite, in this case

$$\sigma^2(X) = \lim_{z\to 1} [g''(z) + g'(z) - (g'(z))^2].$$

Generalize: for integer p, X^p is integrable if and only if

$\lim_{z\to 1} g^{(p)}(z) < \infty$; then this limit equals $E[X(X-1)...(X-p+1)]$.

E.3. *Deviation of a Poisson distribution from its mean.* Let N be an r.v. with distribution $p(\lambda)$. Prove that, for all real u, e^{uN} has a mean, and calculate it. Let h be the function $x \longmapsto (1+x)\text{Log}(1+x) - x$; show that h is positive (strictly positive for $x \neq 0$). Show that, for $\xi > 0$:

$$\inf_{u>0} E[e^{u(N-(1+\xi)\lambda)}] = e^{-\lambda h(\xi)} \quad \text{and}$$

$$P[N \geqslant (1+\xi)\lambda] \leqslant e^{-\lambda h(\xi)}.$$

Show in the same way $P[N \leqslant (1-\xi)\lambda] \leqslant e^{-\lambda h(-\xi)}$.

E.4. Let $X_1, ..., X_n$ be independent integer r.v.'s with generating function g and U an r.v. independent of $X_1, ..., X_n$ taking values in $\{1, ..., n\}$ with generating function f. Calculate the generating function of

$$V = \sum_{i=1}^{U} X_i.$$

E.5. A die is thrown twice. Let X, Y be the results of the two throws, assumed independent. Let $S = X + Y$ and let S' be equal to 1 for S odd and equal to 0 for S even. Show that (X,Y), (X,S'), (Y,S') are pairs of independent r.v.'s and that the three variables X, Y, and S' are not independent.

E.6. *Reliability.* An electrical network comprises n components $C_1, ..., C_n$. Its state is described by $x = (x_1, ..., x_n) \in \{0,1\}^n$, where x_i equals 1 or 0 acording to whether component C_i is working or not ($i = 1, ..., n$). What this implies for the network depends on the connections. A function ϕ is given on $\{0,1\}^n$ into $\{0,1\}$ such that $\phi(x)$ equals 1 or 0 according to whether the system works or not.

(a) If the components are placed in series, $\phi(x) = x_1 ... x_n$; if they are placed in parallel, $\phi(x) = 1 - \prod_{i=1}^{n}(1 - x_i)$. Why? Determine ϕ for the following network where the arrows indicate the direction of the current and the numbers the components:

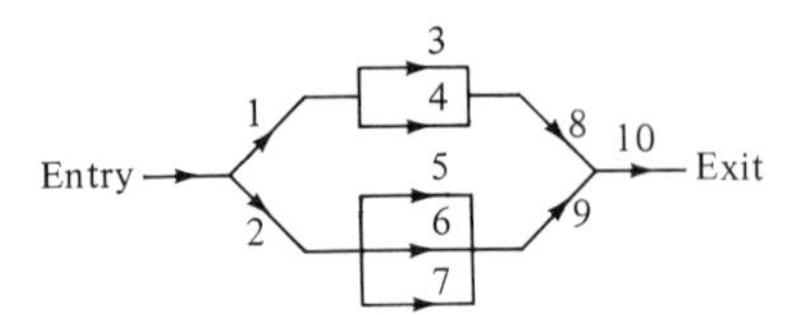

(b) Let $\pi = (p_1, \ldots, p_n) \in]0,1[^n$. Assume that the states of the components $X_1, \ldots, X_n$ are independent Bernoulli r.v.'s with parameters $p_1, \ldots, p_n$. Let $h(\pi)$ be the mean of $\phi(X_1, \ldots, X_n) = \phi(X)$. What is the distribution of $\phi(X)$? Calculate h for the examples in (a). Calculate h for the particular case $p = p_1 = \ldots = p_n$.

(c) For $(x_1, \ldots, x_n) = x \in \{0,1\}^n$ denote by $(1_j,x)$ and $(0_j,x)$ the vector in $\{0,1\}^n$, the components of which are those of x, with the exception of x_j replaced by 1 or 0, respectively. The *utility* of C_j, $u_j(\pi)$, is the mean value of $\phi(1_j,X) - \phi(0_j,X)$. Prove $u_j(\pi) = (\partial h/\partial p_j)(\pi)$. To improve the quality of the network, assume known a cost function: to improve component C_j by changing p_j to $p_j + \Delta_j$ costs

$$\int_{p_j}^{p_j+\Delta_j} \gamma_j(u)du,$$

with γ_j continuous and positive. Show that, at each instant, it is in our interest to improve component C_j for which the ratio $u_j(\pi)/\gamma_j(p_j)$ is the largest.

2.2. A Bernoulli Sample

2.2.1. The Binomial Distribution

An n-sample with a Bernoulli distribution can be constructed by giving $\{0,1\}^n = \Omega$ the following probability P_p. For $(x_1, \ldots, x_n) \in \{0,1\}^n$, set $s_n = x_1 + x_2 + \ldots + x_n$ and

$$P_p(x_1, \ldots, x_n) = P_p(X_1 = x_1, X_2 = x_2, \ldots, X_n = x_n)$$
$$= p^{s_n}(1-p)^{n-s_n}.$$

Denote by E_p the expectation associated with P_p. We will often have in mind playing "heads or tails": X_p equals 1 if the pth throw is heads (or success) and zero if it is tails (failure).

The number of "heads" obtained is $S_n = X_1 + \dots + X_n$; $(S_n = k)$ is the union of the $\binom{n}{k}$ events $(X_1 = x_1, \dots, X_n = x_n)$ with the same probability $p^k(1 - p)^{n-k}$ since S_n equals k. From which

$$P_p(S_n = k) = \binom{n}{k} p^k(1 - p)^{n-k} = b(k;p,n).$$

This distribution of S_n, $b(\cdot;p,n)$ or $b(p,n)$ is called the **binomial distribution with parameter p of order n.** It is the distribution $b(p)^{*n}$. This result is also found by calculating its generating function:

$$(g(z))^n = (p + (1 - p)z)^n = p^n + np^{n-1}(1 - p)z$$
$$+ \dots + \binom{n}{k} p^{n-k}(1 - p)^k z^k + \dots + (1 - p)^n z^n.$$

Assume that S_n is known and is equal to k; for each of the $\binom{n}{k}$ elements $(x_1, \dots, x_n)$ of $\{0,1\}^n$ such that $x_1 + \dots + x_n$ takes the value k, we have

$$P_p(X_1 = x_1, \dots, X_n = x_n | S_n = k) = \frac{1}{\binom{n}{k}}.$$

Thus, *whatever the value of p, the distribution of the sample conditional on $(S_n = k)$ is the uniform distribution on the $\binom{n}{k}$ possible values.*

The line charts shown below represent several binomial distributions. Only the points for which the probability exceeds 0.01 are shown. The mean is shown on the abscissa by a circle.

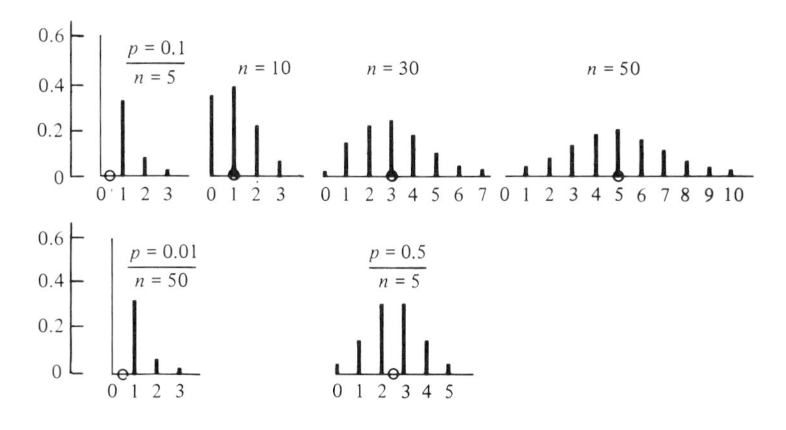

2.2.2. The Weak Law of Large Numbers and Large Deviations

It is common knowledge that if "heads or tails" is played a large number of times, and if at each throw the probability of sucess is p, the frequency of winning will be "without doubt" close to p.

This closeness of S_n/n and p appears first of all through the formulae

$$E_p\left(\frac{S_n}{n}\right) = p; \quad E_p\left(\frac{S_n}{n} - p\right)^2 = \frac{p(1-p)}{n}.$$

The mean of the frequency of winning is p and its variance, the quadratic deviation from p, tends to zero as n tends to ∞. Using Tchebychev's formula, we deduce, for all $\varepsilon > 0$,

$$P_p\left[\left|\frac{S_n}{n} - p\right| \geqslant \varepsilon\right] \leqslant \frac{p(1-p)}{n\varepsilon^2}.$$

This result is known as the **weak law of large numbers (Theorem 2.2.1).** The rate of convergence can be made more precise by writing for $s > 0$:

$$\phi(s) = E_p(\exp(sX_1)) = 1 - p + pe^s;$$

$$E_p(\exp(sS_n)) = [\phi(s)]^n;$$

$$P_p\left(\frac{S_n}{n} \geqslant a\right) = P_p\left[\exp s\left(\frac{S_n}{n} - a\right) \geqslant 1\right)$$

$$\leqslant E_p\left[\exp s\left(\frac{S_n}{n} - a\right)\right] = \left[\phi\left(\frac{s}{n}\right)\right]^n e^{-as};$$

$$P_p\left(\frac{S_n}{n} \geqslant a\right) \leqslant \exp\left[-n\left(\sup_{s>0}(as - \operatorname{Log}\phi(s))\right)\right].$$

Similarly

$$P_p\left(\frac{S_n}{n} \leqslant a\right) \leqslant \exp\left[-n\left(\sup_{s<0}(as - \operatorname{Log}\phi(s))\right)\right].$$

But the function $s \longmapsto as - \operatorname{Log}(1 - p + pe^s)$ is concave; it vanishes at zero and its derivative equals $a - p$. It has a strictly positive maximum $h(a)$ attained for $s > 0$ or for $s < 0$, according to whether we assume $a > p$ or $a < p$.

Theorem 2.2.2 (for large deviations). *Let* $\varepsilon > 0$:

$$P_p\left[\frac{S_n}{n} \geqslant p + \varepsilon\right] \leqslant \exp(-n\, h(p,\varepsilon)),$$

$$P_p\left[\frac{S_n}{n} \leqslant p - \varepsilon\right] \leqslant \exp(-nh(p,-\varepsilon)).$$

where $h(p,\varepsilon)$ *and* $h(p,-\varepsilon)$ *are strictly positive numbers. Setting* $\inf(h(p,\varepsilon),h(p,-\varepsilon)) = H(p,\varepsilon)$ *we then have*

$$P_p\left[\left|\frac{S_n}{n} - p\right| \geqslant \varepsilon\right] \leqslant 2\exp(-nH(p,\varepsilon)).$$

We shall say S_n/n *converges to* p *at an exponential rate.*

2.2.3. Convergence to a Poisson Distribution

A factory produces cloth which sometimes contains faults. The average number of faults per square meter is λ, and the faults are produced at random independently of one another. A square meter of this cloth is cut; let N be the random number of faults in this piece. What is its distribution? Assume that the factory production is regular, i.e., the faults are spread out at random on the cloth. Intuitively speaking, this means that if the observed piece of cloth is cut into n equal pieces the numbers $N_1, ..., N_n$ of observed faults on each of the n pieces will be independent variables with means equal to λ/n. For n large enough, the chances are small of observing two faults on the same piece of material. It therefore seems reasonable to approximate the distribution of the number of observed faults on a piece of material by a Bernoulli distribution with parameter λ/n, and then the distribution of N by the limit distribution of that of a binomial with parameter λ/n and order n, when n tends to ∞.

As shown by the following theorem, this is a Poisson distribution. This theorem is not surprising if the diagrams for $b(0.1;5)$ or for $b(0.01;50)$ are compared with that of $p(0.5)$; or that of $b(0.1;30)$ with $p(3)$.

Theorem 2.2.3. *Let* (p_n) *be a sequence of numbers in* $]0,1[$, *and* $\lambda > 0$ *such that* (np_n) *tends to* λ. *Let* (S_n) *be a sequence of* r.v.'s, S_n *having a binomial distribution with parameter* p_n *and of order* N. *Then*

$$\lim_{n\to\infty} P_{p_n} [S_n = k] = \frac{\lambda^k e^{-\lambda}}{k!}.$$

Proof. Let $q_n = 1 - p_n$. Then

$$P_{p_n}(S_n = 0) = (1 - p_n)^n = \left[1 - \frac{\lambda}{n} + o\left(\frac{\lambda}{n}\right)\right]^n$$

tends to $e^{-\lambda}$ if n tends to ∞;

$$\frac{P_{p_n}(S_n = k+1)}{P_{p_n}(S_n = k)} = \frac{n-k}{k+1}\frac{p_n}{q_n}$$

which tends, for $0 \leqslant k$, to $\lambda/(k+1)$ if n tends to ∞.

Application. *When n is large enough and p is small ($p < 0.1$) we can, approximately, identify the binomial distribution with a Poisson distribution of parameter $\lambda = np$.*

2.2.4. Convergence to a Gaussian Distribution

Consider a sequence (S_n) of r.v.'s, S_n having distribution $b(p,n)$; set $q = 1 - p$. We have studied large deviations of the type $|S_n/n - p| > \varepsilon$. Their probability decreases exponentially. The variance of $\sqrt{n}(S_n/n - p)$ equals pq. It is then natural to be interested in small deviations of the type $|S_n/n - p| > \varepsilon/\sqrt{n}$. We will consider the variable $Z_n = (S_n - np)/\sqrt{npq}$ with mean zero and variance equal to 1. The set I_n of values of Z_n consists of $n + 1$ real numbers at a distance $1/\sqrt{npq}$ apart lying between $-\sqrt{n(p/q)}$, and $\sqrt{n(q/p)}$. We can construct the associated *line chart* by drawing above each $x \in I_n$ a line of height $P[Z_n = x]$. The following theorem gives the asymptotic appearance of this histogram, which is none other than the histogram of $b(p,n)$ with a change of origin and scale.

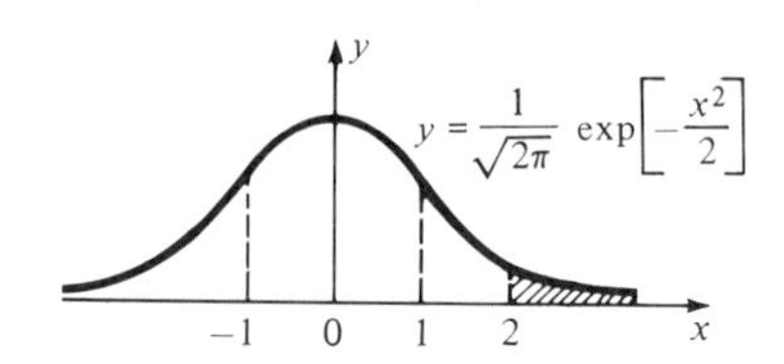

Theorem 2.2.4 (De Moivre-Laplace). *With the preceding notations, for* $-\infty < a < b < \infty$,

$$\sup_{x \in I_n \cap [a,b]} \left| \sqrt{2\pi npq}\, P\left[\frac{S_n - np}{\sqrt{npq}} = x\right] - \exp\left[-\frac{x^2}{2}\right] \right|$$

tends to zero as n tends to ∞.

Proof. Let $x \in I_n$. Set $k = np + x\sqrt{npq}$ and $j = n - k = nq - x\sqrt{npq}$:

$$P[Z_n = x] = P[S_n = k] = \binom{n}{k} p^k q^j.$$

Using *Stirling's formula*,

$$n! = \sqrt{2\pi}\left(\frac{n}{e}\right)^n \sqrt{n}\exp(\theta_n), \quad \text{with} \quad 0 < \theta_n \leqslant \frac{1}{12n};$$

$$P[S_n = k] = \frac{\sqrt{2\pi n}\; n^n e^{-n} e^{\theta_n}}{\sqrt{2\pi k}\; k^k e^{-k} e^{\theta_k} \sqrt{2\pi j}\; j^j e^{-j} e^{\theta_j}}\, p^k q^j$$

$$= \frac{1}{\sqrt{2\pi}} (n/kj)^{1/2} \left(\frac{np}{k}\right)^k \left(\frac{nq}{j}\right)^j e^{\theta}$$

with

$$\theta = \theta_n - \theta_j - \theta_k, \quad |\theta| \leqslant \frac{1}{12}\left(\frac{1}{n} + \frac{1}{j} + \frac{1}{k}\right)$$

$$= \frac{1}{12}\left(\frac{1}{n} + \frac{n}{kj}\right).$$

But $(1/npq)(kj/n)$ tends uniformly to 1 for $x \in [a,b]$ and thus for n large enough we can state

$$|\theta| \leqslant \frac{1}{12}\left(\frac{1}{n} + \frac{2}{pqn}\right).$$

Let us study

$$\text{Log}\left(\frac{np}{k}\right)^k = (x\sqrt{npq} + np)\text{Log}\left[1 - \frac{x\sqrt{npq}}{x\sqrt{npq} + np}\right].$$

But

$$\operatorname{Log}(1+y) = y - \frac{y^2}{2} + \int_0^y \frac{t^2\,dt}{1+t}$$

$$\left|\operatorname{Log}(1+y) - y + \frac{y^2}{2}\right| \leqslant \frac{|y|^3}{3(1-|y|)},$$

$$\left|\operatorname{Log}\left(\frac{np}{k}\right)^k + x\sqrt{npq} + \frac{x^2 npq}{2(np + x\sqrt{npq})}\right| \leqslant \frac{\alpha_n(x)}{n^{3/2}},$$

where $(\alpha_n(x))$ tends uniformly to $(|x|^3/3)(p/q)^{3/2}$ for $x \in [a,b]$;

$$\frac{x^2 npq}{np + x\sqrt{npq}} = \frac{x^2 npq}{np}\left|\frac{1}{1 + x(q/np)^{1/2}}\right|$$

$$= x^2 q\left[1 - x(q/np)^{1/2} + \frac{x^2(q/np)}{1 + x(q/np)^{1/2}}\right],$$

$$\left|\operatorname{Log}\left(\frac{np}{k}\right)^k + x\sqrt{npq} + \frac{x^2 q}{2}\right| \leqslant \frac{\beta_n(x)}{\sqrt{n}},$$

where $(\beta_n(x))$ tends uniformly to $(|x|\sqrt{q})^3/2\sqrt{p}$ for $x \in [a,b]$. Similarly,

$$\left|\operatorname{Log}\left(\frac{nq}{j}\right)^j - x\sqrt{npq} + \frac{x^2 p}{2}\right| \leqslant \frac{\beta'_n(x)}{\sqrt{n}},$$

where $(\beta'_n(x))$ converges uniformly for $x \in [a,b]$,

$$P[S_n = k] = \frac{\gamma_n(x)}{\sqrt{2\pi npq}} \exp\left(-\frac{x^2}{2}\right),$$

where $(\gamma_n(x))$ converges uniformly to 1 for $x \in [a,b]$, which establishes the theorem.

A Study of the Function

$$x \longmapsto \frac{1}{\sqrt{2\pi}} \exp\left(-\frac{x^2}{2}\right).$$

Its graph is the famous *bell-shaped curve*, its points of inflection are ±1 on the abscissa. The area of the plane under the curve equals 1: indeed,

$$\left[\int_{-\infty}^{+\infty}\exp\left(-\frac{x^2}{2}\right)dx\right]^2 = \int_{-\infty}^{+\infty}\int_{-\infty}^{+\infty}\exp\left(-\frac{x^2+y^2}{2}\right)dx\,dy$$

$$= \int_0^{2\pi} d\theta \int_0^{+\infty}\exp\left(-\frac{r^2}{2}\right)r\,dr = 2\pi.$$

$$\frac{1}{\sqrt{2\pi}}\int_{-\infty}^{+\infty}\exp\left(-\frac{x^2}{2}\right)dx = 1.$$

To find out how quickly the curve flattens out, the approximate values of the area below the curve for points on the abscissa greater than 2 or 3 are calculated:

$$\frac{1}{\sqrt{2\pi}}\int_2^{\infty}\exp\left(-\frac{x^2}{2}\right)dx \simeq 0.0228,$$

$$\frac{1}{\sqrt{2\pi}}\int_3^{\infty}\exp\left(-\frac{x^2}{2}\right)dx \simeq 0.0013.$$

Theorem 2.2.5, Central Limit Theorem. *Using the above notations for* $-\infty \leqslant a < b \leqslant +\infty$:

$$\lim_{n\to\infty} P\left[a \leqslant \frac{S_n - np}{\sqrt{npq}} \leqslant b\right] = \frac{1}{\sqrt{2\pi}}\int_a^b \exp\left(-\frac{x^2}{2}\right)dx.$$

Proof. (a) Take, first of all, $-\infty < a < b < \infty$. The DeMoivre-Laplace theorem states that for all $\varepsilon > 0$, there exists an n_0 such that for $n \geqslant n_0$, and $x \in I_n \cap [a,b]$,

$$P[Z_n = x] = \frac{1}{\sqrt{2\pi npq}}[1 + \alpha(x)]\exp\left(-\frac{x^2}{2}\right),$$

$$\text{with } |\alpha(x)| \leqslant \varepsilon;$$

$$P[a \leqslant Z_n \leqslant b]$$
$$= \frac{1}{\sqrt{2\pi npq}} \sum_{x\in[a,b]\cap I_n} [1 + \alpha(x)]\exp\left(-\frac{x^2}{2}\right).$$

However, $[a,b] \cap I_n$ is a partition of $[a,b]$ of width $1/\sqrt{npq}$. By definition of the Riemann integral,

$$\lim_{n\to\infty} \frac{1}{\sqrt{npq}} \sum_{x\in[a,b]\cap I_n} \exp\left(-\frac{x^2}{2}\right) = \int_a^b \exp\left(-\frac{x^2}{2}\right) dx;$$

$$\lim_{n\to\infty} \left| P[a \leqslant Z_n \leqslant b] - \frac{1}{\sqrt{2\pi}} \int_a^b \exp\left(-\frac{x^2}{2}\right) dx \right|$$

$$\leqslant \frac{\varepsilon}{\sqrt{2\pi}} \int_a^b \exp\left(-\frac{x^2}{2}\right) dx.$$

This proves the theorem, for $-\infty < a < b < \infty$, since it is true for all ε.

(b) The above shows that by letting a tend to b, $\lim_{n\to\infty} P[Z_n = b] = 0$. Let $b \in \mathbb{R}$ and λ be a closure point of $(P(Z_n \leqslant b))$; for a sequence of integers $(u(n))$ and $a \leqslant b \leqslant c$:

$$\lambda = \lim_{n\to\infty} P(Z_{u(n)} \leqslant b) \geqslant \lim_{n\to\infty} P(a \leqslant Z_{u(n)} \leqslant b)$$

$$= \frac{1}{\sqrt{2\pi}} \int_a^b \exp\left(-\frac{x^2}{2}\right) dx,$$

$$(1 - \lambda) = \lim_{n\to\infty} P(Z_{u(n)} > b) \geqslant \lim_{n\to\infty} P(b \leqslant Z_{u(n)} \leqslant c)$$

$$= \frac{1}{\sqrt{2\pi}} \int_b^c \exp\left(-\frac{x^2}{2}\right) dx.$$

Therefore, letting a tend to $-\infty$ and c tend to $+\infty$,

$$\lambda \geqslant \frac{1}{\sqrt{2\pi}} \int_{-\infty}^b \exp\left(-\frac{x^2}{2}\right) dx,$$

$$1 - \lambda \geqslant \frac{1}{\sqrt{2\pi}} \int_b^{+\infty} \exp\left(-\frac{x^2}{2}\right) dx.$$

But the sum of the two right hand sides equals 1: thus the above inequalities are equalities.

Consequencces. We thus obtain, for large n, the following very useful approximate values:

$$P[S_n - np \geqslant 2\sqrt{npq}] \simeq P[S_n - np \leqslant -2\sqrt{npq}] \simeq 0.025,$$

$$P[S_n - np \geq 3\sqrt{npq}] \simeq P[S_n - np \leq -3\sqrt{npq}] \simeq 0.0013.$$

These relations will allow a study of the deviation of S_n from its mean.

Note. *The approximation given by the Central Limit Theorem is improved for fairly small n when p is close to* 1/2 (Exercise 2.2.1). *The practice is to accept it for* $np(1 - p) > 18$.

Exercises 2.2.

E.1. Draw the line charts of the distributions $(S_n - np)/\sqrt{npq}$ where S_n is distributed as $b(p,n)$, for $p = 1/2$ and $n = 2,3,5,10$; then for $p = 1/10$ and the same values of n. Notice that they resemble the bell-shaped curve more quickly for $p = 1/2$ than for $p = 1/10$.

E.2. *Weierstrass Approximation Theorem.* For $p \in]0,1[$, consider an r.v. S_n with distribution $b(p,n)$. Show that, for all p and $a > 0$

$$P_p\left[\left|\frac{S_n}{n} - p\right| \geq a\right] \leq \frac{1}{4na^2}.$$

Let u be a continuous function from $[0,1]$ into $\mathbb{R}$. Show that the function $p \longmapsto E_p(u(S_n/n)) = u_n(p)$ extended by $u_n(0) = u(0)$, $u_n(1) = u(1)$ is, for each n, a polynomial. Show that the sequence (u_n) converges uniformly to u.

E.3. *Asymptotic behavior of the hypergeometric distribution.* Consider the hypergeometric distribution defined, for $0 \leq k \leq n$, by

$$F(k) = \frac{\binom{N_1}{k}\binom{N - N_1}{n - k}}{\binom{N}{n}}.$$

Set $N_1/N = \hat{p} = 1 - \hat{q}$. In what follows, assume that N_1 and n are functions of N (and let N tend to ∞).

(b) Prove that if the ratio N_1/N tends to p and N tends to ∞ for n and k fixed, we have

$$F(k) = b(k;p,n)\left[1 + o\left(\frac{1}{N}\right)\right].$$

Interpretation. The hypergeometric distribution corresponds to the number of individuals of type T, if a group of n individuals in a population of size N is chosen at random, when the effective total of type T individuals is N_1. The binomial distribution with parameter $p = N_1/N$ corresponds to n independent choices of individuals in the population with, after each choice, replacement of the chosen individual in order not to change the population. For large N these two procedures are similar.

(b) The variance of the hypergeometric distribution is $n\hat{p}\hat{q}[1 - (\hat{n}-1)/(N-1)]$ (see Exercise 1.3.9). It is the same as that of $b(\hat{p},n)$ to within the correction term $n-1/N-1$. Assume that $\hat{p}$ and $n/N = \hat{r}$ tend, respectively, to p and r if N tends to ∞ $(0 < p < 1,\ 0 < r < 1)$. The analogue of the DeMoivre-Laplace theorem is (prove it), for $-\infty < a < b < \infty$,

$$\sup_{x \in J_N \cap [a,b]} \left| \sqrt{2\pi n\hat{p}\hat{q}(1-\hat{r})}\, F(x\sqrt{n\hat{p}\hat{q}(1-\hat{r})} + n\hat{p} - \exp\left(-\frac{x^2}{2}\right) \right|$$

tends to zero if N tends to ∞, J_N being the set

$$\left\{ \frac{k - n\hat{p}}{\sqrt{n\hat{p}\hat{q}(1-\hat{r})}},\quad k = 0,1,\ \ldots,\ n \right\}.$$

Prove that for $-\infty \leqslant a < b \leqslant \infty$,

$$\sum_{a \leqslant \frac{k-n\hat{p}}{\sqrt{n\hat{p}\hat{q}(1-\hat{r})}} \leqslant b} F(k)$$

tends to

$$\frac{1}{\sqrt{2\pi}} \int_b^a \exp\left(-\frac{x^2}{2}\right) dx,$$

when N tends to ∞.

(c) Assume now, that for $\lambda > 0$, N_1, n, and N tend to ∞ and $n(N_1/N)$ tends to λ. Prove

$$\lim_{N \to \infty} F(k) = \frac{e^{-\lambda}\lambda^k}{k!}.$$

E.4. *Sampling with replacement in a finite population.* In a population of N individuals, r distinct types can be distinguished. Let $N_1, N_2, \ldots, N_r$ be the respective number of individuals of type 1,2, ..., r. A survey is carried out of n individuals: let X_i be the response of the ith individual. Finally, let $Z_j = \Sigma_{i=1}^{r} \mathbf{1}_{(X_i=j)}$ be the number of responses of type j. If the survey is carried out by sampling globally a group of n individuals, this is a case of sampling without replacement studied in Exercise 1.3.9. $(Z_1, \ldots, Z_r)$ has a hypergeometric distribution.

Sampling is now carried out with replacement. For $j = 1, \ldots, r$, set $p_j = N_j/N$. Assume that after each choice the individual questioned is replaced back in the population and that successive choices are independent. The r.v.'s $(X_1, \ldots, X_n)$ are independent and $P(X_i = j) = p_j$, $(i = 1, \ldots, n;\ j = 1, \ldots, r)$. Let $E = \{(i_1, \ldots, i_r) \in \mathbb{N}^r;\ i_1 + \ldots + i_r = n\}$. For $(i_1, \ldots, i_r) \in E$, show

$$P[Z_1 = i_1, \ldots, Z_r = i_r] = \frac{n!}{i_1! \ldots i_r!} p_1^{i_1} \ldots p_r^{i_r}.$$

This distribution concentrated on E is the **multinomial distribution of order n with parameter** $p = (p_1, \ldots, p_r)$. Determine the distribution of Z_1 and the covariance of Z_1 and Z_2. For $(s_1, \ldots, s_r) \in \mathbb{R}^r$, calculate $E(s_1^{Z_1} \ldots s_r^{Z_r})$.

2.3. Estimation

2.3.1. Surveys

Before each advertising campaign or each election, it is now popular to carry out **surveys**, i.e., to choose "at random" individuals from the population and to ask them their views. Assume, for the sake of simplicity, that we are concerned with a referendum or with a question having two answers: "Do you prefer this product or that?".

n people are questioned and we write $X_i = 1$ or 0 according to whether the ith answer is yes or no. If we have made a sufficiently "random" choice of these people (which is not very easy – this is the **sampling** problem, see Exercise 2.4), their n answers follow a hypergeometric distribution. If the people questioned are chosen from a large population, we can consider, which is what we shall do from now on, that we are dealing with a sample from a Bernoulli distribution (Exercise 2.2.3). The survey institute announces the frequency of the "yes" in its enquiry as $\bar{X} = (X_1 + ... + X_n)/n$ and we think that the proportion of "yes" in the entire population, θ, is close to $\bar{X}$. θ is **estimated** by $\bar{X}$.

We observe then an n-sample $(X_1, ..., X_n)$ from a Bernoulli distribution with unknown parameter θ, $\theta \in [0,1]$, which we are trying to estimate. What follows is the description of the ways of doing this, with their advantages and disadvantages. The notations are those of Section 2.2 replacing p by θ.

2.3.2. The Sample (or Empirical) Mean

$$\bar{X} = \frac{X_1 + ... + X_n}{n} = \frac{S_n}{n},$$

the frequency of observed 1's is the *a priori* natural way to evalute θ. This is an **estimator**, i.e., a function of the observations used to estimate θ. What are its good points?

(a) $E_\theta(\bar{X}) = (1/n)E_\theta(S_n) = \theta$: *the average value of* $\bar{X}$ *is* θ, *whatever the value of* θ. We say $\bar{X}$ is **unbiased**. It is the only unbiased estimator which is a function of S_n. Let g be a function from $\{0, ..., n\}$ into $\mathbb{R}$, satisfying

$$E_\theta(g(S_n)) = \sum_{k=0}^{n} g(k) \binom{n}{k} \theta^k (1-\theta)^{n-k} = \theta$$

Then

$$\sum_{k=0}^{n} \left[g(k) - \frac{k}{n}\right] \binom{n}{k} \left[\frac{\theta}{1-\theta}\right]^k = 0.$$

This being true for all $\theta \in [0,1[$, the polynomial $\Sigma_{k=0}^{n}[g(k) - (k/n)]\binom{n}{k}x^k$ is identical to 0 and $g(k)$ equals k/n for all k

$$g(S_n) = \bar{X}.$$

(b) The distribution of the sample $(X_1, ..., X_n)$ has been obtained if the parameter equals θ. The probability of observing $(x_1, ..., x_n)$ is, setting $s_n = x_1 + ... + x_n$,

$$\theta^{s_n}(1-\theta)^{n-s_n}.$$

If, therefore, $(x_1, ..., x_n)$ is observed we can think of attributing to θ the "most likely" value, i.e., the one which maximizes the function $u \longmapsto u^{s_n}(1 - u)^{n-s_n} = L(u,s_n)$. For $s_n = 0$ (resp. $s_n = n$) take $\theta = 0$ (resp. $\theta = 1$). For $s_n = 1, ..., n - 1$, we calculate on $]0,1[$:

$$\frac{\partial}{\partial u} \operatorname{Log} L(u,s_n) = \frac{\partial}{\partial u}(s_n \operatorname{Log} u + (n - s_n)\operatorname{Log}(1 - u))$$

$$= \frac{s_n - nu}{u(1 - u)}.$$

The function $u \longmapsto L(u,S_n)$ is a maximum for $u = S_n/n$. The random variable $L(\theta,S_n)$ is the **likelihood** of the sample if the parameter equals θ. We have

$$L(\bar{X},S_n) = \sup\{L(\theta,S_n);\ \theta \in [0,1]\}.$$

We shall say $\bar{X}$ **is the maximum likelihood estimator.**

(c) When the sample size n increases to $+\infty$, (S_n/n) tends to θ exponentially quickly; we shall say $\bar{\mathbf{X}}$ **is an exponentially consistent estimator.**

(d) The error made in attributing to θ the value $\bar{X}$ can be measured by the **quadratic risk**

$$E_\theta(\bar{X} - \theta)^2 = \frac{\theta(1 - \theta)}{n}.$$

This decreases as $1/n$.

(e) This error can also be measured by the probability of deviating from θ by more than a value a given *a priori*,

$$P_\theta(|\bar{X} - \theta| \geq a) = P_\theta\left(\left|\frac{S_n - n\theta}{\sqrt{n\theta(1 - \theta)}}\right| \geq \frac{a\sqrt{n}}{\sqrt{\theta(1 - \theta)}}\right).$$

Thus from the Central Limit Theorem for $n \geq (4\theta(1-\theta))/a^2$ and n large enough, this probability is less than 0.05. The estimator $\bar{X}$ therefore has many good properties... .

2.3.3. Estimators, Functions of the Sample Mean

Why do we often forget, when we carry out a survey, the list of answers $X_1, ..., X_n$ and why do we summarize this by the sum S_n (or the sample mean $\bar{X} = S_n/n$)? This can be justified by the study of maximum likelihood as above. The probability of observing $(x_1, ..., x_n)$ depends only on the sum $s_n = x_1 + ... + x_n$ (also, therefore, the maximum likelihood estimator). Let us show that the calculation of the quadratic risk also justifies this summary; in other words, that for all functions h from $\{0,1\}^n$ into $\mathbb{R}$, there exists another $\hat{h}$ on $\{0,1, ..., n\}$ such that, for all θ,

$$E_\theta([h(X_1, ..., X_n) - \theta]^2) \geq E_\theta([\hat{h}(S_n) - \theta]^2);$$

i.e., the quadratic risk of $\hat{h}(S_n)$ is smaller than that of $h(X_1, ..., X_n)$.

Proposition 2.3.6. *For an arbitrary function h from $\{0,1\}^n$ into* **R**, *define the function $\hat{h}$ on $\{0,1, ..., n\}$* by

$$\hat{h}(s) = \frac{1}{\binom{n}{s}} \sum_{(x_1, ..., x_n) \in \{0,1\}^n; x_1 + ... + x_n = s} h(x_1, ..., x_n)$$

The estimator $\hat{h}(S_n)$ of θ has a smaller quadratic risk than that of $h(X_1, ..., X_n)$. We say that S_n is sufficient or exhaustive.

Proof. The probability conditional on $(S_n = s)$ is, as we have seen, uniformly distributed on all the possible values of the n-sample. Thus the expectation of $h(X_1, ..., X_n)$ conditional on $S_n = s$ equals $\hat{h}(s)$, for arbitrary θ. The variance of the variable $(h(X_1, ..., X_n) - \theta)$ for this probability conditional on $S_n = s$ is positive:

$$E_\theta[(h(X_1, ..., X_n) - \theta)^2 | S_n = s]$$
$$\geq (E_\theta[h(X_1, ..., X_n) - \theta | S_n = s])^2 = (\hat{h}(s) - \theta)^2.$$

From which

$$E_\theta[(h(X_1, ..., X_n) - \theta)^2 1_{(S_n = s)}]$$
$$\geq (\hat{h}(s) - \theta)^2 P_\theta(S_n = s);$$

and by summing over s,

$$E_\theta(h(X_1, ..., X_n) - \theta)^2 \geqslant E_\theta(\hat{h}(S_n) - \theta)^2.$$

2.3.4. Bayesian Estimators

It therefore seems wise to use the estimator $\bar{X}$, if nothing is known *a priori* about θ. However, suppose that we observe goods produced by one or other of two machines; the first machine does not produce any defective goods and the second produces them on average one in two. Here, θ equals either 0 or 1/2. We try to find out which machine provided the observed sample. Of course, here, as soon as the sample has a defective object ($S_n > 0$) θ will be estimated by 1/2 and not by $\bar{X}$... . *A priori* information on θ changes the problem.

If the two machines produce defective items in proportions θ_1 and θ_2, respectively, and if we know that there is one chance in two that the sample was provided from one machine or the other, we shall look for an estimator $g(S_n)$ which minimizes the *average quadratic risk*

$$\frac{1}{2} E_{\theta_1}[g(S_n) - \theta_1]^2 + \frac{1}{2} E_{\theta_2}[g(S_n) - \theta_2]^2.$$

Thus, suppose that ν is a probability defined on a countable subset D of $[0,1]$. We can try to minimize the average quadratic risk

$$\begin{aligned}
&\sum_{\theta \in D} \nu(\theta) E_\theta[(g(S_n) - \theta)^2] \\
&= \sum_{\theta \in D} \nu(\theta) \left[\sum_{k=0}^{n} (g(k) - \theta)^2 \binom{n}{k} \theta^k (1 - \theta)^{n-k} \right] \\
&= \sum_{k=0}^{n} \binom{n}{k} \left[\sum_{\theta \in D} \nu(\theta)(g(k) - \theta)^2 \theta^k (1 - \theta)^{n-k} \right].
\end{aligned}$$

We therefore take for $g(k)$ the real number which minimizes the function $x \longmapsto \Sigma_{\theta \in D} \nu(\theta)(x - \theta)^2 \theta^k (1 - \theta)^{n-k} = \psi(x)$. The derivative ψ' of ψ is

$$\psi'(x) = 2 \sum_{\theta \in D} \nu(\theta)(x - \theta)\theta^k (1 - \theta)^{n-k}.$$

The case where ν is a Dirac measure on 0 or 1 is excluded, for which we know that θ equals 0 or 1. *The average quadratic risk for the probability* ν, *defined on a countable* *subset D of* [0,1], *called the* **prior distribution** *of* θ, *is a minimum if we choose the* **associated Bayesian estimator** $g(S_n)$ *where g is defined by*

$$g(k) = \frac{\sum_{\theta \in D} \theta^{k+1}(1-\theta)^{n-k}\nu(\theta)}{\sum_{\theta \in D} \theta^{k}(1-\theta)^{n-k}\nu(\theta)}.$$

If ν is a Dirac distribution which charges only the point θ_1, we find (fortunately) $g = \theta_1$. In the preceding example,

$$g(k) = \frac{\theta_1^{k+1}(1-\theta_1)^{n-k} + \theta_2^{k+1}(1-\theta_2)^{n-k}}{\theta_1^{k}(1-\theta_1)^{n-k} + \theta_2^{k}(1-\theta_2)^{n-k}}.$$

Nothing prevents us from taking for ν a positive measure on [0,1] and trying to minimize $\int d\nu(\theta)E_\theta[(g(S_n) - \theta)^2]$ (any reader of this chapter who does not know what a measure is can take a positive continuous function f defined on]0,1[, called the **density** and write $f(\theta)d\theta$ instead of $d\nu(\theta)$). Assume, first of all, that ν is bounded ($\int_0^1 f(\theta)d\theta$ is finite). An identical calculation to the preceding one shows that the maximum is obtained for g defined by

$$g(k) = \frac{\int \theta^{k+1}(1-\theta)^{n-k}d\nu(\theta)}{\int \theta^{k}(1-\theta)^{n-k}d\nu(\theta)}.$$

We shall say that $g(S_n)$ is the **Bayesian estimator for the prior measure** ν. The calculation is easy for the measure $\nu_{\alpha\beta}$, the density of which is $f_{\alpha\beta}$ on [0,1] with $f_{\alpha\beta}(\theta) = \theta^{\alpha-1}(1-\theta)^{\beta-1}$, $\alpha > 0$, $\beta > 0$. Recall the following classical analysis formulas. Define the beta function β by

$$\beta(x,y) = \int_0^1 \theta^{x-1}(1-\theta)^{y-1}d\theta.$$

This equals

$$\frac{\Gamma(x)\Gamma(y)}{\Gamma(x+y)} \quad \text{with} \quad \Gamma(x) = \int_0^\infty e^{-\theta}\theta^{x-1}d\theta;$$

$\Gamma(n+1) = n!$ for $n \in \mathbb{N}$ and $\Gamma(x+1) = x\Gamma(x)$.

The Bayesian estimator is then defined by

$$g(k) = \frac{\int_0^1 \theta^{\alpha+k}(1-\theta)^{n-k+\beta-1}d\theta}{\int_0^1 \theta^{\alpha+k-1}(1-\theta)^{n-k+\beta-1}d\theta}$$

$$= \frac{\Gamma(\alpha+k+1)}{\Gamma(\alpha+\beta+n+1)}\,\frac{\Gamma(\alpha+\beta+n)}{\Gamma(\alpha+k)} = \frac{\alpha+k}{n+\alpha+\beta}.$$

We obtain the estimator $g(S_n) = (a + S_n)(n + \alpha + \beta)$. Its quadratic error is

$$E_\theta([g(S_n) - \theta]^2)$$
$$= \frac{\theta^2[(\alpha+\beta)^2 - n] + \theta[n - 2\alpha(\alpha+\beta)] + \alpha^2}{(n+\alpha+\beta)^2}.$$

Take for example $\alpha + \beta = \sqrt{n}$ and $2\alpha(\alpha + \beta) = n$. Then, $\alpha = \beta = \sqrt{n}/2$. The estimator $(S_n + \sqrt{n}/2)/(n + \sqrt{n})$ has a quadratic risk equal to $1/4(\sqrt{n} + 1)^2$, no matter what the value θ of the parameter.

For the density $\theta \longmapsto 1/\theta(1 - \theta)$, the integral of which is infinite on $[0,1]$, we calculate directly:

$$\int_0^1 \frac{d\theta}{\theta(1-\theta)} E_\theta(g(S_n) - \theta)^2$$

$$= \int_0^1 \frac{d\theta}{\theta(1-\theta)} \sum_{k=0}^{n} (g(k) - \theta)^2\theta^k(1-\theta)^{n-k}\binom{n}{k}.$$

It can be seen that the average risk is infinite, except for $g(0) = 0$ and $g(n) = 1$. Assuming this, we obtain as above that, for $0 < k < n$,

$$g(k) = \frac{\int_0^1 \theta^{k+1}(1-\theta)^{n-k}\dfrac{d\theta}{\theta(1-\theta)}}{\int_0^1 \theta^{k}(1-\theta)^{n-k}\dfrac{d\theta}{\theta(1-\theta)}} = \frac{k}{n}; \quad g(S_n) = \frac{S_n}{n}$$

is the sample mean.

2.3.5. Admissible Estimators

Section 2.3.4 leads us to greater flexibility. The sample mean is perhaps not always the best estimator. The extreme case is, of course, that where θ is certainly equal to θ_0, the best estimator is then θ_0 and its quadratic risk vanishes! However, this estimator would give a quadratic risk of $(\theta - \theta_0)^2$ for another θ.

Let us examine, as a function of θ, the variations in the quadratic risk of the various estimators we have come across. We see that none of these curves always lies below the others. For example, setting $\theta = u + 1/2$;

$$\frac{\theta(1-\theta)}{n} \leqslant \frac{1}{4(\sqrt{n}+1)^2}$$

is equivalent to $(1/4 - u^2) \leqslant n/(\sqrt{n} + 1)^2_4$ or

$$\left[\theta - \frac{1}{2}\right]^2 \geqslant \frac{1}{4} - \frac{n}{4(\sqrt{n}+1)^2} = \frac{2\sqrt{n}+1}{4(\sqrt{n}+1)^2}.$$

The estimator with constant risk has a quadratic risk less than that of the sample mean on an interval centered on 1/2, the length of which decreases as $\sqrt{2}n^{-1/4}$ for large n, which is important for small n. None of these estimators is better than all the others for arbitrary θ. After what follows, this is still true in the family of all estimators.

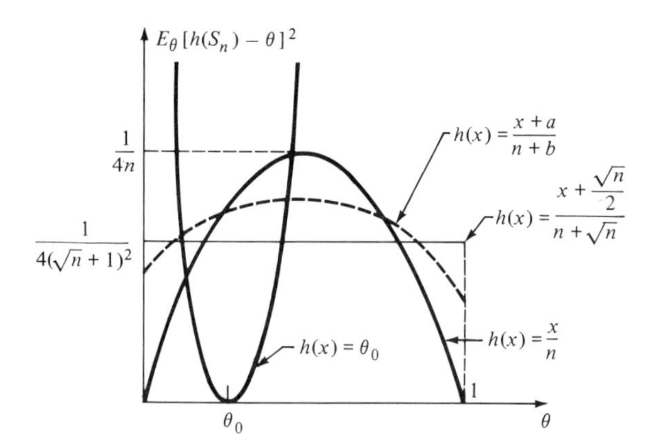

Definition 2.3.7. An estimator $h(S_n)$ is **better** than $g(S_n)$ if, for all θ, the quadratic risk of $h(S_n)$ is less than that of $g(S_n)$ and if there exists a θ where this inequality is strict. The estimator of $g(S_n)$ is **admissible** if no better estimator exists.

Let $h(S_n)$ be a better estimator than $g(S_n)$. Its average risk for a given measure ν on $]0,1[$ is at most equal to that of $g(S_n)$. Thus $g(S_n)$ cannot be the unique Bayesian estimator associated with ν.

Proposition 2.3.8. *All the unique Bayesian estimators of* θ *are admissible; in particular, the sample mean. All the estimators of the form* $(S_n + a)/(n + b)$ *for* $a > 0$ *and* $b > 0$ *and the constant estimator equal to* $\theta_1 \in [0,1]$ *are admissible.*

On the other hand, the estimator $2\overline{X}$ or an estimator equal to a constant > 1 are not admissible. It is quite clear that the notion of admissibility is sometimes associated with uninteresting estimators, such as the constant estimator. However it is rare for interesting estimators to be nonadmissible!

2.3.6. Minimax

A prudent statistician, when concerned with a problem in which the stakes are high (risk of fatal accident in using a medicine ...) tries to minimize the maximum risk. An estimator with this property is called **minimax**. Of course, a minimax estimator is admissible.

Here the estimator

$$\frac{S_n + \sqrt{n}/2}{n + \sqrt{n}}$$

with constant risk is minimax. Since it is admissible, all other estimators have a maximum risk at least equal to the constant risk.

Exercises 2.3

E.1. *An n-sample from a Bernoulli distribution.* We observe X_1 and X_2 independent Bernoulli r.v.'s with parameter $\theta \in]0,1[$. Show that an unbiased estimator of $\theta/(1 - \theta)$ which is a function of X_1 and X_2 cannot be found.

E.2. With the aid of an n-sample from a Bernoulli distribution try to estimate the variance $\theta(1 - \theta)$. $\overline{X}$ being the sample mean, we propose the estimator $T = \overline{X}(1 - \overline{X})$. Verify that this is a biased estimator and give an unbiased estimator of $\theta(1 - \theta)$ which is a multiple of T.

E.3. *An n-sample from a Poisson distribution.* Consider $(X_1, ..., X_n)$ an n-sample from a Poisson distribution with an unknown parameter θ, $\theta \in [0,\infty[$; for example (Section 2.2.3), X_i is the number of observed faults on the ith piece of cloth examined.

(1) We are trying to estimate θ.

(a) Is the estimator $\overline{X} = (X_1 + ... + X_n)/n$ unbiased? What is its quadratic risk?

(b) Show that, for all estimators $h(X_1, ..., X_n)$, there exists an estimator $\hat{h}(X_1 + ... + X_n)$ of θ with a lower quadratic risk: $\overline{X}$ is a sufficient statistic.

(c) What is the maximum likelihood estimator of θ?

(2) We now try to estimate $e^{-\ell\theta}$, the probability that in ℓ future experiments, we always observe 0.

(a) We propose the estimator $e^{-\ell\overline{X}}$. Is it unbiased? What is its quadratic risk?

(b) Determine a function g from $\mathbb{N}$ into $\mathbb{R}$ such that $g(n\overline{X})$ is an unbiased estimator. What happens if you take $n = 1$, $\ell = 2$?

E.4. *An n-sample from a uniform distribution.* We observe an n-sample $X_1, ..., X_n$ from the uniform distribution on $\{1,2,3, ..., \theta\}$ which, for integer θ, gives the weight $1/\theta$ to each of the integers $1, ..., \theta$. We try to estimate θ. For $(x_1, ..., x_n) \in (\mathbb{N} \setminus \{0\})^n$ with probability of $\{(X_1, ..., X_n) = (x_1, ..., x_n)\}$ is therefore $(1/\theta^n)1_{[\sup_{1 \leq i \leq n} x_i \leq \theta]}$.

(a) What is the maximum likelihood estimator M of θ? What is its distribution and that of $(X_1, ..., X_n)$ conditional on $M = k$?

(b) Consider the estimator

$$2\,\frac{X_1 + ... + X_n}{n} - 1$$

of θ. Is it unbiased? What is its quadratic risk?

(c) Study the Bayesian estimator for the measure ν defined by $\nu(\{\theta\}) = \theta^n/2^\theta$ for all θ.

(d) Show that for every estimator $g(X_1, \ldots, X_n)$ of θ, there exists an estimator $\hat{g}(\sup_{1\leqslant i\leqslant n} X_i)$ the quadratic risk of which is lower. It is also said that $\sup_{1\leqslant i\leqslant n} X_i$ is sufficient.

E.5. *A zero sum game. Vocabulary.* Two players, A and B, play (neither knows what the other is going to play). If A chooses to play $x \in X$ and B to play $y \in Y$ then A wins $\phi(x,y)$ and B loses $\phi(x,y)$. The function ϕ of $X \times Y$ into $\mathbb{R}$ is the pay-off matrix. A negative gain is, of course, a loss.

– A plays x: his minimum gain is $\alpha(x) = \inf_y \phi(x,y)$. If he is prudent he will look for a "*maximin*" *strategy*, i.e., he will play $\overline{x}$, if it exists, such that

$$\alpha = \alpha(\overline{x}) = \inf_y \phi(\overline{x},y) = \sup_x \inf_y \phi(x,y).$$

– B plays y: the maximum gain of A is $\beta(y) = \sup_x \phi(x,y)$. If he too is prudent, he will try to find a *minimax strategy* $\overline{y}$, if it exists, such that $\beta = \beta(\overline{y}) = \sup_x \phi(x,\overline{y}) = \inf_y \sup_x \phi(x,y)$. In general, $\alpha \leqslant \beta$; for $\alpha = \beta$ we say that the game has a *value* $V = \alpha = \beta$, and $(\overline{x},\overline{y})$ is the *equilibrium* position

$$\phi(x,\overline{y}) \leqslant \phi(\overline{x},\overline{y}) \leqslant \phi(\overline{x},y) \quad \text{for all } (x,y).$$

A game x for A is *better* than x' if for all $y \in Y$, $\phi(x,y) \geqslant \phi(x',y)$, the inequality being strict for at least one y. The game x for A is *admissible* if no game is better than it. The same definitions are obtained for B by using the gain $\psi(x,y) = -\phi(x,y)$. An element $x \in X$ (resp. $y \in Y$) is called either a *game* or *strategy* of A (resp. B).

Determine the admissible strategies for A and for B, $\overline{x}$, $\overline{y}$, α, and β in the following cases.

(a) $X = \{0,1\} = Y$, ϕ given by the following tables:

$y \backslash x$	0	1
0	5	3
1	−1	−2

$y \backslash x$	0	1
0	5	3
1	−1	2

$y \backslash x$	0	1
0	5	1
1	−1	2

(b) Player B chooses a point in $\mathbb{Z}$ and A tries to get close to it. B's choice is assumed random: the set Y is the set of distributions on $\mathbb{Z}$. If B plays $y \in \mathbb{Z}$ and if A plays $x \in \mathbb{Z}$, the gain of A is $-(x - y)^2$. If B plays in a random manner with distribution F and if A plays x, his mean gain is $-\Sigma_{y \in \mathbf{Z}}(x - y)^2 F(y)$. What is the best choice of $x \in \mathbb{Z}$ (*the Bayesian game associated with the strategy F of B*) for player A, when B plays F with finite mean and variance?

(c) Let $\overline{x}$ be an admissible game for A, the gain $y \longmapsto \phi(\overline{x},y)$ of which is constant. Show that $\overline{x}$ is maximin.

(d) Here is a payoff matrix of type

$y \backslash x$	0	1
0	a	b
1	c	d

with $\begin{cases} a > b, & c < d \\ a > c, & b < d \end{cases}$.

(Why do we make a hypothesis of this kind in order that the strategies of A and B are not evident?)

d_1 – Player A uses a roulette to decide whether he plays 0 or 1. His strategy is the choice of the probability p of playing 1. His average gain is then

$$\begin{cases} \phi(p,0) = bp + a(1 - p) & \text{if } B \text{ plays } 0, \\ \phi(p,1) = dp + c(1 - p) & \text{if } B \text{ plays } 1. \end{cases}$$

Represent graphically the curves $p \longmapsto \phi(p,0)$, $p \longmapsto \phi(p,1)$, $p \longmapsto \alpha(p)$. Calculate $\overline{p}$, $\alpha(\overline{p})$, $\phi(\overline{p},0)$, $\phi(\overline{p},1)$.

d_2 – Player B also makes a draw to decide what to play. He chooses $q \in [0,1]$, the frequency of 1, or the probability of playing 1. If A has chosen p and B has chosen q, A wins on average $\phi(p,q) = q\phi(p,1) + (1 - q)\phi(p,0)$. Here $X = Y = [0,1]$. Show that

$$\inf_{q \in [0,1]} \phi(p,q) = \inf\{\phi(p,0), \phi(p,1)\},$$

$$\sup_{p \in [0,1]} \phi(p,q) = \sup\{\phi(0,q), \phi(1,q)\}.$$

Calculate $\overline{p}$, $\overline{q}$, α, and β, and show that $(\overline{p},\overline{q})$ is an equilibrium position for the game.

(e) Consider the payoff matrix below.

y \ x	0	1
1	1	3
2	2	1.5
3	2.5	3.5
4	4	5
5	3	1

e_1 – Player A, as in d_1, makes a choice of the probability of playing 1. His average gain, if Y plays j ($j = 1, ..., 5$) is $\phi(p,j)$. Draw the 5 curves $p \longmapsto \phi(p,j)$. Determine graphically, then calculate, the admissible strategies for B, the maximin strategy $\overline{p}$ for A, and $\phi(\overline{p},j)$ for $j = 1, ..., 5$.

e_2 – Here $X = \{0,1\}$ but the player B's choice is an r.v. taking values in $\{1,2,3,4,5\}$ with distribution q; q_j is the weight of j ($j = 1, ..., 5$). Thus Y is the set of these distributions. Let $\psi(0,q)$ and $\psi(1,q)$ be the average gains of A when he plays 0 or 1, and $M(q)$ the point in the plane with coordinates $(\psi(0,q), \psi(1,q))$. Draw the set of points $M(q)$. What are B's admissible strategies? Determine graphically, then calculate, an admissible strategy of constant loss for B. Give a minimax strategy for B and the value of the minimax gain. If ϕ, instead of being A's gain, is B's gain, what are his admissible strategies? Give a maximin strategy.

2.4. Tests, Confidence Intervals for a Bernoulli Sample, and Quality Control

2.4.1. Quality Control

A machine produces some objects. If it functions normally, there is a proportion $\theta \leq \theta_0$ of defective objects. The quality controller takes samples of the production in order to control this proportion, i.e., to reveal a possible fault in the machine in the course of production. He is concerned with process

control. The problem is not to become worried too quickly for the slightest overstepping of θ_0, by the observed proportion in the sample, but to worry enough if this proportion is too large.

Another quality controller examines the production on leaving the factory. If he is a client, he does not want to accept the stock unless θ is less than a number θ_0 which he sets himself *a priori*. The quality controller must answer the question "$\theta \leqslant \theta_0$?" This is customer control. We are concerned with problems of **testing**.

The controller may also be asked to give an interval in which θ lies , with a small risk of error fixed in advance. This is an estimation problem, where the estimation of θ itself is replaced by the estimation of the two bounds of a **confidence interval**.

In all these problems, the observation of all objects is impossible (too long, too costly; in the production of jars of jam, for example, it would simply destroy all the objects examined ...). What must be done, therefore, is to take a sample of n objects and make a decision on the population.

2.4.2. Upper Control Limits

In the problem of process quality control, where the greatest proportion of defective goods is θ_0, we will only be worried when the observed proportion $\bar{X}$ in the n-sample is too large. Let $\alpha > 0$ and M_α be the smallest integer such that:

$$P_{\theta_0}[S_n \leqslant M_\alpha] = \sum_{k=0}^{M_\alpha} \binom{n}{k} \theta_0^k(1 - \theta_0)^{n-k} \geqslant 1 - \alpha.$$

We also have (this will be verified in Section 2.4.3) $P_\theta(S_n \leqslant M_\alpha) \geqslant 1 - \alpha$, when production is normal. Thus if $S_n > M_\alpha$ is observed, we see an event with probability $\leqslant \alpha$ if the production is normal; when α is "small" there is reason to be worried. The choice of α depends on the problem. The smaller it is, the fewer unnecessary adjustments will be made, but the more we will allow an abnormal production to build up. Often the quality controller will use the **upper control limit** $L_c = M_{0.001}$ and the **upper surveillance limit** $L_s = M_{0.025}$. From time to time, he takes a sample of n items produced by the machine. If the proportion of defective goods exceeds L_s

he steps up the surveillance (the next sample taken will be larger, for example); if it exceeds L_c he calls in the repair engineer. These values of L_s and L_c are tabulated for various values of θ_0. When n is large enough, we can use the Central Limit Theorem approximation:

$$L_c \simeq \theta_0 + 3(\theta_0(1 - \theta_0)/n)^{1/2},$$

$$L_s \simeq \theta_0 + 2(\theta_0(1 - \theta_0)/n)^{1/2}.$$

We have obtained **tests** in order to answer the question "$\theta \leqslant \theta_0$" or "$\theta > \theta_0$"? or to **test** "$\theta \leqslant \theta_0$" against "$\theta > \theta_0$". We decide to reject the hypothesis "$\theta \leqslant \theta_0$" for $X > L_s$. An error can be made with probability $P_\theta(X > L_s) \leqslant 0.05$ if $\theta \leqslant \theta_0$. We say that $(X > L_s)$ is the **rejection region** of the test and that its *level* is 0.05 (if the **rejection region** $(X > L_c)$ is chosen, the level becomes 0.001). Of course we could always also decide "$\theta = \theta_0$" and the level of such a test is zero ... . However, the other type of error must be taken into account, the error which consists of deciding "$\theta \leqslant \theta_0$" when "$\theta > \theta_0$" is true: its probability is $P_\theta(X \leqslant L_s)$ for the first test, and 1 for the test which always decides "$\theta \leqslant \theta_0$".

2.4.3. A Study of the Function $\theta \longmapsto P_\theta(S_n \leqslant a)$ for $a = 1, ..., n - 1$.

Consider the distribution of S_n:

$$P_\theta(S_n = k) = f_\theta(k) = \binom{n}{k} \theta^k(1 - \theta)^{n-k}$$

$$= \binom{n}{k} \exp\left[k \operatorname{Log} \frac{\theta}{1 - \theta} + n \operatorname{Log}(1 - \theta)\right].$$

For $\theta_0 < \theta_1$,

$$\frac{f_{\theta_1}(k)}{f_{\theta_0}(k)} = \exp\left[k \operatorname{Log} \frac{\theta_1}{\theta_0} \frac{1 - \theta_0}{1 - \theta_1} + n \operatorname{Log} \frac{1 - \theta_1}{1 - \theta_0}\right].$$

Thus, the sequence

$$\left\{\frac{f_{\theta_1}(k)}{f_{\theta_0}(k)}\right\}_{0 \leqslant k \leqslant n}$$

is increasing. Let $C = f_{\theta_1}(a)/f_{\theta_0}(a)$ and $\alpha = P_{\theta_0}(S_n \leqslant a) = \Sigma_{k \leqslant a} f_{\theta_0}(k)$. For $1 \leqslant k \leqslant n-1$, $(f_{\theta_1}(k) - Cf_{\theta_0}(k))(1_{(k \leqslant a)} - \alpha)$ is negative, strictly so for $k \neq a$ and

$$\sum_{k=0}^{a} (f_{\theta_1}(k) - Cf_{\theta_0}(k))(1_{(k \leqslant a)} - \alpha) = \sum_{k=0}^{a} f_{\theta_1}(k) - \alpha < 0.$$

Proposition 2.4.9. *For $\theta_0 < \theta_1$, the sequence*

$$\left\{\frac{f_{\theta_1}(k)}{f_{\theta_0}(k)}\right\}_{k=0,\ldots,n}.$$

is increasing, and for all $a = 1, \ldots, n-1$, the function $\theta \longmapsto P_\theta(S_n \leqslant a)$ decreases strictly from 1 to 0.

Consequences. For θ and α in $]0,1[$, call the numbers $q_\alpha^+(\theta)$ and $q_\alpha^-(\theta)$ such that

$$P_\theta(\bar{X} > q_\alpha^+(\theta)) \leqslant \alpha < P_\theta(\bar{X} \geqslant q_\alpha^+(\theta))$$

and

$$P_\theta(\bar{X} < q_\alpha^-(\theta)) \leqslant \alpha < P_\theta(\bar{X} \leqslant q_\alpha^-(\theta)),$$

quantiles of $\bar{X}$ of order α. Take for q^+ (resp. q^-) the smallest (resp. the largest) possible quantile of order α. The preceding proposition shows that, n being fixed, the quantiles are monotone functions of θ: q_α^+ increases and q_α^- decreases. For $x \in]0,1[$ a multiple of $1/n$, we can find two numbers θ_α^+ and θ_α^- (which depend on α and on n) such that

$$P_\theta(\bar{X} > x) \begin{cases} \leqslant \alpha & \text{for } \theta \leqslant \theta_\alpha^-(x); \\ > \alpha & \text{for } \theta > \theta_\alpha^-(x) \end{cases}$$

$$P_\theta(\bar{X} < x) \begin{cases} \leqslant \alpha & \text{for } \theta \geqslant \theta_\alpha^+(x); \\ > \alpha & \text{for } \theta < \theta_\alpha^+(x) \end{cases}.$$

Thus

$$\begin{cases} x \geqslant q_\alpha^+(\theta) \text{ is equivalent to } \theta \leqslant \theta_\alpha^-(x) \\ \qquad \text{and } P_\theta(\theta < \theta_\alpha^-(\bar{X})) \leqslant \alpha < P_\theta(\theta \leqslant \theta_\alpha^-(\bar{X})) \\ x \leqslant q_\alpha^-(\theta) \text{ is equivalent to } \theta \geqslant \theta_\alpha^+(x) \\ \qquad \text{and } P_\theta(\theta > \theta_\alpha^+(\bar{X})) \leqslant \alpha < P_\theta(\theta \geqslant \theta_\alpha^+(\bar{X})). \end{cases}$$

The safety limits L_s and L_c used in Section 2.3.2 are, respectively, $q_{0.025}^+$ and $q_{0.001}^+$. Nomograms representing the functions θ_α^+ and θ_α^- for the usual values of α and various values of n appear in statistical tables. We are dealing with curves which are a continuous approximation of a discontinuous situation.

For n large the central limit theorem approximation can be used and the number ϕ_α introduced such that

$$\frac{1}{\sqrt{2\pi}} \int_{\phi_\alpha}^{\infty} \exp\left(-\frac{x^2}{2}\right) dx = \frac{1}{\sqrt{2\pi}} \int_{-\infty}^{-\phi_\alpha} \exp\left(-\frac{x^2}{2}\right) dx = \alpha.$$

For n large,

$$q_\alpha^+(\theta) \simeq \theta + \phi_\alpha(\theta(1-\theta)/n)^{1/2},$$

$$q_\alpha^-(\theta) \simeq \theta - \phi_\alpha(\theta(1-\theta)/n)^{1/2}.$$

Statistical tables give to 4 decimal places the values of ϕ_α for the usual values of α. For example, the values which we have already used,

$$\phi_{0.025} = 1.9600 \simeq 2 \quad \text{and} \quad \phi_{0.001} = 3.0902 \simeq 3.$$

2.4.4. Various Tests for a Bernoulli Sample

(a) Reconsider the quality control test, where "$\theta = \theta_0$" is tested against "$\theta > \theta_0$" or "$\theta \leqslant \theta_0$" is tested against "$\theta > \theta_0$".

For a **level** $\alpha \in]0,1[$ given in advance, we could give the following decision rule: decide to accept "$\theta = \theta_0$" (or "$\theta \leqslant \theta_0$") when we observe $X \leqslant q_\alpha^+(\theta_0)$ and reject this hypothesis otherwise. We define thus a **test** of "$\theta = \theta_0$" (or $\theta \leqslant \theta_0$) **against** $\theta > \theta_0$ which satisfies the following properties:

– *Its* **level** *is* α, i.e., *for* $\theta \leqslant \theta_0$, *we decide to reject the hypothesis* $\theta > \theta_0$ *only with a probability less than* α.

– *It is* **unbiased**, i.e., *for* $\theta > \theta_0$ *we have a probability greater than* α *of rejecting* "$\theta \leqslant \theta_0$".

(b) The preceeding test could have been used for the quality control of the reception of stock by the manufacturer, who fixes *a priori* a small probability α of rejecting a batch of stock (deciding that θ is too large) when it was of good quality ($\theta \leqslant \theta_0$).

If the inspection is carried out by a customer, the opposite point of view holds: He fixes *a priori* a low probability $\beta \in]0,1[$ of accepting the stock (deciding that θ is small enough), when it is unacceptable ($\theta \geqslant \theta_1$). He will then use the following test of "$\theta \geqslant \theta_1$" against "$\theta < \theta_1$": he decides to accept "θ_1" when he observes $X \geqslant q_\beta^-(\theta_1)$. This test satisfies analogous properties to the preceding test:

– Its level is β, i.e., for $\theta \geqslant \theta_1$, an error is made only with probability less than β;

– it is unbiased, i.e., for $\theta < \theta_1$, the correct decision is taken with probability greater than β.

(c) A test can also be for testing **the validity of a hypothesis.**

A Genetic Example. To test his laws (see Exercises 1.3.6 and 1.3.7), Mendel crosses two pure types of beans, one colored, the other colorless. According to his model each parent has two genes associated with "colored" C or "colorless" I; an offspring "chooses" at random a gene from each parent. Colored is the dominant gene. If a bean has two different genes, it is colored. Hence the diagrams for two parents, one

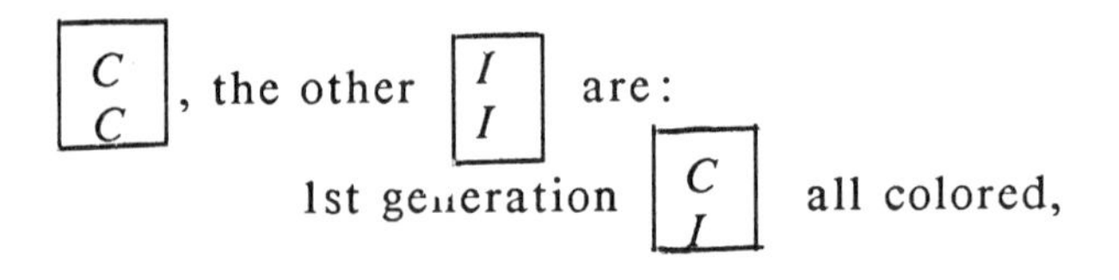

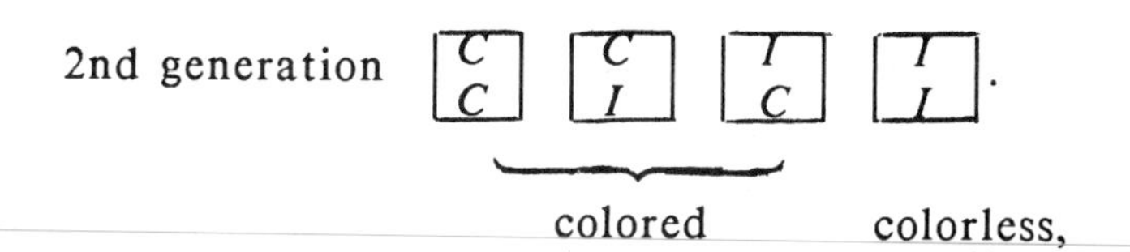

Thus the probability that a bean in the second generation is colored is, if the model is correct, 3/4. n second generation beans are observed and we denote $X_i = 1$ if the ith bean is colored. If Mendel is correct, we are dealing with a sample from a binomial distribution with proportion 3/4. We are concerned with testing a hypothesis of type "$\theta = \theta_0$" against "$\theta \neq \theta_0$". The problem is symmetrical. We could therefore, being given α, use the following method. Decide "$\theta = \theta_0$" if $\bar{X}$ is observed in the interval $[q^-_{\alpha/2}(\theta_0), q^+_{\alpha/2}(\theta_0)]$ (supposing $\alpha < 1$). For such a test, if "$\theta = \theta_0$" is true, the probability of error is less than α.

2.4.5. Confidence Intervals

For an n-sample from a Bernoulli distribution with parameter θ, sample mean $\bar{X}$, Tchebychev's inequality always implies

$$P_\theta[|\bar{X} - \theta| \geqslant \varepsilon] \leqslant \frac{\sigma^2_\theta(\bar{X})}{\varepsilon^2} = \frac{\theta(1-\theta)}{n\varepsilon^2} \leqslant \frac{1}{4n\varepsilon^2}.$$

Thus for all $\alpha > 0$ and all θ

$$P_\theta\left[|\bar{X} - \theta| \geqslant \sqrt{1/4n\alpha}\right] \leqslant \alpha.$$

If we state that "θ is in the interval $]\bar{X} - \sqrt{1/4n\alpha}, \bar{X} + \sqrt{1/4n\alpha}[$" the probability of being mistaken is less than α, no matter what the true value of θ. $[\bar{X} - \sqrt{1/4n\alpha}, \bar{X} + \sqrt{1/4n\alpha}]$ is a **confidence interval, with confidence level $1 - \alpha$.**

We can, thanks to the tables of functions θ^+_α and θ^-_α (which depend on n and α) defined in Section 4.3, obtain a more precise result: *the* **two-sided** *confidence interval* $[\theta^-_{\alpha/2}(\bar{X}), \theta^+_{\alpha/2}(\bar{X})]$ *is a confidence interval with confidence level* $1 - \alpha$, for $0 < \alpha < 1/2$.

The above are the upper and lower bounds used in surveys. However, a quality controller is interested only in an **upper confidence bound** of level $1 - \alpha$, below which he can state that θ will be found, with probability $1 - \alpha$. Thus we define **one-sided** *confidence intervals* $[0, \theta_\alpha^+(X)]$ and $[\theta_\alpha^-(X), 1]$: *these are confidence intervals with confidence level* $1 - \alpha$, i.e., the probability of being in error by stating θ is in one of these intervals is less than α. For example, for $\bar{X} = 0.7$ and $n = 8$, we obtain, thanks to the tables, the confidence intervals [0; 0.95] and [0.30; 1] at the 0.975 level and [0.30; 0.95] at the 0.95 level.

For large n, we can use the normal approximation

$$P_\theta[|\bar{X} - \theta| \geqslant \phi_\alpha\sqrt{\theta(1 - \theta)/n}] \simeq 2\alpha.$$

This means that by accepting a probability of error of 2α, we can consider that $\bar{X} - \theta$ is of the order of $1/\sqrt{n}$ and $[\bar{X}(1 - \bar{X}) - \theta(1 - \theta)]$ is of the same order. We can therefore neglect the error of order $1/n$ made by replacing $\sqrt{\theta(1 - \theta)/n}$ by the estimated value $(\bar{X}(1 - \bar{X})/n)^{1/2}$ and use the following confidence intervals of level $1 - \alpha$

$$[0, \bar{X} + \phi_\alpha(\bar{X}(1 - \bar{X})/n)^{1/2}; \quad [\bar{X} - \phi_\alpha(\bar{X}(1 - \bar{X})/n)^{1/2}, 1]$$

$$[\bar{X} - \phi_{\alpha/2}(\bar{X}(1 - \bar{X})/n)^{1/2}; \quad \bar{X} + \phi_{\alpha/2}(\bar{X}(1 - \bar{X})/n)^{1/2}].$$

Exercises 2.4

Random Numbers. Choosing numbers at random is not straightforward. What is a sequence of 0's and 1's taken at random? Try to choose at random a sequence of 100 figures equal to 0 or 1, writing them as you like; then play at heads or tails (heads = 1, tails = 0); then open (at random) a telephone directory and take the second to last figure of the numbers (even = 0, odd = 1). If your numbers are really chosen at random, you have a 100-sample of a $b(\frac{1}{2})$ distribution.

(a) Assuming that we are dealing with a 100-sample from a $b(p)$ distribution, test "$p = \frac{1}{2}$" against "$p \neq \frac{1}{2}$" at the 0.05 level.

(b) Carry out the same test on each of the following 50-samples: $(X_{2n})_{1 \leqslant n \leqslant 50}$; $(X_{2n-1})_{1 \leqslant n \leqslant 50}$; or $(Y_n)_{1 \leqslant n \leqslant 50}$ by setting $Y_n = 1$ if $X_{2n-1} + X_{2n}$ is even and $Y_n = 0$

otherwise.

(c) Testing the hypothesis "we are dealing with an n-sample from a distribution" is more difficult. The sequence 0...0 1...1 (50 times 0 then 50 times 1) will be accepted by test (a); the sequence 001100110... will be accepted for tests (a) and (b). However, they are obviously not random (refer to Exercise 1.3.10). Think up a test based on this idea for $n = 7$ and apply it to your first 7 figures.

2.5. Observations of Indeterminate Duration

2.5.1. Observations of Random Duration

A sequence of experiments is carried out taking values in a finite set E. If the number of experiments is indeterminate, the set of possible results is $\bigcup_{n \geq 1} E^n$, the set of finite sequences. Of course no one ever carries out an infinite experiment! However it is often advantageous to introduce the space $\Omega = E^{\mathbb{N}}$ of all infinite sequences of elements of E. Each result of n experiments is then considered as the first n components of an element $\omega = (x_p)$ of Ω; denoting by X_n the nth coordinate function from Ω into E, which associates with ω its nth component, the variable X_n appears as the "nth experiment."

After n experiments, events $\{(X_1, ..., X_n) \in \Gamma\}$ for $\Gamma \subset E^n$ can be observed. With such an event we associate the subset of Ω, $\{\omega; (X_1(\omega), ..., X_n(\omega)) \in \Gamma\}$ denoted in the same way as the event. Let $\mathcal{A}_n$ be the set of these subsets of Ω, called the *events prior to n*.

The experimenter or player can decide to stop at a random time ν. The decision to stop at time n will depend on results he has obtained from the first n games, the player cannot predict the result of future games! The random time ν at which he will stop is thus such that $\{\nu = n\}$ is an event prior to n.

We call a **finite stopping time** ν, a function from Ω into $\mathbb{N}$ for which, for arbitrary $n \in \mathbb{N}$ we have $\{\nu = n\} = \{\omega; \nu(\omega) = n\} \in \mathcal{A}_n$. If we stop the observations at the time $\nu = n$, the observable events are of the form $\{\nu = n\} \cap \{(X_1, ..., X_n) \in \Gamma_n\}$ where $\Gamma_n \subset E^n$. An event prior to ν will be of the form $\bigcup_n \{\nu = n\} \cap \{(X_1, ..., X_n) \in \Gamma_n\}$ with $\Gamma_n \subset E^n$ for all n. Let $\mathcal{A}_\nu$ be the set of all these subsets: the traces of $\mathcal{A}_n$ and of $\mathcal{A}_\nu$ on $\{\nu = n\}$

are the same. An element of $\mathcal{A}_\nu$ is called an *event prior to* ν.

The event $\{\nu = n\}$ is an element of $\mathcal{A}_n$. Thus for $(i_1, ..., i_n) \in E^n$, it contains $\{X_1 = i_1, ..., X_n = i_n\}$ or it does not intersect it. Let $J_n \subset E^n$ such that

$$\{\nu = n\} = \bigcup_{(i_1,...,i_n)\in J_n} \{X_1 = i_1, ..., X_n = i_n\}.$$

Designate by X_ν the function which maps ω to $X_{\nu(\omega)}(\omega)$. The observation stopped at the instant ν, $(X_1, ..., X_\nu)$, takes values in $H = \bigcup_n J_n$. An element of $\mathcal{A}_\nu$ is the union of events $\{(X_1, ..., X_\nu) = j\}$, for $j \in H$.

2.5.2. Observation of Unlimited Duration of a Sample with Distribution F Defined on Finite E

With the preceding framework, it is natural to define for all n and $(i_1, ..., i_n)$ the probability $P[X_1 = i_1, X_2 = i_2, ..., X_n = i_n] = F(i_1) \ ... \ F(i_n)$. We shall assume here that there exists a probability defined on a set $\mathcal{A}$ of subsets of Ω, called events, containing $\mathcal{A}_n$ and satisfying the following properties.

– for all n and $(i_1, ..., i_n) \in E^n$, $P(X_1 = i_1, ..., X_n = i_n) = F(i_1) \ ... \ F(i_n)$.

– for all $A \in \mathcal{A}$, $P(A)$ is a number between 0 and 1.

– for $A \in \mathcal{A}$ and $B \in \mathcal{A}$ with $A \subset B$, $P(A) \leqslant P(B)$.

– if (A_n) is a sequence of elements of $\mathcal{A}$, ${}_nA_n$ is in A and $P({}_nA_n) \leqslant \Sigma_n P(A_n)$.

– if A is in $\mathcal{A}$, its complement A^c is in $\mathcal{A}$ and $P(A^c) = 1 - P(A)$.

We shall return in Chapter 3 to these properties of a probability, and in Volume 2 to its existence.

Certain events will have zero probability. This is the case, for example, for $\{X_n = i$ for all $n\}$ when $F(i) \neq 1$. It is inconvenient to use a space Ω which is too large. We consider that an event of zero probability never occurs, or rather, *almost never*. An *almost sure* (a.s.) *property* is true except on an event of zero probability.

We shall say that the sequence (X_n) is a sequence of independent r.v.'s with distribution F. All the calculations made in Sections 2.1 and 2.2 remain valid. Here, however, we can vary n, randomly stopping the experiments, and obtain asymptotic theorems.

Let ν be the finite stopping time studied in Section 2.5.1. For $j = (i_1, ..., i_n) \in H$, the probability of observing $\{(X_1, ..., X_\nu) = j\}$ is $P[(X_1, ..., X_\nu) = j] = F(i_1) ... F(i_n)$. If ϕ is a bounded function from H into $\mathbb{R}$, the expectation of $\phi(X_1, ..., X_\nu)$ is thus

$$E(\phi(X_1, ..., X_\nu)) = \sum_n \bigcup_{(i_1,...,i_n) \in J_n} \phi(i_1, ..., i_n)F(i_1)...F(i_n).$$

2.5.3. The Strong Law of Large Numbers

The notations are as above: $E = \{0,1\}$, F is a Bernoulli distribution with parameter θ, and P_θ is the associated probability on (Ω, A).

Theorem 2.5.10. *For a sequence of independent Bernoulli variables with parameter* θ, *and for* $S_n = X_1 + ... + X_n$,

$$P_\theta\left[\omega;\ \lim_{n\to\infty} \frac{S_n(\omega)}{n} = \theta\right] = 1.$$

We say that S_n/n *converges almost surely to* θ.

Proof. *Let* $\omega \in \Omega$; $S_n(\omega)/n$ does not tend to θ if, and only if, there exists a $k \in \mathbb{N}$, such that for all n, there exists $n' \geq n$ satisfying

$$\left|\frac{S_{n'}(\omega)}{n'} - \theta\right| \geq \frac{1}{k}.$$

But, according to Section 2.2.2,

$$P_\theta\left[\bigcup_{n' \geq n} \left\{\left|\frac{S_{n'}(\omega)}{n'} - \theta\right| \geq \frac{1}{k}\right\}\right] \leq \sum_{n' \geq n} 2e^{-n'H(\theta,1/k)}$$

$$= \frac{2\exp(-nH(\theta,\frac{1}{k}))}{1 - \exp(-H(\theta,\frac{1}{k}))}.$$

The event $A_k = \{$for all n there exists $n' \geq n$ such that $\{|(S_{n'}/n - \theta| \geq 1/k\}$, included in $\bigcup_{n' \geq n}\{|(S_{n'}(\omega)/n') - \theta| \geq 1/k\}$, has a smaller probability; this being true for all n, it has zero probability. Let N be the set of ω's such that $S_n(\omega)/n$ does not tend to θ:

$$N = \bigcup_{k \in \mathbf{N}} A_k \quad \text{and} \quad P_\theta(N) \leqslant \sum_{k \in \mathbf{N}} P_\theta(A_k) = 0.$$

Corollary 2.5.11. *If, in the preceding model, the parameter* θ *has an unknown value in* $[0,1]$, *the law of large numbers indicates that the sample mean is an estimator which converges, almost surely, to the parameter* θ: *it is a strongly consistent estimator of* θ.

2.5.4. Random Walks

The fluctuations in a game are often described by a sequence $(Y_n)_{n \geqslant 1}$ of independent r.v.'s equal to +1 or −1 (+1 for a success, −1 for a failure). The number of points won after n games is thus $S_n = Y_1 + \dots + Y_n$. Assume $P(Y_n = +1) = p = 1 - P(Y_n = -1)$. For $X_n = (1 + Y_n)/2$, the distribution of X_n is $b(p)$. Thus $S_n = 2(X_1 + \dots + X_n) - n$, and (S_n/n) converges a.s. to $2p - 1 = E(Y_1)$. For $p > 1/2$, (S_n) tends to $+\infty$; for $p < 1/2$, (S_n) tends to $-\infty$.

A player will stop playing at a time n when he is ruined or when he judges his gain to be sufficient. Thus for $S_n \leqslant -a$ or for $S_n \geqslant b$, with $a > 0$ (his fortune) and $b > 0$ (the gain with which he is satisfied), the random time at which he will stop is thus

$$\nu_{a,b} = \inf\{n;\ S_n \leqslant -a, \text{ or } S_n \geqslant b\}$$

(the infimum of an empty set of $\mathbf{N}$ being $+\infty$).

This is not a finite stopping time. It equals, for example, $+\infty$ for the sequence which equals alternately +1 and −1 and $a = b = 2$. However, it is an *almost surely finite time* (the game will almost surely have an end). For $p \neq 1/2$, this results from the fact that $(|S_n|)$ tends a.s. to ∞. For $p = 1/2$ this results from the following more general lemma. We will say further that $\nu_{a,b}$ is a "*stopping time*" since, for all n

$$\{\nu_{a,b} = n\} = \bigcap_{p=1}^{n-1} \{S_p \in]-a,b[\} \cap \{S_n \notin]-a,b[\}$$

is an event prior to n.

Lemma 2.5.12. *Let* (Z_n) *be a sequence of independent* *identically distributed* r.v.'s, *taking values in a finite subset E of $\mathbb{R}$, not identically zero. Let* $S_n = Z_1 + \ldots + Z_n$, $-a < 0 < b$ *and* $\nu_{a,b} = \inf\{n;\ S_n \geq b$ or $S_n \leq -a\}$. *Then* $\nu_{a,b}$ *is almost surely a finite stopping time. A number* $\delta < 1$ *exists such that* $P[\nu_{a,b} \geq n]$, *after a certain point, is majorized by* δ^n ; $\nu_{a,b}$ *has moments of all orders.*

Proof. The problem remains the same when (Z_n) is replaced by $(-Z_n)$. We can thus assume that there exists a $c > 0$, $P(Z_n > c) = \eta > 0$. Let k be an integer $\geq (b+a)/c$:

$$P[S_k > kc] \geq P[Z_1 > c\ ,\ \ldots,\ Z_k > c] = \eta^k = 1 - \delta_1.$$

Thus

$$P[(S_k \leq b+a), (S_{2k} - S_k \leq b+a), \ldots, (S_{nk} - S_{(n-1)k} \leq b+a)]$$
$$\leq (1 - \eta^k)^n = \delta_1^n.$$

The event $\{\nu_{a,b} \geq nk\}$ contains $\bigcap_{p=1}^{n-1}(S_{(p+1)k} - S_{pk} \leq b + a)$. From this, $P[\nu_{a,b} \geq nk] \leq \delta_1^n$; $P[\nu_{a,b} = \infty] \leq P[\nu_{a,b} \geq nk] \leq \delta_1^n$. This being true for all n, $P[\nu_{a,b} = \infty]$ is zero and the r.v. $\nu_{a,b}$ is a.s. finite. For $p \geq 1$, we have

$$E(\nu_{a,b}^p) \geq \sum_{n \in \mathbf{N}} ((n + 1)k)^p P[nk < \nu_{a,b} \leq (n + 1)k]$$
$$\leq \sum_{n \in \mathbf{N}} ((n + 1)k)^p \delta_1^n < \infty.$$

Denote by δ_2 the kth root of δ_1 and $[\cdot]$ the integer part. Once n is greater than k we have

$$P(\nu_{a,b} \geq n) \leq P(\nu_{a,b} \geq [n/k]k) \leq \delta_1^{[n/k]} \leq \delta_2^{n-k}.$$

From which the lemma follows, by taking $\delta_2 < \delta < 1$.

2.5.5. Sequential Statistics for a Bernoulli Distribution

Independent observations (X_n) with distribution $b(\theta)$ are taken. Of course we want to stop after a finite number of observations, but this number can depend on the observed results: we can stop at a finite stopping time ν. We have seen in Section 2.5.1:

$$\{\nu = n\} = \bigcup_{(i_1,\ldots,i_n)\in J_n} \{X_1 = i_1, \ldots, X_n = i_n\}$$

for $J_n \subset \{0,1\}^n$. For $J \in H = \bigcup_n J_n$, $j = (i_1, \ldots, i_n)$, and $s(j) = i_1 + \ldots + i_n$, we have, for all θ,

$$P_\theta[(X_1, \ldots, X_\nu) = j] = \theta^{s(j)}(1 - \theta)^{n-s(j)}\begin{bmatrix} n \\ s(j) \end{bmatrix}.$$

The value of θ for which the observation is most probable is that which maximizes this probability: it is $s(j)/n$. Thus the **maximum likelihood estimator** of θ after the observation $(X_1, \ldots, X_\nu)$ is again the sample mean $S_\nu/\nu = \bar{X}_\nu$ (setting $S_\nu(\omega) = X_1(\omega) + \ldots + X_{\nu(\omega)}(\omega)$).

We call the **likelihood** of the experiment $(X_1, \ldots, X_\nu)$, the variable defined on Ω,

$$L_\nu(\theta) = \theta^{S_\nu}(1 - \theta)^{\nu - S_\nu}\begin{bmatrix} \nu \\ S_\nu \end{bmatrix}.$$

Let ϕ be a positive function from H into $\mathbb{R}$ and θ_0, θ_1 two values of the parameter. We have

$$E_{\theta_1}[\phi(X_1, \ldots, X_\nu)] = \sum_n \sum_{j\in J_n} \begin{bmatrix} n \\ s(j) \end{bmatrix} \theta_1^{s(j)}(1 - \theta_1)^{n-s(j)}\phi(j)$$

$$= \sum_n \sum_{j\in J_n} \begin{bmatrix} n \\ s(j) \end{bmatrix} \theta_0^{s(j)}(1 - \theta_0)^{n-s(j)}$$

$$\times \left[\phi(j)\frac{\begin{bmatrix} n \\ s(j) \end{bmatrix} \theta_1^{s(j)}(1 - \theta_1)^{n-s(j)}}{\begin{bmatrix} n \\ s(j) \end{bmatrix} \theta_0^{s(j)}(1 - \theta_0)^{n-s(j)}}\right]$$

$$= E_{\theta_0}\left[\frac{L_\nu(\theta_1)}{L_\nu(\theta_0)}\,\phi(X_1, \ldots, X_\nu)\right].$$

2.5.6. The Choice of the Length of an Experiment

Recall the problem of quality control described in Section 2.4.4 where $\theta_0 < \theta_1$ are the proportions fixed by the manufacturer and by the customer. A quality controller who wishes to satisfy both will try to accept a stock such as $\theta \geq \theta_1$ only with a probability $\leq \beta$ and to reject a stock such as $\theta \leq \theta_0$ only with a probability ω. If the sample size is fixed, we have seen how to satisfy one or the other of these demands, but not both. It is possible, though, to choose a random duration ν of the experiment. The likelihood ratio

$$\frac{L_\nu(\theta_1)}{L_\nu(\theta_0)} = \exp\left[S_\nu \mathrm{Log}\,\frac{\theta_1}{\theta_0} + (\nu - S_\nu)\,\mathrm{Log}\,\frac{1-\theta_1}{1-\theta_0}\right]$$

is, for ν fixed, an increasing function of S_ν. Thus for all M, there exists A such that

$$\{S_\nu \geq M\} = \left\{\frac{L_\nu(\theta_1)}{L_\nu(\theta_0)} \geq A\right\}.$$

If this set is chosen as the rejection region, the errors are then, for $\theta = \theta_1$ or $\theta = \theta_0$,

$$P_{\theta_1}\left[\frac{L_\nu(\theta_1)}{L_\nu(\theta_0)} < A\right] = E_{\theta_0}\left[\frac{L_\nu(\theta_1)}{L_\nu(\theta_0)} 1_{\left\{\frac{L_\nu(\theta_1)}{L_\nu(\theta_0)} < A\right\}}\right] \leq A$$

$$P_{\theta_0}\left[\frac{L_\nu(\theta_1)}{L_\nu(\theta_0)} \geq A\right] = E_{\theta_1}\left[\frac{L_\nu(\theta_0)}{L_\nu(\theta_1)} 1_{\left\{\frac{L_\nu(\theta_0)}{L_\nu(\theta_1)} \leq 1/A\right\}}\right] \leq \frac{1}{A}.$$

Setting

$$Z_p = X_p \mathrm{Log}\,\frac{\theta_1}{\theta_0} + (1 - X_p)\mathrm{Log}\,\frac{1-\theta_1}{1-\theta_0}\ (p = 1, \ldots, n),$$

we have

$$\mathrm{Log}\,\frac{L_n(\theta_1)}{L_n(\theta_0)} = Z_1 + \ldots + Z_n.$$

The random time

$$\nu = \inf\left\{ n: \frac{L_n(\theta_1)}{L_n(\theta_0)} \leqslant \frac{1}{\alpha} \text{ or } \frac{L_n(\theta_1)}{L_n(\theta_0)} \leqslant \beta \right\}$$

is equal to $\inf\{n; Z_1 + \ldots + Z_n \notin]\text{Log } \beta, \text{Log } (1/\alpha)[\}$, for $0 < \alpha < 1$, and $0 < \beta < 1$, this is an almost surely finite stopping time (Section 2.5.4). We can thus decide to stop the experiment at this time ν. Then

$$P_{\theta_1}\left[\frac{L_\nu(\theta_1)}{L_\nu(\theta_0)} \leqslant \beta\right] \leqslant \beta, \quad P_{\theta_0}\left[\frac{L_\nu(\theta_1)}{L_\nu(\theta_0)} \geqslant \frac{1}{\alpha}\right] \leqslant \alpha.$$

Thus, given α and β, we obtain a **sequential likelihood ratio test** for quality control: *the stock is examined successively. At time n the choice will lie among three decisions:*

- decide to continue examine the stock, for $Z_1 + \ldots + Z_n \in]\text{Log } \beta, \text{Log } 1/\alpha[$;
- decide to reject the stock, for $Z_1 + \ldots + Z_n \geqslant \text{Log } (1/\alpha)$,
- decide to accept the stock, for $Z_1 + \ldots + Z_n \leqslant \text{Log } \beta$.

The random time of the end of the experiments is an integrable stopping time for arbitrary θ. For $\theta \leqslant \theta_0$, the probability of rejecting the stock is less than α and, for $\theta \geqslant \theta_1$, the probability of accepting it is less than β.

Exercises 2.5.

E.1. *The negative binomial distribution.* Within the framework of Section 2.5.3, with $0 < p < 1$, we are given $m \geqslant 1$ and consider the random time $\nu = \inf\{n \geqslant 1; S_n = m\}$. What is the distribution of $\nu - m$? Determine its generating function, mean, and variance.

E.2. *A sequential test.* We are trying to detect a fault amongst objects produced by a machine. Let θ be the probability that an object is faulty ($0 \leqslant \theta \leqslant 1$). We test the hypothesis "$\theta = 0$" against "$\theta > 0$". The notation is that of Section 2.5.5. Let

$$\nu = \begin{cases} \inf\{n; X_n = 1\} & \text{on } \bigcup_n \{X_n = 1\} \\ \infty & \text{on } \bigcap_n \{X_n = 0\} \end{cases}.$$

(a) The following sequential test is suggested. Sampling is continued until the time $\inf(\nu,n)$ and we decide "$\theta > 0$" for $\nu \leqslant n$ and "$\theta = 0$" otherwise. Show that $\inf(\nu,n)$ is a finite stopping time. Determine, as a function of θ, the average duration of sampling $E_\theta(\inf(\nu,n))$ and the power of the test $P_\theta(\nu \leqslant n)$, for $\theta > 0$. What is the probability of error when θ is zero?

(b) We consider another test, of which the duration of sampling is a stopping time ν' less than or equal to n and such that the probability of error, for $\theta = 0$, is zero. Show that the test studied in (a) is at least as economical, i.e., that ν' is greater than ν.

E.3. *Gambler's ruin.* For the sequence (Y_n) studied in Section 2.5.4:

(a) Verify that $E((q/p)^{S_n}) = 1$ and $E(S_n) = n(2p - 1)$. Prove that, for $k < n$ and $\Gamma \in \mathcal{A}_k$,

$$E\left[\left(\frac{q}{p}\right)^{S_n} 1_\Gamma\right] = E\left[\left(\frac{q}{p}\right)^{S_k} 1_\Gamma\right];$$

$$E[(S_n - n(2p - 1))1_\Gamma] = E[(S_k - k(2p - 1))1_\Gamma].$$

Let ν be a stopping time less than n. Prove

$$E\left[\left(\frac{q}{p}\right)^{S_\nu}\right] = 1, \qquad E[S_\nu] = (2p - 1)E(\nu).$$

(b) a and b being strictly positive integers and $\nu_{a,b}$ the stopping time studied in Section 2.5.4, prove

$$E\left[\left(\frac{q}{p}\right)^{S_{\nu_{a,b}}}\right] = 1, \quad E(S_{\nu_{a,b}}) = (2p - 1)E(\nu_{a,b}).$$

(Notice that $\inf(\nu_{a,b},n)$ is a stopping time. Deduce from this the probability that the gambler finishes up ruined, $P(S_{\nu_{a,b}} = -a)$, and, for $p \neq 1/2$, the mean value of the duration of the game $E(\nu_{a,b})$.

(c) For $p = 1/2$, prove $E(S_\nu^2) = E(\nu)E(Y_1^2)$ when ν is an integer n, then when ν is a stopping time less than n, then for $\nu = \nu_{a,b}$. Calculate $E(\nu_{a,b})$.

E.4. *Prediction.* Let F be a distribution defined on $\mathbb{Z}$. We call the *mode* of F one of the values k, such that $F(\{k\}) = \sup_{h \in \mathbb{Z}} F(\{h\})$.

If a distribution F defined on $\mathbb{Z}$ is known, and if Y is an unobservable r.v. with distribution F, Y is often predicted by a mode of F (a similar principle to that of maximum likelihood (Section 2.3.2)).

(a) Calculate the modes of $b(\theta,n)$. In which case is the mode unique?

(b) Assume now that Y, unobservable, follows a $b(\theta,n)$ distribution, θ unknown, $\theta \in [0,1]$. We can observe an r.v. X with distribution $b(\theta,m)$ independent of Y. Having observed X, we want to predict Y. Calculate and compare the following three predictions:

(1) $\hat{Y}_1$ is the mode of $b(\hat{\theta}_m(X),n)$, $\hat{\theta}_m(X)$ the maximum likelihood estimator of θ when X has been observed.

(2) If $L(\theta,X)$ and $L(\theta,Y)$ are the respective likelihoods of X and Y, $\hat{Y}_2$ is defined by

$$\sup_\theta L(\theta,X)L(\theta,\hat{Y}_2) = \sup_y \sup_\theta L(\theta,X)L(\theta,y).$$

(3) $\hat{Y}_3$ is defined, when X equals x, as a y which maximizes the probability (independent of θ) of $\{Y = y\}$ conditional on $\{X + Y = x + y\}$.

(c) We are given, at each instant n, two observations U_n and V_n with independent binomial distributions $b(\theta)$, thus an observation taking values in $\{0,1\}^2$. Assume that the variables (U_n,V_n) are independent. For the r.v.'s τ and ν taking values in $\mathbb{N}$, denote $X = U_1 + \dots + U_\tau$ and $Y = V_1 + \dots + V_\nu$ and try to predict Y with the aid of X. Generalize procedures (1) and (2) described in (b) and compare the four predictions of Y in the two following situations:

(α) $\nu = 14$, $\tau = 21$: we observe $X = 3$

(β) $\nu = 14$, $\tau = \inf\{m; U_1 + \dots + U_m = 3\}$

and we observe $\tau = 21$ (and thus $X = 3$).

Bibliographic Notes

Feller (vol. 1) contains, in an elementary form, all the probabilistic aspects of the game of heads or tails and many interesting applications to random walks (samples of a distribution taking values in $\{-1, 1\}$). All the probability books cover this in depth (e.g., Hodges and Lehmann, Renyi, Metivier, Reeb, and Fuchs.

The statistical part features in general books which will be mentioned later. A simple approach can be found in books on quality control.

Grant and Leavenworth is a good introduction to statistical quality control and contains many tables for use in sampling inspection plans.

Exercise 2.1.6 is an introduction to reliability inspired by Barlow and Proschan (Chapter 1). Exercise 2.2.5 is an introduction to the theory of games. Ventsel's very elementary text can be read with interest; also Vajda's book. If you want to go further in this theory, which is very useful in economics, you can read, for example, Ekeland.

Chapter 3
PROBABILISTIC VOCABULARY OF MEASURE THEORY

INVENTORY OF THE MOST USEFUL TOOLS

Objectives

We are led to use the tools of measure theory to describe a random phenomenon. In this section we do not claim to give a course on measure theory and integration, which is indispensible.

All we are doing here is establishing a dictionary translating the geometric intuition of the measure of surfaces into a probabilistic intuition and also giving details of certain ideas in current usage in statistics and probability such as the manipulation of distributions. In order to arrive there by the shortest possible route, we shall give the essential technical "tricks" which make measure theory proofs work. This rather nasty section can be omitted, and consulted when reference is made to it.

3.1. Probabilistic Models

3.1.1. Intuitive Foundation

Let us study a random phenomenon. In fact, various realizations can be observed. Often a complete description does not interest us and it is sufficient to consider whether certain particular "events" take place. For example, if we observe a shot at a target, each realization of this phenomenon can be described by many characteristics, such as the trajectory and speed of the bullet, the type of gun, the position of the firer... ; but we are interested only in the events which describe the bullet's final position, "the bullet scores 100," "the bullet misses the target"... .

If Ω is the set of all realizations of a random phenomenon, to each event A corresponds the subset $\hat{A}$ of Ω of realizations which result in event A. Let us develop this correspondence between sets and events. Ω corresponds to the certain event, the empty set 0 corresponds to the impossible event. If we can observe two events A and B, the same applies to the event "A or B" and the event "A and B." If $\hat{A}$ and $\hat{B}$ are the subsets of Ω associated with A and B, then $\hat{A} \cup \hat{B}$ is associated with "A or B" and $\hat{A} \cap \hat{B}$ is associated with "A and B". We can also observe the event "not A" i.e., "A does not take place" with which the complement of $\hat{A}$, $\hat{A}^c$ is associated. Thus, to the set of events associated with a random phenomenon corresponds, intuitively, a family of subsets of the space of realizations, which is closed under union, intersection, and complementation, and which contains the empty set.

To describe the preceding random phenomenon, we are also interested in the "probability" of each "event." The probability of the certain event must be 1 and the probability of the impossible event must be 0. If A and B are two mutually exclusive events, the probability of the event "A or B" is the sum of the probabilities of A and of B. We are now going to see how measure theory provides an acceptable mathematical model for the preceding intuitions.

3.1.2. Measure Spaces

Measure theory provides a similar framework, which rests on the idea of the measure of surfaces. We are given a set Ω and a family $\mathcal{A}$ of subsets of Ω closed under the taking of complements and countable unions. These properties imply that $\mathcal{A}$ is closed under countable intersection and contains Ω and ϕ. We say that $\mathcal{A}$ is a **σ-algebra** of Ω, and $(\Omega,\mathcal{A})$ is a **measure space**. (Intuitively the elements of $\mathcal{A}$ are the subsets of Ω of which the surface can be measured.)

The pair $(\Omega,\mathcal{A})$ corresponds to the intuitive structure associated with a random phenomenon: Ω corresponds to the set of realizations and $\mathcal{A}$ corresponds to the set of subsets of Ω associated with the events... . This is apart from the fact that intuition leads to assuming $\mathcal{A}$ closed under *finite* intersection and union. The introduction of countable operations in the probabilistic model (as in the geometric model, moreover) is purely technical; without it hardly any results are obtainable.

Our basic model will therefore be a measurable space $(\Omega,\mathcal{A})$; we can establish a first dictionary. For $A \subset \Omega$ denote by $\mathbf{1}_A$ the **indicator function** of A, which equals 1 on A and 0 on A^c. For $\mathcal{C}$ a set of subsets of Ω, denote by $\sigma(\mathcal{C})$ the **σ-algebra generated** by $\mathcal{C}$, the smallest σ-algebra containing $\mathcal{C}$. If Ω is a topological space, its **Borel σ-algebra** B_R is the σ-algebra generated by its open sets.

Dictionary 3.1.1. $(\Omega,\mathcal{A})$ Is a Measurable Set

Probabilistic language	Geometrical or set language	Notation
Ω: **event** space	Ω: the set with which we work	
Ω: **certain** event		
ϕ: **impossible** event	ϕ: empty set	
ω is a trial or a realization	ω is an element of Ω	$\omega \in \Omega$
A is an **event**	A is a measurable set	$A \in \mathcal{A}$
A or B	Union of A and B	$A \cup B$
A and B	Intersection of A and B	$A \cap B$
A and B are **incompatible**	A and B are disjoint	$A \cap B = \phi$
B **implies** A	A contains B	$A \supset B$
The **opposite** of A	The complement of A	$A^c = \Omega \backslash A$

For $\Omega = \mathbb{R}$ or $\Omega = \overline{\mathbb{R}} = [-\infty,+\infty]$, the Borel σ-algebra is also (Exercise 3.1.7)

$$\sigma(]-\infty,x];\ x \in \mathbb{R}) = \sigma(]-\infty,x[;\ x \in \mathbb{R}) = \sigma(]a,b[;\ -\infty < a < b < \infty).$$

Likewise,

$$B_{\mathbb{R}^k} = \sigma\left[\prod_{i=1}^{k}\ ]-\infty,x_i]\ ;\ (x_i)_{1 \leq i \leq k} \in \mathbb{R}^k\right].$$

If $(\mathcal{B}_i)_{i \in I}$ is a family of σ-algebras contained in $\mathcal{A}$ (sub σ-algebras of $\mathcal{A}$), denote $\bigvee_{i \in I} \mathcal{B}_i = \sigma(\cup_{i \in I} \mathcal{B}_i)$; denote $\mathcal{B}_1 \vee \mathcal{B}_2 = \sigma(\mathcal{B}_1 \cup \mathcal{B}_2)$.

3.1.3. Measurable Functions, Observations

Let $(E,\mathcal{E})$ be a measurable space and X a function from Ω into E; for $\Gamma \subset E$ denote $(X \in \Gamma) = \{\omega;\ X(\omega) \in \Gamma\}$. If Γ is measurable we can by observing the values of X, observe the event $(X \in \Gamma)$. The observation of X signifies the observation of the **σ-algebra generated by X**, $\sigma(X) = \{(X \in \Gamma);\ \Gamma \in \mathcal{E}\} = X^{-1}(\mathcal{E})$. The function X from $(\Omega,\mathcal{A})$ into $(E,\mathcal{E})$ is **measurable** when $\sigma(X)$ is contained in $\mathcal{A}$.

We often use real observations, $E = \mathbb{R}$ or $E = \overline{\mathbb{R}}$. The σ-algebra used is then, except where otherwise stated, $B_{\mathbb{R}}$ or $B_{\overline{\mathbb{R}}}$ and we often omit mention of it. A measurable X is said to be a **random variable**; abbreviated r.v. For $E = \mathbb{R}^k$, X is said to be a **random vector of dimension** k.

The set of finite random variables is a vector space over $\mathbb{R}$ which contains the indicator functions of events, thus their linear combinations, called **step** r.v.'s. Each positive r.v. X is the increasing limit of a sequence $(X^{(n)})$ of step r.v.'s, defined, for example, by setting

$$X^{(n)} = \sum_{k \leq 2^n n} \frac{k}{2^n}\, 1_{\{k/2^n \leq X < (k+1)/2^n\}}\ .$$

Each r.v. X is the difference between two positive r.v.'s, its positive part $X_+ = \sup(0,X)$ and its negative part $X_- = -\inf(0,X)$.

3.1.4. Probability Spaces

We have seen in Chapter 1 an introduction to the idea of probability starting from that of frequency, and the construction of examples of particular probability spaes. Other intuitive notions of probability are of geometric origin. If a needle is dropped from a great height on to a floor Ω, for $A \subset \Omega$, the probability that the needle falls in A is $s(A)/s(\Omega)$, $s(A)$ being the area of A. Probability merges here with the measure of surfaces. In the case of the throw of a die, it is the idea of symmetry which gives a probability of 1/6 of a 3 coming up.

According to the intuitive rules of Section 3.1.1, P must almost be a measure, apart from the fact that the σ-additivity is not intuitively useful, only additivity appears to be neessary. There again (and as in geometry...) for technical reasons, we shall assume σ-additivity to be satisfied in the mathematical model used. From which comes the definition: a **probability** P on a measurable space $(\Omega,\mathcal{A})$ is a positive measure on $(\Omega,\mathcal{A})$ such that $P(\Omega)$ equals 1. The measure space $(\Omega,\mathcal{A},P)$ will be called a **probability space.**

Dictionary 3.1.2.

Probabilities	Measure Theory	Notation
Probability	Positive measure with total mass equal to 1	P
Probability space	Measure space provided with a measure of mass 1	$(\Omega,\mathcal{A},P)$
An **almost sure** event A	A^c has **measure zero**	$P(A) = 1$ or $\omega \in A$ a.s.
An **almost sure** property P (the set of ω which do not satisfy P is negligible.)	Property P holds **almost everywhere**	"$\omega \in$ P" a.s.

Vocabulary on Measures. *In this book, a "measure" on the measurable space* $(\Omega,\mathcal{A})$ *will always be a positive measure.* A set A is **charged** by μ if its measure is nonzero (if it is not negligible). The measure is **bounded** if $\mu(\Omega)$ is finite; it is

σ-**finite** if Ω is the union of a countable family of measurable sets of finite measure. For a family of measures $(\mu)_{i \in I}$ on (Ω, A) and a family $(\alpha_i)_{i \in I}$ of positive real numbers, the **mixed measure** $\Sigma_{i \in I} \alpha_i \mu_i$ is $A \longmapsto \Sigma_{i \in I} \alpha_i \mu_i(A)$. The **Dirac measure** δ_ω is $A \longmapsto 1_A(\omega)$.

3.1.5. Convergence of r.v.'s

Let (Ω, A, μ) be a measure space and (A_n) a sequence of events. The event "an infinite number of the events A_n takes place" called **lim sup**(A_n) is

$$\overline{\lim}\ A_n = \left\{\omega;\ \sum_n 1_{A_n}(\omega) = \infty\right\} = \lim_p \downarrow \bigcup_{n \geq p} A_n.$$

Let (X_n) be a sequence of r.v.'s. The event "(X_n) converges to the r.v. X" may be written $\{\lim X_n = X\} = \{\overline{\lim}|X_n - X| = 0\}$. For each sequence (α_p) which decreases to 0, staying strictly positive, it is the complement of $\cup_p[\overline{\lim}_n|X_n - X| \geq \alpha_p\}]$.

Definition 3.1.3. (X_n) **converges almost surely to** X, when $N = \{\omega;\ \lim_n X_n(\omega) = X(\omega)\}^c$ has measure zero.

The following lemma gives a practical condition of a.s. convergence to zero of a sequence of set indicators.

Borel-Cantelli Lemma 3.1.4. *Let (A_n) be a sequence of events of a measure space (Ω, A, μ). Assume that $\Sigma\mu(A_n) < \infty$. Then the event $\overline{\lim}\ A_n$ has measure zero.*

Proof.

$$\mu(\overline{\lim}\ A_n) = \lim_{p\to\infty} \downarrow \mu\left\{\bigcup_{n \geq p} A_n\right\} \leq \lim_{p\to\infty} \downarrow \sum_{n \geq p} \mu(A_n) = 0.$$

Definition 3.1.5. Let (Ω, A, μ) be a measure space. Let (X_n) be a sequence of a.s. finite r.v.'s on (Ω, A, μ). We say that (X_n) **converges in measure** to an r.v. X if for all $\varepsilon > 0$, we have

$$\lim_{n\to\infty} \mu(|X_n - X| \geq \varepsilon) = 0.$$

Proposition 3.1.6. *For each sequence of r.v.'s (X_n) which converges in measure to an r.v. X, we can extract a sub-sequence* *which converges almost surely to X.*

Proof. For all $k \in \mathbb{N}$, $\lim_{n\to\infty}\mu\{|X_n - X| \geqslant 1/k\}$ equals 0. Thus there exists an integer n_k such that, for all $n \geqslant n_k$,

$$\mu\left\{|X_n - X| \geqslant \frac{1}{k}\right\} \leqslant \frac{1}{k^2},$$

From which $\Sigma_k\mu\{|X_{n_k} - X| \geqslant 1/k\} < \infty$ and the Borel-Cantelli lemma then gives

$$\left\{\lim_{k\to\infty} X_{n_k} = X\right\}^c = \overline{\lim_{k\to\infty}}\left\{|X_{n_k} - X| \geqslant \frac{1}{k}\right\}$$

has measure zero.

Definition 3.1.7. If the measure μ is a probability P, convergence in measure is called **convergence in probability.**

Proposition 3.1.8. *In a probability space $(\Omega, \mathcal{A}, P)$ a.s. convergence implies convergence in probability.*

Proof. If (X_n) is a sequence of r.v.'s which converges a.s. to an r.v. X we have, for all $\varepsilon > 0$ (following from the fact that P is bounded).

$$P\left\{\overline{\lim_{n\to\infty}}\, |X_n - X| \geqslant \varepsilon\right\} = 0,$$

$$0 = \lim_{n\to\infty} \downarrow P\left\{\bigcup_{p\geqslant n} \{|X_p - X| \geqslant \varepsilon)\right\} \geqslant \overline{\lim_{n\to\infty}}\, P\{|X_n - X| \geqslant \varepsilon\};$$

from which $\lim_{n\to\infty}P\{|X_n - X| \geqslant \varepsilon\} = 0$.

Proposition 3.1.9. *In a probability space every Cauchy sequence converges in probability.*

Proof. Let (X_n) be a Cauchy sequence, i.e., such that for all $\varepsilon > 0$,

$$\lim_{n,m\to\infty} P\{|X_n - X_m| \geqslant \varepsilon\} = 0.$$

For each $k \in \mathbb{N}$, there exists an integer n_k such that, if n and m are integers greater than n_k, we have

$$P\left\{|X_n - X_m| \geq \frac{1}{k^2}\right\} \leq \frac{1}{k^2}.$$

Take the sequence (n_k) to be increasing (and $n_0 = 0$). The sum

$$\sum_k P\left\{|X_{n_{k+1}} - X_{n_k}| \geq \frac{1}{k^2}\right\}$$

is finite. The Borel-Cantelli lemma then implies that

$$N = \overline{\lim_{k\to\infty}} \left\{|X_{n_{k+1}} - X_{n_k}| \geq \frac{1}{k^2}\right\}$$

has measure zero. Let $\omega \in N^c$; the sum $\Sigma_k |X_{n_{k+1}}(\omega) - X_{n_k}(\omega)|$ is finite, thus $\Sigma_k (X_{n_{k+1}}(\omega) - X_{n_k}(\omega))$ converges absolutely. Its sum is

$$\lim_{p\to\infty} \sum_{k \leq p} (X_{n_{k+1}}(\omega) - X_{n_k}(\omega)) = \lim_{p\to\infty} (X_{n_{p+1}}(\omega) - X_0(\omega)).$$

Thus (X_{n_k}) converges on N^c to a finite r.v. X, i.e., (X_{n_k}) converges P a.s. and in probability to X. For all $\varepsilon > 0$ and all $\eta > 0$, there exists a $k \in \mathbf{N}$ such that

$$P\left\{|X_{n_k} - X| \geq \frac{\varepsilon}{2}\right\} \leq \frac{\eta}{2} \quad \text{and} \quad \frac{1}{k^2} \leq \inf\left[\frac{\varepsilon}{2}, \frac{\eta}{2}\right].$$

Then, for all $n \geq n_k$,

$$P\{|X_n - X| \geq \varepsilon\} \leq P\left\{|X_n - X_{n_k}| \geq \frac{\varepsilon}{2}\right\} + P\left\{|X_{n_k} - X| \geq \frac{\varepsilon}{2}\right\} \leq \eta.$$

The sequence (X_n) converges in probability to X.

3.1.6. Monotone Classes

Definition 3.1.10. A nonempty family $\mathcal{B}$ of subsets of Ω is a **monotone class** if the following properties are satisfied:

(a) $\Omega \in \mathcal{B}$;
(b) for $A_1 \in \mathcal{B}$, and $A_2 \in \mathcal{B}$ included in A_1, $A_1 \backslash A_2$ is in $\mathcal{B}$;
(c) if (A_n) is an increasing sequence of elements of $\mathcal{B}$, then $\lim \uparrow A_n$ belongs to $\mathcal{B}$.

Trick 3.1.11. *Every monotone class which contains a set* $\mathcal{B}$ *of subsets of* Ω *closed under intersection of a finite number of elements, contains the* σ*-algebra generated by* $\mathcal{B}$.

Proof. (1) The family of monotone classes which contain $\mathcal{B}$ is nonempty, since it contains the set of subsets of Ω. Its intersection is a monotone class which contains $\mathcal{B}$, and which we denote by $\lambda(\mathcal{B})$.

(2) Let us show that $\lambda(\mathcal{B})$ is closed under finite intersections. We consider the set γ_1 of elements A of $\lambda(\mathcal{B})$ such that $A \cap B$ belongs to $\lambda(\mathcal{B})$, for all $B \in \mathcal{B}$; γ_1 is monotone and contains $\mathcal{B}$ (verify this), and thus contains $\lambda(\mathcal{B})$. Let then γ_2 be the set of elements A of $\lambda(\mathcal{B})$, such that $A \cap B$ is in $\lambda(\mathcal{B})$ for all $B \in \lambda(\mathcal{B})$; γ_2 is monotone and contains $\mathcal{B}$ (verify this), and thus contains $\lambda(\mathcal{B})$.

(3) Let us show that a monotone class λ closed under finite intersection is a σ-algebra. Let $A \in \gamma$; A^c is in γ, since A^c equals $\Omega\backslash A$. If A and B are in γ, $A \cup B = (A^c \cap B^c)^c$ is in γ, and γ is closed under finite unions. Thus if (A_n) is a sequence of elements of γ, $\cup A_n$ belongs to γ_1 since it is the increasing limit of $(\bigcup_{p<n} A_p)$ which is contained in γ.

(4) $\lambda(\mathcal{B})$ is a σ-algebra which contains $\mathcal{B}$, thus $\sigma(\mathcal{B})$. $\sigma(\mathcal{B})$ is monotone and contains $\mathcal{B}$, thus $\lambda(\mathcal{B})$. As a result, $\lambda(\mathcal{B})$ coincides with $\sigma(\mathcal{B})$.

Corollary 3.1.12. *Let* $(\Omega, \mathcal{A})$ *be a measurable space,* $\mathcal{A}$ *being generated by* $\mathcal{B}$ *closed under intersection of a finite number of elements. Then two probabilities* P *and* Q *which coincide on* $\mathcal{B}$ *coincide on* $\mathcal{A}$.

Proof. The set of $A \in \mathcal{A}$ such that $P(A)$ and $Q(A)$ are equal is a monotone class containing $\mathcal{B}$, thus it contains $\mathcal{A}$.

Trick 3.1.13. *Let* $(\Omega, \mathcal{A})$ *be a measurable set and let* $\mathcal{C}$ *be closed under intersection of a finite number of elements which generate* $\mathcal{A}$. *A vector space* $\mathcal{F}$ *of functions from* Ω *into* $\mathbb{R}$ *contains all finite* r.v.'s (*respectively, bounded*) *on* $(\Omega, \mathcal{A})$, *when the following hypotheses hold*:

(1) $1_\Omega \in \mathcal{F}$;
(2) *for all* $A \in \mathcal{C}$, 1_A *is in* $\mathcal{F}$;
(3) *if* (f_n) *is an increasing sequence of positive functions of* $\mathcal{F}$

and if $\lim_{n\to\infty} \uparrow f_n$ *is finite (respectively bounded) then* $\lim_{n\to\infty} \uparrow f_n$ *is in* $\mathcal{F}$.

Proof. Let γ be the set of subsets A of Ω such that 1_A is in $\mathcal{F}$. It is easy to verify that γ is monotone and contains $\mathcal{C}$, thus γ contains $\mathcal{A}$. Since $\mathcal{F}$ is a vector space it contains all the step r.v.'s. If f is a positive r.v., there exists a sequence (f_n) of step r.v.'s which increases to f. Thus from (3), f is in $\mathcal{F}$ if it is finite (respectively bounded). Let f be an arbitrary r.v.. f is the difference of two positive r.v.'s, f_+ *and* f_- ; thus if f is finite (respectively bounded) f is in $\mathcal{F}$.

3.1.7. Distribution Functions

Let P be a probability on $\mathbb{R}^k$. It is characterized by its value on

$$\left\{ \prod_{i=1}^{k}]-\infty, x_i];\ \ (x_i)_{1 \leqslant i \leqslant k} \in \mathbb{R}^k \right\},$$

thus by its **distribution function**

$$F: (x_i)_{1 \leqslant i \leqslant k} \longmapsto P\left[\prod_{i=1}^{k}]-\infty, x_i] \right].$$

If P *is a probability on* $\mathbb{R}$, *its distribution function,* $x \longmapsto F(x) = P([-\infty, x])$ *is an increasing function, continuous on the right, from* $\mathbb{R}$ *into* $\mathbb{R}$ *and such that*

$$\lim_{x \to -\infty} F(x) = 0 \quad \text{and} \quad \lim_{x \to +\infty} F(x) = 1.$$

Continuity on the right results from

$$\lim_{\substack{h \to 0 \\ h > 0}} F(x+h) = \lim_{\substack{h \to 0 \\ h > 0}} P([-\infty, x+h]) = P(]-\infty, x]).$$

Likewise,

$$\lim_{\substack{h \to 0 \\ h > 0}} F(x-h) = F(x-) = P(]-\infty, x[\ ).$$

The jump of F *at* x *is thus* $P(\{x\})$, denoted also by $P(x)$.

N.B. In certain books, the function $x \longmapsto P(]-\infty, x[)$ is used, which is continuous on the left.

It follows from an extension theorem in measure theory that every function F from $\mathbb{R}$ into $\mathbb{R}$ which satisfies the above properties is the distribution function of a probability on $\mathbb{R}$: this will be assumed. (We shall come back to this result in corollary 3.4.35.)

A probability on $\mathbb{R}$ *can thus be given by a distribution function* F. We also often denote this probability by F. This probability will be said to be **continuous** or **diffuse**, if its distribution function is continuous, i.e., if it does not charge the points of $\mathbb{R}$.

Exercises 3.1. *Recap of Operations on Sets and Events*

E.1. Let A, B, and C be three events. Explain with the aid of operations on events (or sets) the following events and calculate their indicators starting from those of A, B, and C: "only A occurs", "A and B occur, but not C", "one of the three events occurs," "two events at most occur," "two events occur, but not three," "none of the events occurs."

E.2. Let (A_n) be a sequence of events. Write down with the aid of the operations $\cup$ and $\cap$ the following event:

$$\underline{\lim}\, A_n = \text{"All the } A_n \text{ from a certain point onwards are realized."}$$

Calculate the indicators of $\underline{\lim}\, A_n$ and of $\overline{\lim}\, A_n$ as a function of those of A_n.

E.3. Let f be a function from Ω_1 into Ω_2, $(A_i)_{i\in I}$ a family of subsets of Ω_2, and A and B subsets of Ω_2. Show

$$1_{f^{-1}(A)} = 1_A \circ f;\quad f^{-1}(A\backslash B) = f^{-1}(A)\backslash f^{-1}(B),$$

$$f^{-1}\left[\bigcup_{i\in I} A_i\right] = \bigcup_{i\in I} f^{-1}(A_i),\quad f^{-1}\left[\bigcap_{i\in I} A_i\right] = \bigcap_{i\in I} f^{-1}(A_i).$$

E.4. Let A and B be subsets of Ω_r. Determine $\sigma(A)$, $\sigma(A) \cap \sigma(B)$, $\sigma(A) \cup \sigma(B)$, $\sigma(A,B)$, $\sigma(A) \vee \sigma(B)$. Let $(A_n)_{n\in N}$ be a partition of Ω_r. Describe $\sigma\{A_n;\ n \in \mathbb{N}\}$. Describe $\sigma(\{\omega\};\ \omega \in \Omega)$.

E.5. Let $(C_i)_{i\in I}$, $(C_j)_{j\in J}$ be two families of subsets of Ω. Show

$$\sigma\left[\bigcup_{i\in I} C_i\right] \vee \sigma\left[\bigcup_{j\in J} C_j\right] = \sigma\left[\bigcup_{\alpha\in I\cup J} C_\alpha\right].$$

E.6. *Image of a σ-algebra.* Let $f\colon \Omega_1 \longmapsto \Omega_2$, and A_2 be a σ-algebra of Ω_2. Verify that $f^{-1}(A_2)$ is a σ-algebra. If C_2 is a set of subsets of Ω_2, prove $\sigma(f^{-1}(C_2)) = f^{-1}(\sigma(C_2))$.

E.7. *Borel σ-algebras.* Let Ω be a topological space, O the set of open sets, and F the set of closed sets. The Borel σ-algebra on Ω is $B_\Omega = \sigma(O)$.

(a) Prove $\sigma(O) = \sigma(F)$.
(b) Show $B_{\mathbb{R}} = \sigma\{]-\infty,a];\ a \in \mathbb{R})\} = \sigma\{]-\infty,a[\ ;\ a \in \mathbb{R}\}$.
(c) Show that $B_{\mathbb{R}^k}$ is generated by cubes with vertices, the coordinates of which are rational.

E.8. Let (X_n) be a sequence of r.v.'s. Prove that $\sup X_n$, $\inf X_n$, $\overline{\lim}\, X_n$ and $\underline{\lim}\, X_n$ are r.v.'s and that $\{\omega;\ \lim X_n(\omega)\ \text{exists}\}$ is measurable.

E.9. *Trace of a σ-algebra.* Let A be a σ-algebra of Ω and $\Gamma \subset \Omega$; set $A|_\Gamma = \{A \cap \Gamma;\ A \in A\}$. Show that $A|_\Gamma$ is a σ-algebra of Γ and that if Ω is a topological space the trace on Γ of the Borel σ-algebra is the Borel σ-algebra of Γ together with the trace topology.

E.10. *Measurable functions.* Let (Ω,A), (Ω',A') be two measurable spaces, $P(\Omega)$ the set of subsets of Ω.

(a) For $A = P(\Omega)$ or $A' = (\phi,\Omega')$ what are the measurable functions from (Ω,A) into (Ω',A')?
(b) If Ω and Ω' are topological spaces together with their Borel σ-algebras, continuous functions are measurable. Are there any others?

E.11. Let, on $\mathbb{Z}$, $A = \{A;$ for all $n > 0$, $2n \in A$ is equivalent to $(2n + 1) \in A\}$. Verify that A is a σ-algebra and that the mapping from $(\mathbb{Z},A)$ into $(\mathbb{Z},A)$ defined by $\omega \longmapsto \omega + 2$ is measurable; it is a bijection, but its inverse is not measurable.

E.12. *Composition of measurable functions.* Let X be measurable from $(\Omega, \mathcal{A})$ into $(E, \mathcal{E})$, and f measurable from $(E, \mathcal{E})$ into $(F, \mathcal{F})$. Show that $f \circ X$ is measurable and that $\sigma(X)$ contains $\sigma(f \circ X)$.

E.13. Let f be the function $(x,y) \longmapsto x^2 + y^2$ from $\mathbb{R}^2$ into $\mathbb{R}$ and $\mathcal{A}$ a σ-algebra of $\mathbb{R}$. Describe $f^{-1}(\mathcal{A})$.

E.14. *Symmetric sets.* Let ζ be the family of symmetric sets of $\mathbb{R}$, i.e., such that "$x \in \Gamma$" implies "$(-x) \in \Gamma$." Show that ζ is a σ-algebra. In the case $k = 1$, show that it is the inverse image of the set of subsets of $\mathbb{R}$ by one of the functions $x \longmapsto x^2$ or $x \longmapsto |x|$.

E.15. Let σ be a permutation of $(1, \ldots, n)$, and ϕ_σ the mapping $(x_1, \ldots, x_n) \longmapsto (x_{\sigma(1)}, \ldots, x_{\sigma(n)})$ of $\mathbb{R}^n$ on to $\mathbb{R}^n$. Let B be the set of Borel subsets of $\mathbb{R}^n$ invariant under ϕ_σ, for all permutations σ and $X_1, \ldots, X_n$ the coordinate functions on $\mathbb{R}^n$. Show

$$B = \sigma\left[\sum_{i=1}^{n} X_i, \sum_{i \neq j} X_i X_j, \ldots, \prod_{i=1}^{n} X_i\right]$$

$$= \sigma\left[\sum_{i=1}^{n} X_i, \sum_{i=1}^{n} X_i^2, \ldots, \sum_{i=1}^{n} X_i^n\right] = \sigma(X_{(1)}, \ldots, X_{(n)}),$$

where $(X_{(1)}(x), \ldots, X_{(n)}(x))$ is for $x = (x_1, \ldots, x_n) \in \mathbb{R}^n$ the sequence x reordered in increasing order. (We shall prove that $(X^{(1)}, \ldots, X^{(n)})$, the *order statistic*, is a random vector on $\mathbb{R}^n$.)

E.16. Let E be a metric space together with its Borel σ-algebra B_E and P a probability on E. Show that, for all $A \in B_E$,

$$P(A) = \inf\{P(G); \quad G \text{ open}, \quad G \supset A\}$$

$$= \sup\{P(F); \quad F \text{ closed}, \quad F \subset A\}.$$

E.17. (a) Show that a distribution function on $\mathbb{R}$ has at most a countable infinity of discontinuities. It is characterized by its values on a well chosen countable set.

(b) Let F be the distribution function of a probability on $\mathbb{R}^k$ and $F_1, \ldots, F_k$ its "marginal distribution functions": for

$1 \leq i \leq k$, and $x_i \in \mathbb{R}$, $F_i(x_i)$ is the probability of $\{(y_1, ..., y_k); y_i \leq x_i\}$. Show that, if F is discontinuous at $x = (x_1, ..., x_k)$, one of the marginal distribution functions is discontinuous at x_i. Show that F has at most a countable infinity of discontinuities and that it is characterized by its values at its continuity points.

E.18. *A.s. continuous functions.* Let h be a measurable function from one metric space E into another. Let D_h be the set of points of E where h is not continuous.

(a) Show that D_h is a Borel set.
(b) Assume E is equipped with the Borel σ-algebra and a measure μ. Show that there is no implication between "h is μ a.s. continuous" and "h is μ a.s. equal to a continuous function," the first assertion signifying "$\mu(D_h) = 0$" and the second, "there exists h' continuous with $\mu(h \neq h') = 0$."
(c) Let A be a Borel set of E, ∂A its boundary. Show that "$\mathbf{1}_A$ is μ a.s. continuous" is equivalent to "$\mu(\partial A) = 0$."
(d) What is a μ a.s. continuous function f when μ is a Dirac measure at $a \in E$? or when μ is a mixture of a countable family of Dirac measures?

E.19. *Completion of a σ-algebra.* Let $(\Omega,\mathcal{A},\mu)$ be a measure space and $\mathcal{N}$ the set of "negligible subsets of Ω" i.e., the subsets contained in a set $A \in$ of measure zero. Show that $\bar{\mathcal{A}} = \{A \cup N; A \in \mathcal{A}, N \in \mathcal{N}\}$ is a σ-algebra ("μ-completion σ-algebra of $\mathcal{A}$"). Show that if we set $\bar{\mu}(A \cup N) = \mu(A)$ for $A \in \mathcal{A}$ and $N \in \mathcal{N}$, a measure $\bar{\mu}$ is defined on $(\Omega,\bar{\mathcal{A}})$ which extends μ.

3.2. Integration

3.2.1. Mechanism of the Theory of Integration

Let $(\Omega,\mathcal{A},\mu)$ be a measure space. The integral of an r.v. X defined on $(\Omega,\mathcal{A})$ associated with μ, denoted by $\int X \, d\mu$ or $\mu(X)$, is defined in the following manner:

– For $X = \Sigma_{i=1}^k \alpha_i \mathbf{1}_{A_i}$ a step r.v.:

$$\int X \, d\mu = \sum_{i=1}^{k} \alpha_i \mu(A_i);$$

– For X a positive increasing limit of a sequence of step r.v.'s $(X^{(n)})$:

$$\int X\,d\mu = \lim_n \uparrow \int X^{(n)} d\mu;$$

– For X **quasi-integrable**, i.e., such that $\int X_+ d\mu$ or $\int X_- d\mu$ is finite:

$$\int X\,d\mu = \int X_+ d\mu - \int X_- d\mu.$$

The common method of demonstrating a property which is verified for all quasi-integrable r.v.'s by following the above three stages will be called **trick 3.2.14.**

The r.v. X is integrable if $\int |X| d\mu = \int X_+ d\mu + \int X_- d\mu$ is finite. For integrable r.v.'s X and Y, the r.v. $X + iY$ taking values in $\mathbb{C}$ is said to be integrable and is denoted by $\int (X + iY) d\mu = \int X\,d\mu + i \int Y\,d\mu$.

Let us take for μ a probability P, and $X = \Sigma_{i=1}^{k} \alpha_i 1_{A_i}$: $\int X\,dP$ is the barycenter of the points $\alpha_1, \ldots, \alpha_k$, weighted by the probabilities that X takes these values. As in Chapter 1, $\int X\,dP$ is a **mean value** of X, or its **expectation**, denoted by $E(X)$. A random vector $X = (X_1, \ldots, X_n)$ is integrable if its components are, and its mean is then $E(X) = (E(X_1), \ldots, E(X_n))$.

We shall denote by L^p the space of r.v.'s X such that $\int |X|^p d\mu$ is finite ($p > 0$), and L^∞ the space of bounded r.v.'s. These spaces indexed by + are the sets of their positive functions. Trick 3.2.14 easily leads to:

3.2.15. Properties of the Integral. (a) $X \longmapsto \int X\,d\mu$ *is linear on* ^{1} *and increasing.* (b) Beppo-Levi's Theorem. *Let* (X_n) *be an increasing sequence of positive* r.v.'s:

$$\lim \uparrow \int X_n d\mu = \int \lim \uparrow X_n d\mu.$$

(c) Fatou's Theorem. *Let* (X_n) *be a sequence of* r.v.'s *minorized by an integrable* r.v.:

$$\underline{\lim}_{n \to \infty} \int X_n d\mu \geq \int \underline{\lim}_{n\to\infty} X_n d\mu.$$

(d) Lebesgue's Theorem. *If* (X_n) *is a sequence of* r.v.'s *such that* $(|X_n|)$ *is majorized by an integrable* r.v. *and if the sequence* (X_n) *converges almost surely to* X, *then* $\lim_{n\to\infty} \int X_n d\mu = \int X\,d\mu$.

Inequalities 3.2.16. (a) The function $x \longmapsto e^x$ is strictly convex, i.e., for $t \in]0,1[$: $e^{tx+(1-t)y} \leqslant te^x + (1 - t)e^y$, the inequality being strict for $x \neq y$.

Let U and V be two r.v.'s such that e^U and e^V are integrable with respect to μ:

$$\left[\frac{e^U}{\mu(e^U)}\right]^t \left[\frac{e^V}{\mu(e^V)}\right]^{1-t} \leqslant t\,\frac{e^U}{\mu(e^U)} + (1-t)\,\frac{e^V}{\mu(e^V)}.$$

Integrating

$$\mu\,(e^{tU}e^{(1-t)V}) \leqslant [\mu(\epsilon^U)]^t[\mu(e^V)]^{1-t}.$$

The inequality is strict unless U and V are proportional.

(b) For $p \in]0,1[$, set

$$\|X\|_p = \left(\int |X|^p d\mu\right)^{1/p}.$$

We also define the **essential bound** of X:

$$\|X\|_\infty = \sup[a;\ \mu\{|X| \geqslant a\} > 0].$$

The following inequalities can be shown for $p \in [1,\infty]$ and $1/p + 1/q = 1$:

$$\|X \cdot Y\|_1 \leqslant \|X\|_p \|Y\|_q \quad \textbf{(Hölder)};$$

$$\|X + Y\|_p \leqslant \|X\|_p + \|Y\|_p \quad \textbf{(Minkowski)}.$$

For $p < \infty$, Hölder results from (a) by taking $U = p \operatorname{Log} |X|$, $V = q \operatorname{Log} |Y|$ and $t = 1/p$. Minkowski results from Hölder by writing $(X + Y)^p = X(X + Y)^{p-1} + Y(X + Y)^{p-1}$. For $p = \infty$, the proofs are easy.

For r, p, q in $]0,\infty]$ and $1/p + 1/q = 1/r$, take in (a), $U = p \operatorname{Log} |X|$, $V = q \operatorname{Log} |Y|$ and $t = r/p$. The relation becomes $\|X \cdot Y\|_r \leqslant \|X\|_p \|Y\|_q$. If μ is a probability, taking $Y = 1$ and $r < p$, we obtain $\|X\|_r \leqslant \|X\|_p$.

(c) **Jensen's Inequality.** *Let X be an integrable random vector, defined on a probability space $(\Omega, \mathcal{A}, P)$ taking values in A, which is convex in $\mathbb{R}^k$. Let ϕ be a convex function from A into $\mathbb{R}$, the* r.v. *$\phi(X)$ is quasi-integrable and $E[\phi(X)] \geqslant \phi[E(X)]$ if ϕ is strictly convex, the inequality is strict when X is not constant* a.s.

Proof. At each point $x \in A$, ϕ is the upper bound of the affine functions (called support functions) which are less than ϕ. A linear function ℓ from $\mathbb{R}^k$ into $\mathbb{R}$ can always be found, such that for all $y \in A$, $\phi(y) \geqslant \phi(x) + \ell(y - x)$. If ϕ is strictly convex, the inequality is strict for $y \neq x$.

Taking $x = E(X)$ and $y = X$, we obtain

$$\phi(X) \geqslant \phi(E(X)) + \ell(X - E(X)),$$

but $E(\ell(X)) = \ell(E(X))$, from which the result follows.

3.2.2. Equivalence Classes of r.v.'s and L^p Spaces

Almost sure equality on $(\Omega, \mathcal{A}, \mu)$ is an equivalence relation between r.v.'s. The integral, convergence in measure, and a.s. convergence are properties closed under this equivalence relation, which can be defined for equivalence classes. We will talk about these notions either for r.v.'s or equivalence classes without being precise. When an element is chosen from an equivalence class X, we say that a **version** of X has been chosen. X will designate, according to the case, its class or a version of the class. The space L^p is the space of equivalence classes of the r.v.'s of L^p. Together with $\|\cdot\|_p$, L^p is a Banach space.

When μ is a probability and $p < q$, L^p contains L^q. From Tchebychev's inequality, convergence in L^p implies convergence in measure:

$$\varepsilon^p \mu\{|X_n - X| \geqslant \varepsilon\} \leqslant \int_{\{|X_n - X| \geqslant \varepsilon\}} |X_n - X|^p d\mu$$

$$\leqslant (\|X_n - X\|_p)^p.$$

Proposition 3.2.17. *In a probability space we have*

(a) *for* $0 < p < q \leqslant \infty$; $L^p \supset L^q$ *and* $\| \ \|_p \leqslant \| \ \|_q$.

(b) $(X_n) \xrightarrow{\text{a.s.}} X \Rightarrow (X_n) \xrightarrow{P} X$

$$(X_n) \xrightarrow{L^p} X \Leftarrow (X_n) \xrightarrow{L^p} X.$$

Dictionary 3.2.18

$(\Omega,\mathcal{A},P)$ probability space	$(\Omega,\mathcal{A},\mu)$ measure space
X is **integrable**: $E(\lvert X\rvert) < \infty$	$\int \lvert X\rvert d\mu < \infty$
$E(X)$ the **expectation** of X	$\int X\, d\mu$ or $\int X(\omega)d\mu(\omega)$ Integral of X
X is **centered** if $E(X)$ is zero	
$E(\lvert X\rvert^p) < \infty$: X **has a moment of order** p	$\lVert X\rVert_p = (\int \lvert X\rvert^p d\mu)^{1/p} < \infty$ or $X \in L^p$
$E(X^p)$ is the **moment of order** p **of** X	
(X_n) converges **in probability** to X: $(X_n) \xrightarrow{P} X$	(X_n) converges **in measure** to X: $(X_n) \xrightarrow{\mu} X$

For a probability space $(\Omega,\mathcal{A},P)$, a particular vocabulary is used in L^2. Let $X \in L^2$, $E(X^2)$ is its **quadratic mean**, $\sigma^2(X) = E(X - E(X))^2 = E(X^2) - [E(X)]^2$ its **variance**, and $\sigma(X)$ the positive square root of $\sigma^2(X)$, its **standard deviation**. Convergence in L^2 is called **convergence in quadratic mean**. For X and Y in L^2, the **covariance** of X and Y is defined as

$$\Gamma(X,Y) = E[(X - E(X))(Y - E(Y))] = E(XY) - E(X)E(Y),$$

their **correlation coefficient** $\rho(X,Y) = \Gamma(X,Y)/\sigma(X)\sigma(Y)$. Hölder's inequality then becomes $\rho^2(X,Y) \leq 1$. When $\rho(X,Y)$ is zero the variables X and Y are said to be **uncorrelated**. This is the vocabulary which we have already used for the discrete distributions in Sections 1.3.2 and 1.3.3. The covariance is a scalar product on the subspace of centered r.v.'s, from which

$$\sigma^2(X + Y) = \sigma^2(X) + \sigma^2(Y) + 2\Gamma(X,Y).$$

3.2.3. Densities

Consider a measure μ on the measure space $(\Omega,\mathcal{A})$ and a positive r.v. f. The measure $f\cdot\mu$, of which the **density** with respect to μ is f, is the measure which associates to $A \in \mathcal{A}$, $(f\cdot\mu)(A) = \int_A f\,d\mu = \int 1_A f\,d\mu$. However, be careful; *the density of $f\cdot\mu$ is only defined up to an equality μ a.s.!* We will often refer to "the" density of f by designating a version of this density -- but this is again an abuse of language. For $(\Omega,\mathcal{A}) = (\mathbb{R}^k, B_{\mathbb{R}^k})$ and μ Lebesgue measure λ, we will say the *measure of the density f*, λ being understood. Let X be a quasi-integrable r.v. for $f\cdot\mu$; then (Trick 3.2.16) $X\cdot f$ is quasi-integrable for μ, and

$$\int X\,d(f\cdot\mu) = \int Xf\,d\mu.$$

3.2.4. Bounded Measures on $\mathbb{R}^k$

Consider the sets of continuous functions, from $\mathbb{R}^k$ into $\mathbb{R}$, C_K, C_0, and C_b with compact support, which tend to 0 at infinity, and which are bounded respectively. Let $\mathcal{M}$ be the set of positive measures F, bounded on $\mathbb{R}^k$ ($F(\mathbb{R}^k)$, denoted $\|F\|$, is finite). A measure F of $\mathcal{M}$ is characterized by its valued on compact sets (apply Corollary 3.1.12 to its restriction to a given compact set K, then consider a sequence (K_n) of compact sets, the union of which is $\mathbb{R}^k$). It is also characterized by the integrals $F(f)$ of functions $f \in C_K$. In fact, each compact indicator function of $\mathbb{R}^k$ is the decreasing limit of a sequence of functions of C_K. The function $f \mapsto F(f)$ is a positive linear from bounded on C_K. We shall assume that such a linear from is a bounded measure on $\mathbb{R}^k$. This is an *extension theorem*, which will be found in every measure theory course.

The set of functions $\{\exp i\langle\cdot,y\rangle;\ y \in \mathbb{R}^k\}$ separates the points of $\mathbb{R}^k$ and contains the identity function 1. Let A be the algebra formed by linear combinations of such functions. From the Stone-Weierstrass Theorem, each function f of C_K is the uniform limit of a sequence of functions of A on a compact set containing its support.

To $F \in \mathcal{M}$, associate its **Fourier transform** ϕ_F, defined on $\mathbb{R}^k$ by $\phi_F(t) = \int \exp[i\langle t,x\rangle]dF(x)$. This function ϕ_F charcterizes the integrals with respect to F of functions of A. Let F_1 and

F_2 in M have the same Fourier transform. For $f \in C_K$ and $\varepsilon > 0$, take a compact set Γ containing the support of f and such that $F_1(\Gamma^c)$ and $F_2(\Gamma^c)$ are less than ε. Let $a \in A$ be such that $\sup\{|f(x) - a(x)|;\ x \in \Gamma\}$ is majorized by η:

$$|F_1(f) - F_2(f)| \leqslant |F_1(a) - F_2(a)|$$

$$+ 2(\|a\| + \|f\|)\varepsilon + 2\eta$$

$$\leqslant 2(\|a\| + \|f\|)\varepsilon + 2\eta.$$

Let η, then ε, tend to 0: $F_1(f) = F_2(f)$. *The Fourier transform characterizes F.* Likewise a positive bounded measure F on $[0,\infty[$ is characterized by its **Laplace transform** ψ_F defined on $[0,\infty[$ by

$$\psi_F(t) = \int \exp(-tx)dF(x).$$

In fact, the Stone-Weierstrass Theorem also applies on $[0,\infty[$ to the algebra generated by the functions $\{\exp(-t\cdot);\ t > 0\}$.

Exercises 3.2. *Convergence of* r.v.'s (Exercises 1 – 8)

E.1. Give counterexamples to show that almost sure convergence does not imply convergence in measure on the measure space $(\Omega,\mathcal{A},\mu)$ when μ is not bounded.

E.2. Let P be Lebesgue measure on $[0,1[$ together with its Borel σ-algebra $B_{[0,1[}$. Let $n \in \mathbb{N}$; there are two unique integers p and k such that $n = 2^k + p$ and $0 \leqslant p < 2^k$. Let $X_n = \mathbf{1}_{[p/2^k,(p+1)/2^k[}$. Prove that (X_n) converges in probability, but not P a.s.

E.3. Let (X_n) be a sequence of r.v.'s on a probability space $(\Omega,\mathcal{A},P)$ which converges in probability to a constant α. Let f be a real function differentiable at α; let $f'(\alpha)$ be this derivative. For all $n \in \mathbb{N}$, set

$$Z_n(X_n - \alpha) = f(X_n) - f(\alpha) - f'(\alpha)(X_n - \alpha);$$

$Z_n = 0$ on $\{X_n = \alpha\}$. Prove that (Z_n) tends to 0 in probability.

E.4. Let (Ω, A, P) be a probability space. Give counter examples proving that between the different types of convergence there are no other implications than those indicated by Proposition 3.2.17 (see Exercise 3.2.2).

E.5. Let (X_n) and X be positive, integrable r.v.'s on a measure space (Ω, A, μ). Assume that (X_n) converges a.s. to X and that $\lim_{n\to\infty} \int X_n d\mu$ equals $\int X\, d\mu$. Prove that (X_n) converges to X in $L^1(\Omega, A, \mu)$ (first of all verify that $(X - X_n)_+$ tends to 0 in $L^1(\Omega, A, \mu)$).

E.6. Let Ω be a space together with the σ-algebra $P(\Omega)$ of its subsets and a Dirac measure δ at a point $a \in \Omega$. Describe $L^p(\Omega, P(\Omega), \delta)$ for $p \in]0, +\infty]$. Compare, in $(\Omega, P(\Omega), \delta)$ convergence almost surely, in probability and in L^p.

E.7. Let (X_n) be a sequence of r.v.'s on a measure space (Ω, A, μ) such that the series $\Sigma\mu\{X_{n+1} \neq X_n\}$ converges. Prove that (X_n) converges a.s.

E.8. In the following examples (f_n) is a sequence of r.v.'s on $\mathbb{R}$ equipped with Lebesgue measure. Study whether or not the theorems of Fatou and Lebesgue apply:

$$f_n = \frac{1}{n}\mathbf{1}_{[n,\infty[}; \; f_n = n\mathbf{1}_{[1/n,2/n]}\; ; \; f_n = \frac{1}{n}\mathbf{1}_{[0,n]}\; ;$$

$$f_n = n(-1)^n\mathbf{1}_{[n,n+1]}; \; f_{2n} = \mathbf{1}_{[0,1]}\; ; \; f_{2n+1} = \mathbf{1}_{[1,2]}\; .$$

E.9. *Regression.* For X and Y in $L^2(\Omega, A, P)$, generalize 1.1 and 1.3 and determine the **regression line and the orthogonal regression line.**

E.10. *Linear or affine dependence.* (a) Let H be a Hilbert space with scalar product $<,>$, and $x_1, \ldots, x_n$ n vectors in H. Prove that linear dependence of these vectors is equivalent to the Gram determinant vanishing

$$[<x_i, x_j>]_{1 \leq j,j \leq n}\; .$$

(b) Interpret this by taking $H = L^2(\Omega, A, P)$ and $X_1, \ldots, X_n$ n square integrable r.v.'s; then by taking $(X_1 - E(X_1), \ldots, X_n - E(X_n))$.

E.12. *Essential upper bound.* Let $(\Omega,\mathcal{A},P)$ be a probability space and $(X_i)_{i\in I}$ a family of integrable r.v.'s, majorized by an integrable r.v.. X is said to be an essential upper bound of $(X_i)_{i\in I}$, denoted by $X = \text{ess sup}_I X_i$ if "for all i, $X_i \leqslant Y$, P-a.s." is equivalent to "$X \leqslant Y$, P-a.s."

(a) If I is countable, show: $\text{ess sup}_I X_i = \sup_{i\in I} X_i$ (P-a.s.).
(b) Show that if an r.v. $X = \text{ess sup}_I X_i$ is known, then X is unique, up to equivalence P a.s.
(c) Consider the set $\mathcal{I}$ of countable subsets of I. Set $\sigma = \sup_{J\in\mathcal{I}} E[\sup_{i\in J} X_i]$. Show that σ is attained for at least one J_0. Deduce from this the existence of $\text{ess sup}_{i\in I} X_i$. Show that for a family of sets, $(A_i)_{i\in I}$, $\text{ess sup}_I \mathbf{1}_{A_i}$ is a.s.

the indicator function of a set. This set (defined up to a set of measure zero) is called the **essential upper bound** of $(A_i)_{i\in I}$ and denoted by $\text{ess sup}_I A_i$.

E.13. *Atom.* (a) An atom A of $(\Omega,\mathcal{A},\mu)$ is a non-negligible event such that for $B \subset A$, B or $A\backslash B$ is negligible. Show that each r.v. X is constant μ a.s. on A. Show that if μ is σ-finite, $\mathcal{A}$ can only contain a countable number of atoms.

Let μ be a σ-finite measure on $\mathbb{R}^k$ and A an atom. Show that there exists a compact set $K \subset A$ such that $A\backslash K$ is negligible (see Exercise 3.1.16). Show that there exists a point $x \in K$ *such that* $K\backslash\{x\}$ is negligible. Deduce from this that *an atom in* $\mathbb{R}^k$ *can only be a point* (up to a negligible set).

A measure without atoms is said to be **continuous** or **diffuse**. On $\mathbb{R}$ this notion coincides with that of Section 3.1.7. Prove that a distribution with continuous distribution function on $\mathbb{R}^k$ is **diffuse**: give a counter example to the converse, for $k > 1$.

(b) Let $(\Omega,\mathcal{A},P)$ be an atomless probability space. Show that, for $a \in [0,1]$, there exists at least one $A \in \mathcal{A}$, with probability equal to a. (Show, with the help of Exercise 12, that $\{B;\ P(B) \leqslant a\}$ is inductive for a.s. inclusion and consider a maximal element.)

E.14. Show that, for a measure which is not a probability, Jensen's inequality may not hold.

3.3. The Distribution of a Measurable Function

3.3.1. Image Measures

When a random phenomenon is studied, the space of realizations $(\Omega,\mathcal{A},P)$ can have several definitions according to the description of the phenomenon which is given (think of the descriptions, more or less detailed, of firing on a target). However, if a measurable function X from $(\Omega,\mathcal{A})$ into $(E,\mathcal{E})$ is observed, we know the events of the σ-algebra $\sigma(X) = X^{-1}(\mathcal{E})$ generated by X and the range space. Using the image under X of P will often allow us to have situations independent of the exact form of $(\Omega,\mathcal{A},P)$.

Recall that if μ is a measure and f is a measurable function taking values in $(E,\mathcal{E})$, the image $f(\mu)$ of μ by f is the measure defined on $(E,\mathcal{E})$ by $f(\mu)(\Gamma) = \mu(f^{-1}(\Gamma))$. The probability interpretation is the following.

Interpretation 3.3.19. The **distribution** of a measurable function X from $(\Omega,\mathcal{A},P)$ into $(E,\mathcal{E})$ is the image measure under X, denoted by F_X, defined for $\Gamma \in \mathcal{E}$ by $F_X(\Gamma) = P(X \in \Gamma)$. This is also denoted by $F_X = X(P)$.

Proposition 3.3.20. *The* r.v.'s *on* $(\Omega,\sigma(X))$ *are the compositions of* r.v.'s *on* $(\mathcal{E},P)$ *and of* X.

Proof. For $A = X^{-1}(\Gamma)$ in $\sigma(X)$, we have $\mathbf{1}_A = \mathbf{1}_\Gamma o X$. The proposition holds for a step r.v.

Let f be a positive r.v. on $(\Omega,\sigma(X))$; f is the increasing limit of a sequence of step r.v.'s (f_p) on $(\Omega,\sigma(X))$. For each p, $f_p = g_p o X$ where g_p is a stepped r.v. on $(E,\mathcal{E})$. Let Γ be the set of those $x \in E$ such that $(g_p(x))$ converges. The sequence (g_p) converges on $X(\Omega)$. Thus if we define $g = \lim_{p\to\infty} g_p$ on Γ and g constant on Γ^c, g is an r.v. on $(E,\mathcal{E})$ and $f = g o X$. For an arbitrary f, the result follows by considering its positive part and its negative part.

Passage to the Quotient for a.s. Equality. Let X_1 and X_2 be two measurable functions from $(\Omega,\mathcal{A},P)$ into $(E,\mathcal{E})$, equal a.s. . Their distributions, denoted by F_X, are the same. Let f_1 and f_2 be two r.v.'s defined on $(E,\mathcal{E})$ and equal a.s. for this distribution: $f_1(X_1)$ and $f_2(X_2)$ are equal P-a.s. We can thus consider X and f, equivalence classes of X_1 for equality P-a.s.

and of f_1 for equality F_X-a.s.; F_X is the "distribution of X" and $f(X)$ an equivalence class of r.v.'s for equality P a.s. This allows us, in what follows, to identify an equivalence class of r.v.'s and a version of this class.

Proposition 3.3.21. (a) *For f a positive* r.v. *on* $(E,\mathcal{E})$, $\int f\, dF_X = E(f(X))$.

(b) *f is F_X-integrable if and only if $f(X)$ is P-integrable, the same equality then holding.*

(c) *For $0 < p \leqslant \infty$, f is in $L^p(E,\mathcal{E},F_X)$ if, and only if, $f o X$ is in $L^p(\Omega,\sigma(X),P)$.*

This proposition is shown with the help of Trick 3.2.14.

3.3.2. Distributions Defined on $\mathbb{R}$ or on $\mathbb{R}^k$

Studying the distribution of a finite r.v. or of a random vector of dimension k amounts to studying a probability on $\mathbb{R}$ or $\mathbb{R}^k$. The term "distribution" will often be used to designate a probability on $\mathbb{R}$ or $\mathbb{R}^k$. Such a distribution F is characterized by its distribution function, which we will denote most often by F.

We will denote by either $\int f\, dF$ or $\int f(x) dF(x)$ the integral of an r.v. defined on $\mathbb{R}^k$, if this integral exists. Likewise, if λ is Lebesgue measure on $\mathbb{R}^k$, we write $\int f\, d\lambda$ or $\int f(x) dx$.

Let X be a finite r.v. with distribution F. The mean, variance, and moment of order n of X and F, if they exist, are

$$E(X) = \int x\, dF(x); \quad \sigma^2(X) = \int (x - E(X))^2 dF(x)$$

$$= \int x^2 dF(x) - \left(\int x\, dF(x)\right)^2; \quad E(X^n) = \int x^n dF(x).$$

When X is a random vector of dimension k, denoted $X(\omega) = (X_1(\omega), ..., X_k(\omega))$, the k r.v.'s $X_1, ..., X_k$ are the **components** of X. The **marginal distributions** of F_X are the distributions $F_{X_1}, ..., F_{X_k}$ of $X_1, ..., X_k$. They are not sufficient to characterize F_X (Exercise 3.3.7).

On the other hand, the **Fourier transform** ϕ of a probability F on $\mathbb{R}^k$, characterizes F (Section 3.2.4). If F is the distribution of a random vector X, we have $\phi(u) =$

$E(e^{i<u,X>})$. ϕ is called the **characteristic function of** X. Likewise the distribution F of a positive r.v. is characterized by its **Laplace transform,** $t \mapsto E(e^{-tX})$, defined on $\mathbb{R}_+$. If X takes values in $\mathbb{N}$, it is characterized by its generating function g: $z \mapsto g(z) = \Sigma F(n)z^n = E[z^X]$, defined on $[-1,1]$.
The following is an important extension:

Proposition 3.3.22. *For a σ-finite measure μ on $\mathbb{R}^k$, consider the set J of elements Z of $\mathbb{C}^k$, such that the* r.v. $x \mapsto e^{<z,x>}$ *is integrable; define the function $\hat{\mu}$ on J by $\hat{\mu}(z) = \int e^{<z,x>} d\mu(x)$.*

(a) *J is convex.*
(b) *If z is in the interior of J, $\hat{\mu}$ is infinitely differentiable at z and for $(i_1, i_2, ..., i_k) \in \mathbb{N}^k$:*

$$\int |x_1^{i_1} x_2^{i_2} \dots x_k^{i_k}| \, |e^{<z,x>}| d\mu(x) < \infty,$$

$$\frac{\partial^{i_1 + \dots + i_k} \hat{\mu}}{\partial z_1^{i_1} \partial z_2^{i_2} \dots \partial z_k^{i_k}}(z) = \int x_1^{i_1} x_2^{i_2} \dots x_k^{i_k} e^{<z,x>} d\mu(x).$$

(c) Log $\hat{\mu}$ *is a convex function on the interior of $I = J \cap \mathbb{R}^k$; it is strictly convex if μ is not concentrated on a hyperplane of $\mathbb{R}^k$.*

Proof. The convexity of J follows from Inequality 3.2.16(a). For $0 < t < 1$ and z and z' in J we have

$$|e^{t<z,x>+(1-t)<z',x>}| \leq t|e^{<z,x>}| + (1-t)|e^{<z',x>}| \, .$$

On I, the inequality is strict unless $e^{<z,\cdot>}$ and $e^{<z',\cdot>}$ are proportional μ a.s., which is not possible when z and z' are distinct, unless μ is a Dirac measure at a point of $\mathbb{R}^k$. To prove (b), consider z in the interior of J, such that the ball with center z and radius r is in J. Let $|h_i| < r$ and $H_i = h_i e_i$, e_i being the ith element of the canonical basis of $\mathbb{C}^k$:

$$\left|\frac{e^{<z,x>} - e^{<z+H_i,x>}}{h_i} - x_i e^{<z,x>}\right|$$

$$= \left|e^{<z,x>} \frac{e^{h_i x_i} - 1 - h_i x_i}{h_i}\right|$$

$$\leqslant |e^{<z,x>}| \sum_{k \geqslant 2} \frac{|h_i^{k-1} x_i^k|}{k!} \leqslant |e^{<z,x>}| \frac{e^{|rx_i|}}{r};$$

$$r \left| \frac{e^{<z,x>} - e^{<z+H_i,x>}}{h_i} - x_i e^{<z,x>} \right|$$

$$\leqslant |e^{<z+re_i,x>}| + |e^{<z-re_i,x>}| .$$

From this the integrability of $x \longmapsto x_i \exp(<z,x>)$ follows, and by Lebesgue's Theorem the stated value for $\partial\hat{\mu}/\partial z_i(z)$. The successive derivatives are obtained likewise by recurrence.

We now prove (c). Let z be in the interior $\mathring{I}$ of I, F the probability on $\mathbb{R}^k$ of density

$$x \longmapsto \frac{e^{<z,x>}}{\hat{\mu}(z)}$$

with respect to μ and $X = (X_1, ..., X_k)$ a random vector with distribution F. *For* $1 \leqslant i, j \leqslant k$,

$$\frac{\partial}{\partial z_i} \operatorname{Log} \hat{\mu}(z) = \frac{1}{\hat{\mu}(z)} \int x_i e^{<z,x>} d\mu(x)$$

$$\frac{\partial^2}{\partial z_i \partial z_j} \operatorname{Log} \hat{\mu}(z) = \frac{1}{\hat{\mu}(z)} \int x_i x_j e^{<z,x>} d\mu(x)$$

$$- \frac{1}{\hat{\mu}^2(z)} \int x_i e^{<z,x>} d\mu(x) \int x_j e^{<z,x>} d\mu(x).$$

Then $\partial^2/\partial z_i \partial z_j \operatorname{Log} \hat{\mu}(z)$ is the covariance $\Gamma(X_i, X_j)$.

We show that the second derivative matrix of $\operatorname{Log} \hat{\mu}$ is positive at the point z. Let $\lambda = (\lambda_1, ..., \lambda_k) \in \mathbb{R}^k$:

$$\sum_{i=1}^{k} \sum_{j=1}^{k} \lambda_i \lambda_j \frac{\partial^2}{\partial z_i \partial z_j} \operatorname{Log} \hat{\mu}(z) = \sum_{i=1}^{k} \sum_{j=1}^{k} \lambda_i \lambda_j \Gamma(X_i, X_j)$$

$$= \sigma^2 \left[\sum_{i=1}^{k} \lambda_i X_i \right].$$

Suppose that $\lambda \neq 0$. If μ is not concentrated on the hyperplane orthogonal to λ, then

$$\mu\{x; <\lambda,x> \neq 0\} = F\{x; <\lambda,x> \neq 0\} > 0$$

and

$$\sigma^2(<\lambda,X>) \neq 0.$$

Thus if μ is not concentrated on any hyperplane, the second derivative of Log $\hat{\mu}$ is positive definite at every point of $\mathring{I}$: Log $\hat{\mu}$ is strictly convex on $\mathring{I}$. If μ is concentrated on the hyperplane orthogonal to λ, then for all $t \in \mathbb{R}$, Log $\hat{\mu}(\lambda t) = 0$: the function is not strictly convex.

Note. *Let* $I = J \cap \mathbb{R}^k$. *Saying that* z *is in* J *is equivalent to saying that its real part* Re(z) *is in* I. *What is important is the convex set* I *in* $\mathbb{R}^k$. *Part* (b) *of the proposition applies at each point in the interior of* I. *If* I *is a neighborhood of* 0, *it proves that* μ *has moments of all orders and allows them to be calculated with the aid of the derivatives at* 0 *of* ϕ. *This result can be stated more precisely* (*see the important Exercise* 3.3.9).

Application. *Exponential model.* For a σ-finite measure μ on a measurable space (Ω, A) and $T = (T_1, \ldots, T_k)$ a random vector defined on (Ω, A) taking values in $\mathbb{R}^k$, consider the set of θ's of $\mathbb{R}^k$, Θ such that $\exp\langle\theta, T\rangle$ is μ-integrable. For $\theta \in \Theta$ and $\omega \in \Omega$, define

$$\psi(\theta) = \text{Log} \int \exp\langle\theta, T\rangle d\mu,$$

$$p_\theta(\omega) = \exp(-\psi(\theta) + \langle\theta, T(\omega)\rangle).$$

The integral of p_θ with respect to μ equals 1: the measure $P_\theta = p_\theta \cdot \mu$ is a probability on (Ω, A). Denote by E_θ the expectation with respect to P_θ. The preceding theorem can be applied to the image measure of μ by T: Θ is a convex set in $\mathbb{R}^k$, ψ is a convex function. For θ in the interior of Θ we can differentiate indefinitely. We obtain

$$\text{grad } \psi(\theta) \cdot \exp \psi(\theta) = \int T \exp\langle\theta, T\rangle d\mu$$

$$\left[\frac{\partial \psi}{\partial \theta_i}(\theta)\frac{\partial \psi}{\partial \theta_j}(\theta) + \frac{\partial^2 \psi(\theta)}{\partial \theta_i \partial \theta_j}\right] \exp \psi(\theta) = \int T_i T_j \exp\langle\theta, T\rangle d\mu \quad (1 \leq i, j \leq k).$$

Otherwise stated,

$$\text{grad } \psi(\theta) = E_\theta(T)$$

$$\frac{\partial^2 \psi}{\partial \theta_i \partial \theta_j} = \Gamma_\theta(T_i, T_j),$$

denoting by Γ_θ the covariance of P_θ. These formulas will be important in what follows. Meanwhile we shall see various examples of exponential models on $\mathbb{R}$ or **exponential families of distributions.**

3.3.3. Examples of Distributions

(A) We have already tackled, in Chapter 2, the principal **discrete distributions.** Certain of these form exponential families (with the preceding notation):

– The family of *binomial distributions of order n* $\{b(p,n);\ 0 < p < 1\}$. If X is the unit r.v. on $\{0,1, ..., n\}$, the density of $b(n,p)$ with respect to the measure $\Sigma_{k=0}^{n}\binom{n}{k}\delta_k$ is

$$\exp\left\{\left[\operatorname{Log}\left(\frac{p}{1-p}\right)\right]X + n \operatorname{Log}(1 - p)\right\}.$$

We have an exponential family with $\theta = \operatorname{Log}(p/(1-p))$, $T = X$, and $\Theta = \mathbb{R}$. Hence we have

$$\psi'(\theta) = np = E_\theta(X); \quad \psi''(\theta) = np(1 - p) = \sigma_\theta^2(X).$$

– The family of *Poisson distributions* $\{p(\lambda);\ \lambda > 0\}$. If X is the identity r.v. on $\mathbb{N}$, the density of $p(\lambda)$ with respect to $\Sigma_{k\geq 0}(1/k!)\delta_k$ is

$$\exp(X \operatorname{Log} \lambda - \lambda).$$

Here $\Theta = \mathbb{R}$, $\theta = \operatorname{Log} \lambda$, and $T = X$; $E_\theta(T) = \psi'(\theta) = \lambda$; $\sigma_\theta^2(T) = \psi''(\theta) = \lambda$.

– On the other hand, the family of uniform distributions on a finite subset of $\mathbb{N}$ is not exponential (verify this).

(B) **Distributions with density** (assumed to be "with respect to Lebesgue measure"). "The" density is only defined a.s. with respect to Lebesgue measure, but the most regular possible version f is chosen. If, for example, f is a continuous version on the interval $]a,b[$ of the density of a distribution defined on $]a,b[$, it is the derivative of the distribution function F on $]a,b[$.

(a) *Uniform distribution* $\mathcal{U}_{(a,b)}$ on the interval (a,b) of $\mathbb{R}$ with density $1/(b - a)\, 1_{(a,b)}$. Its mean is $(a + b)/2$ and its variance

$(b-a)^2/12$. More generally, let k be an integer ($k \geq 1$); let λ_k be Lebesgue measure on $\mathbb{R}^k$. If A is a bounded Borel set of $\mathbb{R}^k$ charged by λ_k, the **uniform distribution** on A is by definition the quotient by $\lambda_k(A)$ of the trace measure λ_k on A.

(b) *Exponential distribution*, $\mathcal{E}(\lambda)$ for $\lambda \in]0,+\infty[$:

$$f(x) = \lambda e^{-\lambda x} 1_{]0,+\infty[}(x).$$

Its mean is $1/\lambda$, its variance $1/\lambda^2$, and its Laplace transform $t \longmapsto \lambda/(\lambda+t)$. The family $\{\mathcal{E}(\lambda);\ \lambda > 0\}$ is, as its name suggests, the simplest exponential family. If X is the identity r.v. on $]0,\infty[$, the density of $\mathcal{E}(\lambda)$ with respect to Lebesgue measure on $]0,\infty[$ is $\exp(-\lambda x + \mathrm{Log}\ \lambda)$. Here, $\Theta =]-\infty,0[$, $\theta = -\lambda$, $T = X$, and $\psi'(\theta) = E_\theta(T) = 1/\lambda$, $\psi''(\theta) = \sigma_\theta^2(T) = 1/\lambda^2$.

(c) *Gaussian distribution*, $\mathcal{N}(m,\sigma^2)$ for $m \in \mathbb{R}$ and $\sigma \in]0,+\infty[$:

$$f(x) = \frac{1}{\sigma\sqrt{2\pi}} \exp\left[-\frac{(x-m)^2}{2\sigma^2}\right].$$

Let us call $\mathcal{N}(0,1)$ the (standard) normal distribution. We have seen in Section 2.2.4 that the integral of

$$x \longmapsto \frac{1}{\sqrt{2\pi}} \exp\left(-\frac{x^2}{2}\right)$$

equals 1: it is definitely the density of a distribution.

For all real z, we have (setting $y = z - x$):

$$\frac{1}{\sqrt{2\pi}} \int \exp\left[zx - \frac{x^2}{2}\right] dx = \frac{1}{\sqrt{2\pi}} \int \exp\left[\frac{z^2}{2} - \frac{y^2}{2}\right] dy$$

$$= \exp\left[\frac{z^2}{2}\right].$$

The function

$$z \longmapsto \frac{1}{\sqrt{2\pi}} \int \exp\left[zx - \frac{x^2}{2}\right] dx$$

is defined on $\mathbb{C}$, and holomorphic following Proposition 3.3.22. It thus coincides on $\mathbb{C}$ with $z \longmapsto \exp[z^2/2]$. Its Taylor expansion is $\exp[z^2/2] = \sum_{n \geq 0} z^{2n}/n!2^n$. This gives the successive derivatives at 0, thus the moments m_n of order n of $\mathcal{N}(0,1)$:

$$m_{2n+1} = 0, \quad m_{2n} = \frac{(2n)!}{2^n n!}.$$

Let X be an r.v. on a probability space $(\Omega,\mathcal{A},P)$ which has a normal distribution; consider the r.v. $\sigma X + m$. Its distribution is $N(m,\sigma^2)$. In fact, if ϕ is a positive r.v. on $(\mathbb{R}, \mathcal{B}_{\mathbb{R}})$, we have

$$E[\phi(\sigma X + m)] = \int \phi(\sigma x + m)\frac{1}{\sqrt{2\pi}} \exp\left[-\frac{x^2}{2}\right]dx$$

$$= \frac{1}{\sigma\sqrt{2\pi}} \int \phi(x)\exp\left[-\frac{(x-m)^2}{2\sigma^2}\right]dx.$$

Its mean is thus m and its variance σ^2; $N(m,0)$ is the Dirac measure at m. For $z \in \mathbb{C}$

$$E[e^{z(m+\sigma X)}] = e^{zm}E[e^{z\sigma X}] = \exp\left[zm + \frac{z^2\sigma^2}{2}\right].$$

An $N(m,\sigma^2)$ *distribution thus has mean* m *and variance* σ^2. *For all* $z \in \mathbb{C}$, *we have*

$$\int e^{zx}dN(m,\sigma^2)(x) = \exp\left[zm + \frac{z^2\sigma^2}{2}\right].$$

In particular, for $z = iu$ and u real, we obtain the *Fourier transform* of $N(m,\sigma^2)$,

$$u \longmapsto \exp\left[ium - \frac{u^2\sigma^2}{2}\right].$$

The family of distributions $\{N(m,\sigma^2);\ m \in \mathbb{R},\ \sigma^2 > 0\}$ is an exponential family. The density of $N(m,\sigma^2)$ may be written

$$f(x) = \exp\left[-\frac{x^2}{2\sigma^2} + \frac{xm}{\sigma^2} - \frac{m^2}{2\sigma^2}\right].$$

We can take $\theta = (m/\sigma^2, -1/2\sigma^2)$ and $T = (X,X^2)$, X being the identity r.v. on $\mathbb{R}$. Then, grad $\psi(\theta) = (E_\theta(X), \ E_\theta(X^2)) = (m,\sigma^2 + m^2)$.

(d) *The Cauchy distribution* with parameter $\alpha > 0$:

$$f(x) = \frac{1}{\pi}\frac{\alpha}{\alpha^2 + x^2}.$$

It does not have a mean value; in fact,

$$-\int_{-\infty}^{0} \frac{\alpha x\, dx}{\alpha^2 + x^2} = \int_{0}^{+\infty} \frac{\alpha x\, dx}{\alpha^2 + x^2} = \infty.$$

Its Fourier transform is $t \longmapsto e^{-\alpha|t|}$: it can be calculated by residues or we can wait until Exercise 3.4.9.

The distributions on $\mathbb{R}$ with a density are the most frequent continuous distributions. However, these are not the only ones (see Exercise 3.3.16).

3.3.4. The Images of Distributions with a Density

Let X be a random vector taking values in an open set U of $\mathbb{R}^k$. Let ϕ be a diffeomorphism (a bijective function of class C^1) of U into an open set V of $\mathbb{R}^k$. Let $\phi_1, \ldots, \phi_k$ denote the components and J_ϕ its Jacobian

$$J_\phi(x_1, \ldots, x_k) = \begin{vmatrix} \frac{\partial \phi_1}{\partial x_1} & \cdots & \frac{\partial \phi_1}{\partial x_k} \\ \cdots & \cdots & \cdots \\ \frac{\partial \phi_k}{\partial x_1} & \cdots & \frac{\partial \phi_k}{\partial x_k} \end{vmatrix}.$$

Let $Y = \phi(X)$; assume that X has a density f_X with respect to Lebesgue measure and Y has a density f_Y. For g, a positive r.v. on V, we have

$$\begin{aligned} E[g(Y)] &= \int_V g(y) f_Y(y) dy = E[g \circ \phi(X)] \\ &= \int_U g \circ \phi(x) f_X(x) dx. \end{aligned}$$

Applying the change of variables formula:

$$\int_V g(y) f_Y(y) dy = \int_U (g \circ \phi(x))(f_Y \circ \phi(x)) |J_\phi(x)| dx.$$

Thus

$$f_{\phi(X)}(x) = \frac{1}{|J_\phi(\phi^{-1}(x))|} f_X(\phi^{-1}(x)).$$

Notice that, in order to apply this formula, it is sufficient to assume that the distribution of X does not charge the complement of U.

Examples. (a) The *Log-normal distribution.* A positive r.v. X is said to have a Log-normal distribution with parameters (m,σ^2) when Log X has distribution $\mathcal{N}(m,\sigma^2)$. Applying the above to the function Log, the density of the Log-normal is obtained, $x \longmapsto (1/|x|)\exp(-1/2\sigma^2[\text{Log } x - m]^2)$.

(b) The mapping $(x,y) \longmapsto (x,y/x)$ is a diffeomorphism of $\mathbb{R}^2\backslash(\{0\} \times \mathbb{R})$ into itself; its Jacobian at the point (x,y) is $1/x$. If (X,Y) has a distribution with density $f_{X,Y}$, this distribution does not charge $\{0\} \times \mathbb{R}$. If $f_{X,Y/X}$ is the density of $(X,Y/X)$, a random vector defined a.s., we have

$$f_{X,Y/X}(u,v) = |u| f_{X,Y}(u,uv).$$

(c) For an affine mapping $x \longmapsto ax + b$ from $\mathbb{R}$ into $\mathbb{R}$, we obtain the formula (which we have used to deduce the density of $\mathcal{N}(m,\sigma^2)$ from that of $\mathcal{N}(0,1)$):

$$f_{aX+b}(x) = \frac{1}{|a|} f_X\left[\frac{x-b}{a}\right].$$

(d) The change of variable formula holds by assuming only that ϕ is a local diffeomorphism. We can, for example, use it for $x \longmapsto x^2$, a diffeomorphism of $\{x > 0\}$ into $\{x > 0\}$, and of $\{x < 0\}$ into $\{x > 0\}$. We have, a.s. on $\mathbb{R}\backslash\{0\}$,

$$f_{X^2}(u) = \frac{1}{2\sqrt{u}} [f_X(\sqrt{u}) + f_X(-\sqrt{u})] 1_{(u>0)}.$$

3.3.5. The Young Transform of a Convex Function; The Cramer Transform of a Distribution

Let ψ be a convex function defined on an interval I of $\mathbb{R}$, taking values in $\mathbb{R}$. We call the **Young transform** h of ψ the function defined on $\mathbb{R}$: $a \longmapsto h(a) = \sup_{u \in I}[ua - \psi(u)]$. It can be seen that h itself is convex (it is the upper bound of affine functions).

Let us asume now that ψ is strictly convex (its graph does not contain any line segments) and differentiable on the interior $\mathring{I}$ of I; its derivative ψ' is strictly increasing. Let $a \in \psi'(\mathring{I})$, then the derivative $a - \psi'(u)$ of $u \longmapsto ua - \psi(u)$ vanishes at $g(a) = \psi'^{-1}(a)$, and $h(a) = ag(a) - \psi(g(a))$. If ψ is twice differentiable, $h'(a) = g(a)$: h and ψ, considered respectively as functions on $\psi'(\mathring{I})$ and $\mathring{I}$, thus have reciprocal derivatives. Note that h is then also strictly convex, and that on $\mathring{I}$, $\psi(u) = \sup_{a \in \psi'(\mathring{I})}[ua - h(a)]$. Hence the relation is one of duality.

Example. $\psi(u) = (1/p)|u|^p$ and $h(a) = (1/q)|a|^q$, with $p > 1$, $1/p + 1/q = 1$.

Note. *If we replace* $\mathring{I}$ *by* I *(which could be different from* $\mathring{I}$*), we can always define* h *for all* a*, but the above properties only hold for* a *in* $\psi(\mathring{I})$*. The above situation can be extended to* $\mathbb{R}^k$ *and the following can be stated.*

Definition 3.3.23 and Properties. Let ψ be a strictly convex real valued function defined on a convex set Θ of $\mathbb{R}^k$, differentiable in the interior $\mathring{\Theta}$ of Θ. The Young transform of ψ is defined, for $a \in \mathbb{R}^k$, by $h(a) = \sup_{u \in \Theta}[\langle u,a \rangle - \psi(u)]$. For $a \in \text{grad } \psi(\mathring{\Theta})$.

$$h(a) = \langle a,g(a) \rangle - \psi(g(a)) \quad \text{with} \quad \text{grad } \psi(g(a)) = a.$$

If ψ is twice differentiable on $\mathring{\Theta}$, h is strictly convex on $\psi'(\mathring{\Theta})$ and

$$\text{grad } h(a) = g(a)$$

$$\psi(u) = \langle u, \text{grad } \psi(u) \rangle - h(\text{grad } \psi(u)), \quad \text{for } u \in \mathring{\Theta}.$$

Let X then be an r.v. with distribution F and I_F the set of real numbers such that $E(e^{tX}) = \int e^{tx}dF(x)$ is finite. Following Proposition 3.3.22, I_F is an interval (A_F,B_F) (the end points may or may not be included) and the function $t \longmapsto \psi_F(t) = \text{Log } E(e^{tX})$ defined on I_F is convex.

Definition 3.3.24. The **Cramer transform** h_F of a distribution F defined on $\mathbb{R}$ is the Young transform of the function ψ_F.

Properties of a Distribution F on $\mathbb{R}$ for which I_F is a Neighborhood of 0. From Proposition 3.3.22, such a distribution F has moments of all orders and ψ_F can be differentiated at 0:

$$\psi_F(0) = 0,\ \psi_F'(0) = m,\ h_F'(m) = 0,\ h_F(m) = -\psi_F(0) = 0.$$

Let us extend ψ_F by $+\infty$ outside I_F.

The function h_F has a minimum, equal to zero, at the point m. It is strictly convex on $\psi(\mathring{I}_F)$, thus strictly positive at each point distinct from m. Let $a > m$; we have $h_F(a) = \sup_{u>0}[ua - \psi_F(u)]$. This can be seen when a is in $\psi'(\mathring{I}_F)$, since the maximum of $ua - \psi_F(u)$ is attained for $u - h_F'(a)$ (see Exercise 3.3.12) for the case when $a \notin \psi'(\mathring{I}_F)$. For arbitrary $u > 0$, we have $1_{(x \geq a)} \leq e^{u(x-a)}$, from which

$$P(X \geq a) \leq \inf_{u>0} E[e^{u(X-a)}] = \inf_{u>0} \{\exp - (ua - \psi_F(u))\}$$
$$\leq \exp[-h_F(a)].$$

Likewise for $a < m$, $h_F(a) = \sup_{u<0}[ua - \psi_F(u)]$;

$$P(X \leq a) \leq \inf_{u<0} E[e^{u(X-a)}] \leq \exp[-h_F(a)].$$

Proposition 3.3.25. *If $I_F = \{t;\ E(e^{tX}) < \infty\}$ is a neighborhood of* 0, *then*

$$\text{for} \quad a > m,\ P(X \geq a) \leq \exp[-h_F(a)],$$

$$\text{and for} \quad a < m,\ P(X \leq a) \leq \exp[-h_F(a)],$$

where h_F is the Cramer transform of the distribution X, strictly positive except for $a = m$.

Exercises 3.3.

All the random vectors are defined on a probability space $(\Omega, \mathcal{A}, P)$.

E.1. Let X be an r.v. Find as a function of the distribution function of X, the distribution functions of the following r.v.'s: X^n (for $n \in \mathbb{Z}$), $aX + b$ (for a and b in $\mathbb{R}$), $[X]$, and $X - [X]$ (where $[X]$ is the integer part of X), Arctan X, e^X.

E.2. Let X be an r.v., the density of which is

$$x \longmapsto \frac{1}{\text{Log } 2} \frac{1}{1 + x} 1_{[0<x<1]} .$$

Verify that the distribution of $1/(X - [1/X])$, where [ ] signifies "the integer part," is the same as that of X.

E.3. *Cauchy Distribution.* Let θ be a uniform r.v. on $[-\pi/2, \pi/2]$. Determine the distribution function and density of $\tan \theta$; verify that this is a Cauchy distribution.

E.4. *Stock Control.* A businessman holds stock equal to s; to order a quantity q cost him cq. The demand which he forecasts is an exponential r.v. D with parameter 1. He thus sells $\inf(D, s + q)$ and will earn $v \inf(D, s + q)$. Knowing s and the buying and selling prices c and v $(c < v)$, what is the order q which maximizes the profit?

E.5. (a) Let F be a distribution on $\mathbb{R}$; define for $t \in]0,1[$: $F^{-1}(t) = \inf\{x; F(x) \geq t\}$. Show

$$\begin{cases} F[F^{-1}(t)] \geq t & \text{with equality for continuous } F; \\ F^{-1}[F(x)] \leq x & \text{with equality for } F \text{ strictly increasing.} \end{cases}$$

(b) Let X be an r.v. with distribution F; prove that $F^{-1}(F(x))$ is equal to X a.s.

(c) Show that, if F is continuous, $F(X)$ has a uniform distribution $[0,1]$. Calculate in this case $\int F^n dF$ for $n \geq 0$.

(d) Show that, if U is uniform on $[0,1]$, $F^{-1}(U)$ has distribution F.

E.6. *A partial order on the distributions on* $\mathbb{R}$. Let X and Y be two r.v.'s with distributions F and G. Assume that for each $s \in \mathbb{R}$, $F(s)$ is less than $G(s)$. We say that the distribution F is larger than G (for once, the confusion of notation is unlucky!). Prove that, for every increasing function f from $\mathbb{R}$ into $\mathbb{R}$, $E(f(X))$ is larger than $E(f(Y))$. Give an example of distributions F and G, F larger than G. For these distributions, construct a pair of r.v.'s X and Y with distributions F and G, satisfying $X \geq Y$; construct another pair of r.v.'s X and Y with distributions F and G for which this property does not hold.

E.7. Let (X,Y) be a random vector with distribution $F_{X,Y}$. Find in the following cases the marginal distributions of (X,Y): the distribution functions of (X,Y), X, Y and $X + Y$; the covariance of (X,Y) (if P and Q are points in $\mathbb{R}^2$, λ_{PQ} is Lebesgue measure on the segment PQ).

(a) $A = (0,1)$, $B = (1,0)$, $C = (0,-1)$, $F_{X,Y} = (1/2\sqrt{2})(\lambda_{AB} + \lambda_{BC})$.
(b) $D = (1,1)$, $E = (-1,1)$, $F = (-1,-1)$, $G = (1,-1)$, $F_{X,Y} = (1/8)(\lambda_{DE} + \lambda_{EF} + \lambda_{FG} + \lambda_{GD})$.
(c) $F_{X,Y}$ the uniform distribution on the triangle ABC.
(d) $F_{X,Y}$ the uniform distribution on the square $DEFG$.
(e) $F_{X,Y} = (1/3)\Sigma_{n\geqslant 1, m\geqslant 1}(1/2^n 3^m)\varepsilon_{(n,m)}$.
(f) $F_{X,Y}$ the uniform distribution on the circumference with center 0 and radius 1:

$$F_{X,Y}(\Gamma) = \frac{1}{2\pi}\int_0^{2\pi} 1_\Gamma(\cos\theta, \sin\theta)d\theta.$$

E.8. Let X,Y be two finite r.v.'s on $(\Omega,\mathcal{A},P)$. Let $F_{X,Y}$, F_X, F_Y be the distributions of (X,Y), X, and Y, respectively. Let f be a finite r.v. on $(\mathbb{R},\mathcal{B}_{\mathbb{R}})$. Prove that $\Gamma_f = \{(x,y); y = f(x)\}$ is a Borel set in $\mathbb{R}^2$. Show that if $F_{X,Y}$ is concentrated on Γ_f, then $F_Y = f(F_X)$. Is the converse true?

E.9. *The relationship between the Fourier transform ϕ and the moments of a distribution F on $\mathbb{R}$.* (a) Show that if $\int|x|^n dF(x)$ is finite, then ϕ is differentiable n times and

$$\phi^{(n)}(t) = i^n\int x^n e^{itx}dF(x).$$

Show that if t and h are two real numbers then we have

$$\left|\ \phi(t+h) - \sum_{k=0}^{n-1}\phi^{(k)}(t)\frac{h^k}{k!}\right| \leqslant \frac{h^n}{n!}\int|x|^n dF(x).$$

(b) Show that if ϕ is twice differentiable at 0, F has a moment of order 2. However, it is useless to look for a converse to (a) when n is odd. Examine the following example: let

$$F = \sum_{n\geqslant 2}\frac{c}{n^2 \operatorname{Log} n}(\varepsilon_n + \varepsilon_{-n}).$$

Verify that ϕ is differentiable, although $\int|x|dF(x)$ is infinite.

E.10. We say that a probability F on $\mathbb{R}$ is arithmetic if it is concentrated on a set of the form $a + h\mathbb{Z}$ ($a \in \mathbb{R}$, $h > 0$). Show that a distribution F on $\mathbb{R}$ is arithmetic if, and only if, its Fourier transform ϕ takes a value ofmodulus 1 at a point different from 0. (Show first of all that if ϕ equals 1 at a point $t \neq 0$, F is concentrated on $(2\pi/t)\mathbb{Z} + a$, a being a real number.)

E.11. Let F be a distribution concentrated on $\mathbb{R}_+$ and ϕ its Laplace transform. Show that ϕ is infinitely differentiable on $]0,\infty[$ and that for each integer n,

$$\int x^n dF(x) = \lim_{t\to 0} |\phi^{(n)}(t)| \leqslant \infty.$$

E.12. *Cramer transform.* Let F be a probability on $\mathbb{R}$, ϕ_F its Laplace transform on the interval I_F where it is defined, ψ_F = Log ϕ_F, and h_F the Cramer transform of F.

(a) Determine I_F, ϕ_F, and h_F for the following distributions F: $N(0,\sigma^2)$, $b(p)$, $p(\lambda)$, $U_{[-1,1]}$. In each case specify $\psi'(I_F)$ and the behavior of h_F at the boundaries of $\psi'(I_F)$.

(b) Let $\overline{F}$ and $\underline{F}$ be the essential upper and lower bounds of F; $\overline{F} = \inf\{a; F(a) = 1\}$, $\underline{F} = \sup\{a; F(a) = 0\}$. Show

$$\lim_{u\to\infty} e^{-ua}\phi_F(u) = \begin{cases} 0 & \text{for } a > \overline{F} \\ F(\{\overline{F}\}), & \text{for } a = \overline{F} \\ \infty & \text{for } a < \overline{F} \end{cases} \quad \text{if } \overline{F} < \infty.$$

Give the analagous result when u tends to $-\infty$.

(c) Show that h_F is finite on $]\underline{F},\overline{F}[$, infinite on $[\underline{F},\overline{F}]^c$, and that $h_F(\overline{F}) = -\text{Log } F(\{\overline{F}\})$ and $h_F(\underline{F}) = -\text{Log } F(\{\underline{F}\})$ if $\overline{F}$ and $\underline{F}$ are finite.

(d) Assume that I_F is a neighborhood of 0; let m be the mean of F. Verify that, for each $a > m$, we have

$$h_F(a) = \sup_{u>0} (ua - \psi_F(u)).$$

(e) Show that, if I_F is nonempty, the Cramer transform h_F characterizes F.

E.13. *The Gamma distribution.* For $a > 0$ and $b > 0$, the distribution $\gamma(a,b)$ with density

$$x \longmapsto \frac{1}{\Gamma(a)} b^a e^{-bx} x^{a-1} 1_{(x>0)}$$

is called the Gamma distribution. Calculate its Laplace transform. Prove that its mean is a/b and its variance a/b^2 (see Section 3.3.4 for the definition of Γ).

Show that the family $\{\gamma(a,b); a > 0, b > 0\}$ is an exponential family. Let X be an r.v. with distribution $\gamma(a,b)$; calculate with the aid of the derivatives of the function Γ the expectation and variance of Log X and the covariance of X and Log X.

E.14. *The Beta distribution.* For $a > 0$ and $b > 0$, we call the Beta distribution $\beta(a,b)$ the distribution concentrated on $[0,1]$ with density

$$x \longmapsto \frac{\Gamma(a+b)}{\Gamma(a)\Gamma(b)} x^{a-1}(1-x)^{b-1}$$

on $[0,1]$, (see Section 2.3.4). Verify that this function is certainly the density of a distribution. Prove that its mean is $a/(a+b)$and its variance $ab/[(a+b)^2(a+b+1)]$.

Show that $\{\beta(a,b);\ a > 0,\ b > 0\}$ is an exponential family. Let X be an r.v. with distribution $\beta(a,b)$; calculate the means and variances of Log X and Log$(1-X)$, as well as their covariance, as a function of the derivatives of Γ.

E.15. *Dyadic numbers and singular distributions on* $\mathbb{R}$ (a) In the model described in Section 2.5.3, Ω is $\{0,1\}^{\mathbb{N}}$ and $\mathcal{A}$ is the σ-algebra generated by coordinate functions $(X_m)_{m\geq 1}$. Let ϕ be the mapping

$$\omega \longmapsto \sum_{n\geq 1} \frac{X_n(\omega)}{2^n}$$

from Ω into $[0,1[$. Let Γ be the set of $\omega = (X_n(\omega))_{n\geq 1}$, each of the terms of which takes the value 1 from a certain point onwards; $\phi(\Gamma) = \mathbb{Q} \cap]0,1[$. Show that ϕ is bijective bimeasurable from $\Omega\backslash\Gamma$, equipped with the trace of $\mathcal{A}$, into $]0,1[$ equipped with the Borel σ-algebra.

(b) Deduce from the existence of non-Borel subsets of $]0,1[$ the fact that there exist subsets of Ω not in $\mathcal{A}$.

(c) Assume that, on $(\Omega,\mathcal{A},P_\theta)$, the r.v.'s X_n are independent and have a Bernoulli distribution with parameter θ. Let F_θ be the distribution of ϕ on $(\Omega,\mathcal{A},P_\theta)$. What is $F_{1/2}$? Prove that, for $\theta \neq \theta'$, the measures F_θ and $F_{\theta'}$ are concentrated on disjoint sets. Prove that for all $\theta \in]0,1[$, the distribution F_θ is continuous but that it only has a density for $\theta = 1/2$.

E.16. *Pareto distributions.* (a) For $r > 0$, $\alpha > 0$, consider the one-sided Pareto distribution with density $f_{r,\alpha}$:

$$x \longmapsto \frac{\alpha r^\alpha}{x^{\alpha+1}} 1_{[x>r]}\,.$$

Calculate its mean and variance if they exist.

(b) For $r < s$ and $\alpha > 0$, consider the two-sided Pareto distribution with density $f_{r,s,\alpha}$:

$$(x,y) \longmapsto \frac{\alpha(\alpha+1)(s-r)^{\alpha}}{(y-x)^{\alpha+2}} 1_{(x<r<s<y)} .$$

Let (X,Y) be a pair of r.v.'s having this distribution. Show that $s - X$ and $Y - r$ have the same distribution and determine it. Calculate $E[Y - X]^2$ and $\Gamma(X,Y)$ for $\alpha > 2$.

3.4. Convergence in Distribution

The notations used here are those of Section 3.2.4.

3.4.1. Weak Convergence of a Sequence of Bounded Measures of $\mathbb{R}^k$

Definition 3.4.26. Let (F_n) be a sequence in M and $f \in M$. The sequence (F_n) converges to F:

(a) **Vaguely** ($(F_n) \xrightarrow{v} F$) if, for all $f \in C_K$, $(F_n(f))$ tends to $F(f)$;

(b) **Weakly** ($(F_n) \xrightarrow{w} F$) if, for all $f \in C_0$, $(F_n(f))$ tends to $F(f)$;

(c) **Narrowly** ($(F_n) \xrightarrow{n} F$) if, for all $f \in C_b$, $(F_n(f))$ tends to $F(f)$.

Theorem 3.4.27. *From each bounded sequence (F_n) of M (i.e., such that $(F_n(\mathbb{R}^k))$ is bounded), we can extract a subsequence which converges weakly. If the sequence (F_n) does not converge weakly we can extract two sequences which converge weakly to distinct limits.*

Proof. The set A introduced in Section 3.2.4 can be replaced by the countable set of linear combinations with rational coefficients of the functions $\exp i\langle u,x\rangle$, $u \in \mathbb{Q}^k$. The set $\{\phi_F(u);\ u \in \mathbb{Q}^k\}$ determines ϕ_F by continuity. If we take the products of these functions by a sequence of continuous functions with compact support increasing towards 1, a countable family A' of C_K is constructed $\{g_p;\ p \geqslant 1\}$ which is dense in C_K. The sequence $(F_n(g_1))$ being bounded, we can extract a convergent sequence $(F_{n,1}(g_1))$; the sequence $F_{n,1}(g_2)$ being bounded, we can extract a convergent sequence $F_{n,2}(g_2)\dots$. Thus by recurrence a nested sequence is constructed $(F_n) \supset (F_{n,1}) \supset (F_{n,2}) \dots \supset (F_{n,p}) \supset \dots$, of such a kind that $(F_{n,p}(g_i))$ converges if n tends to ∞, for $1 \leqslant i \leqslant p$. Then the sequence $(F_{p,p}(g_i))$ converges if p tends to ∞, irrespective of i, $i \geqslant 1$ (this is the "diagonal procedure"). Thus for each $g \in A'$, $(F_{p,p}(g))$ converges to a number $F_0(g)$. For $f \in C_0$ and > 0, we can find $g \in A'$ such that $\sup|f(x) - g(x)|$ is less than $\varepsilon/2M$, with $M = \sup_n F_n(\mathbb{R}^k)$. For $p \geqslant 1$ and $n \geqslant 1$,

$$
\begin{aligned}
&|F_{p,p}(f) - F_{p+n,p+n}(f)| \\
&\qquad \leqslant \frac{\varepsilon}{2} + |F_{p,p}(g) - F_{p+n,p+n}(g)|.
\end{aligned}
$$

For p large enough, this expression is less than ε, and $(F_{p,p}(f))$ is a Cauchy sequence; it converges to a number $\ell(f)$. This function $f \longmapsto \ell(f)$ is a linear form of C_0; it is positive, thus it is a measure bounded by M.

If (F_n) does not converge weakly, we can take $g_1 \in A'$, such that $(F_n(g_1))$ does not converge. Let $\ell(g_1)$ and $\ell'(g_1)$ be two distinct closure points. The sequence $(F_{n,1}(g_1))$ can be constructed in such a way as to converge to $\ell(g_1)$ or to $\ell'(g_1)$ from which two distinct values ℓ and ℓ' are possible by the process described above.

Theorem 3.4.28. *Let (F_n) be a bounded sequence of $\mathcal{M}$. Assume that the sequence of its Fourier transforms (ϕ_{F_n}) converges a.s. for Lebesgue measure to a function ϕ, then there exists $\tilde{\phi}$, $\tilde{\phi} = \phi$* a.s., *which is the Fourier transform of a measure F of $\mathcal{M}$, and (F_n) converges weakly to F.*

Proof. The function

$$
x \longmapsto \exp\left(-\frac{|x-y|^2}{2}\right)
$$

is, for all y, in C_0. However, we have, for $x = (x_j)_{1 \leqslant j \leqslant k}$:

$$\exp\left[-\frac{|x|^2}{2}\right] = \prod_{j=1}^{k} \exp\left(-\frac{x_j^2}{2}\right)$$

$$= \prod_{j=1}^{k} \frac{1}{\sqrt{2\pi}} \int \exp\left(ix_ju_j - \frac{u_j^2}{2}\right) du_j$$

$$= (2\pi)^{-k/2} \int \exp\left(i<x,u> - \frac{|u|^2}{2}\right) du.$$

Thus for $F \in M$,

$$\int \exp\left(-\frac{|x-y|^2}{2}\right) dF(x)$$
$$= (2\pi)^{-k/2} \int dF(x) \int \exp\left(i<x-y,u> - \frac{|u|^2}{2}\right) du.$$

By applying Fubini's Theorem (see the statement of Theorem 4.1.4) we obtain

$$(2\pi)^{k/2} \int \exp\left(-\frac{|x-y|^2}{2}\right) dF(x)$$

$$= \int du\ \phi_F(u) \exp\left[-i<y,u> - \frac{|u^2|}{2}\right].$$

(b) Let us assume that (F_n) converges weakly to F. Taking in these expressions F_n instead of F and letting n tend to ∞, we obtain (under the hypotheses of the theorem)

$$(2\pi)^{k/2} \int \exp\left(-\frac{|x-y|^2}{2}\right) dF(x)$$
$$= \int du\ \phi(u) \exp\left[-i<y,u> - \frac{|u|^2}{2}\right].$$

Measures having densities ϕ and ϕ_F with respect to the measure $\nu(du) = \exp(-|u|^2/2)du$ have the same Fourier transform: they are equal. Thus ϕ and ϕ_F coincide a.s. for Lebesgue measure.

(c) If two sequences extracted from (F_n) converge to the limits F and F', ϕ_F and $\phi_{F'}$ are equal a.s. for Lebesgue measure, to ϕ. Since ϕ_F and $\phi_{F'}$ are continuous, they are equal and F coincides with F'. The sequence (F_n) is weakly convergent following Theorem 3.4.27.

3.4.2. Narrow Convergence of a Sequence of Bounded Measures on $\mathbb{R}^k$

Let (δ_n) be the sequence of Dirac measures on the integers. It is clear that it converges weakly to the null measure, but not narrowly because $(\delta_n(1))$ does not tend to zero. There is a loss of mass.

Theorem 3.4.29. *Let (F_n) be a sequence in $\mathcal{M}$. To say that (F_n) converges narrowly to F is equivalent to saying that it converges weakly to F and that $(F_n(\mathbb{R}^k))$ tends to $F(\mathbb{R}^k)$.*

Proof. Narrow convergence certainly implies weak convergence and the convergence of $(F_n(1))$ to $F(1)$. Conversely we can, for $\varepsilon > 0$, find a compact set Γ such that $F(\Gamma^c)$ is majorized by ε. Thus, also a function $g \in C_K$, majorizing 1_Γ, lying between 0 and 1, such that $F(1 - g)$ is majorized by 2ε.

Let $f \in C_b$:

$$|F_n(f) - F(f)| \leq |F_n(fg) - F(fg)| + \|f\|(F_n(1 - g) + F(1 - g)).$$

But $(F_n(1)) \to F(1)$ and $(F_n(g)) \to F(g)$, from which

$$\overline{\lim_n} |F_n(f) - F(f)| \leq 4\varepsilon\|f\|.$$

This being true for all $\varepsilon > 0$, $(F_n(f))$ tends to $F(f)$.

Corollary 3.4.30. (Levy's Theorem). *Let (F_n) be a bounded sequence in $\mathcal{M}$ such that (ϕ_{F_n}) converges to a function ϕ continuous at zero. Then, ϕ is the Fourier transform of a measure F and (F_n) converges narrowly to F.*

Proof of the Corollary. Following Theorem 3.4.28, (F_n) converges weakly to F and ϕ coincides with ϕ_F a.s. for Lebesgue measure. There exists therefore a sequence (x_k) which tends to zero, such that $\phi_F(x_k) = \phi(x_k)$. If ϕ is continuous at zero, this means that $\phi(0) = \lim F_n(\mathbb{R}^k)$ coincides with $\phi_F(0) = F(\mathbb{R}^k)$.

3.4.3. Narrow Convergence of a Sequence of Probabilities on a Metric Space

Let E be a metric space together with its Borel σ-algebra, (P_n) a sequence of probabilities on E and P a probability on E.

Definition 3.4.31. The sequence (P_n) converges **narrowly** to P ($(P_n) \xrightarrow{n} P$) if, for all continuous bounded functions from E into $\mathbb{R}$, $(P_n(f))$ converges to $P(f)$.

Proposition 3.4.32. *"$(P_n) \xrightarrow{n} P$" is equivalent to saying that, for Γ closed in E, we have*

$$\overline{\lim_n}\, P_n(\Gamma) \leqslant P(\Gamma).$$

Proof. (a) Let Γ be a closed set in E and $d(\cdot,\Gamma)$ the distance to Γ. For $n \in \mathbb{N}$ consider the continuous function $\phi(\cdot) = (1 - nd(\cdot,\Gamma))_+$; it equals 1 on Γ and decreases to 1_Γ if n tends to ∞. For all $\varepsilon > 0$, n can be large enough such that $P(\phi) \leqslant P(\Gamma) + \varepsilon$. If (P_n) converges narrowly to P we have

$$\overline{\lim_n}\, P_n(\Gamma) \leqslant \overline{\lim_n}\, P_n(\phi) = P(\phi) \leqslant P(\Gamma) + \varepsilon.$$

This is true for all $\varepsilon > 0$, thus the stated condition is necessary.

(b) Let us prove the converse. Notice first of all that here, to each bounded measurable $\hat{f}$ the modulus of which is $< M$, we can associate

$$f = \frac{M - \hat{f}}{2M},$$

and the study of the sequence $(P_n(\hat{f}))$ amounts to the study of $(P_n(f))$ for $0 \leqslant f < 1$.

Let f be continuous, $0 \leqslant f < 1$. For each probability P on E, we have

$$\sum_{i=1}^{k} \frac{i-1}{k} P\left[\frac{i-1}{k} \leqslant f < \frac{i}{k}\right]$$

$$\leqslant P(f) < \sum_{i=1}^{k} \frac{i}{k} P\left[\frac{i-1}{k} \leqslant f < \frac{i}{k}\right],$$

$$\frac{1}{k}\sum_{i=1}^{k} P\left[f \geq \frac{i}{k}\right] \leq P(f) < \frac{1}{k} + \frac{1}{k}\sum_{i=1}^{k} P\left[f \geq \frac{i}{k}\right].$$

Let us assume that for each closed set Γ of E we have $\overline{\lim}\, P_n(\Gamma) \leq P(\Gamma)$. Then

$$\overline{\lim}\, P_n(f) \leq \frac{1}{k} + \frac{1}{k}\sum_{i=1}^{k} \overline{\lim}\, P_n\left[f \geq \frac{i}{k}\right]$$

$$\leq \frac{1}{k} + \frac{1}{k}\sum_{i=1}^{k} P\left[f \geq \frac{i}{k}\right] \leq \frac{1}{k} + P(f).$$

This being true for all k, $\overline{\lim}\, P_n(f) \leq P(f)$. Applying this result to $1 - f$, $\underline{\lim}\, P_n(f) \geq P(f)$. The sequence $(P_n(f))$ thus tends to $P(f)$.

Theorem 3.4.33. *Let (P_n) be a sequence of probabilities on a metric space E which converges narrowly to P.*

(a) *Let E' be another metric space and h a measurable function of E into E', P almost surely continuous. The sequence of image measures $(h(P_n))$ converges narrowly to $h(P)$.*

(b) *Let h be a bounded measurable function from E into $\mathbb{R}$, P almost surely continuous, the sequence $(P_n(h))$ tends to $P(h)$.*

Proof. (a) The set D_h of points where h is discontinuous is measurable (Exercise 3.3.18) and $P(D_h)$ is zero. For each closed set Γ of E', the closure $\overline{h^{-1}(\Gamma)}$ of $h^{-1}(\Gamma)$ is contained in $h^{-1}(\Gamma) \cup D_h$;

$$\overline{\lim}\, h(P_n)(\Gamma) = \overline{\lim}\, P_n(h^{-1}(\Gamma)) \leq \overline{\lim}\, P_n[\overline{h^{-1}(\Gamma)}]$$

$$\leq P[\overline{h^{-1}(\Gamma)}] = P(h^{-1}(\Gamma)) = h(P)(\Gamma).$$

Thus $(h(P_n))$ converges narrowly to $h(P)$.

(b) This follows from (a). Let $f \in C_b$, $f(x) = x$ for $|x| \leq \|h\|$. Then $(\int f(h)dP_n)$ tends to $\int f(h)dP$, since $(h(P_n))$ tends to $h(P)$. Thus $(\int h\, dP_n) \to \int h\, dP$.

Corollary 3.4.34. *A sequence (F_n) of distributions on $\mathbb{R}^k$ converges narrowly to the distribution F if, and only if, the sequence $(F_n(x))$ converges to $F(x)$ at each point $x \in \mathbb{R}^k$ where F is continuous.*

Proof. If F is continuous at $x = (x_i)_{1 \leq i \leq k}$, the function

$1_{\Pi_{i=1}^{k}([-\infty,x_i[)}$ is F-almost surely continuous (Exercise 3.1.18),

from which the "only if" part of the corollary follows.

Conversely, if $(F_n(x))$ tends to $F(x)$ at each point x where F is continuous, $\underline{\lim}\ F_n(x)$ is greater than $F(x)$ at each of these points, $\mathbb{R}^k$ thus at 1 because there is at most a countable infinity of discontinuity points, $(F_n(\mathbb{R}^k))$ tends to 1. If F_1 and F_2 are two closure points of (F_n) for weak convergence, these are limits for narrow convergence. The distributions F_1 and F_2 are two closure points of (F_n) for weak convergence, these are limits for narrow convergence. The distributions F_1 and F_2 coincide with F at the continuity points of F; thus they coincide everywhere (Exercise 3.1.17).

Corollary 3.4.35. (*Stieltjes' Integral*). *Let F be a function from $\mathbb{R}$ into* $[0,1]$, *increasing, continuous on the right, and which tends to* 0 (resp. 1) *if the variable tends to* $-\infty$ (resp. $+\infty$). *It is the distribution function of a probability, again denoted by F, on $\mathbb{R}$. The Stieltjes' integral of each function f, continuous F* a.s., *exists and coincides with $F(f)$.*

Proof. Consider for each integer n, the measure

$$F_n = \sum_{p\in\mathbb{Z}} \left[F\left(\frac{p}{n}\right) - F\left(\frac{p-1}{n}\right)\right]\delta_{p/n}.$$

The Stieltjes' integral $I(f)$ is, by definition, the limit, if it exists, of $F_n(f)$. For all bounded functions f the Stieltjes' integral of which exists, in particular for $f \in C_0$, the sequence $(F_n(f))$ tends to $I(f)$. The functional $f \longmapsto I(f)$ is linear and positive on C_0, thus there exists a measure $\hat{F}$ on $\mathbb{R}$ for which $I(f)$ coincides with the integral $\hat{F}(f)$. The sequence (F_n) converges narrowly to $\hat{F}$ and, for every function f, $\hat{F}$ a.s. continuous, the Stieltjes' integral $I(f)$ exists and coincides with $\hat{F}(f)$. At each point x where F is continuous $(F_n(x))$ converges to $F(x)$: the probability $\hat{F}$ converges to the distribution function F. Thus the extension theorem assumed in Section 3.2.4 implies the existence of a probability with distribution function F, assumed in Section 3.1.7.

3.4.4. Probabilistic Interpretation

Definition 3.4.36. Let (X_n) be a sequence of random vectors defined on $(\Omega,\mathcal{A},P)$ taking values in $\mathbb{R}^k$. Let F_{X_n} be the distribution of X_n. Consider a distribution F on $\mathbb{R}^k$ and another random vector X taking values in $\mathbb{R}^k$, with distribution F_X. We say that X_n **converges in distribution to** X (or to F), denoted by $(X_n) \xrightarrow{\mathcal{D}} X$ (or $(X_n \xrightarrow{\mathcal{D}} F)$, when the sequence (F_{X_n}) converges narrowly to F_X (or to F).

The most useful criteria are Levy's Theorem (Corollary 3.4.30) using characteristic functions, and Corollary 3.4.34 using distribution functions. *If (X_n) converges in distribution to X (or F) and if f is an* r.v. *on $\mathbb{R}^k$, F_X (or F) a.s. continuous and bounded, $(E(f(X_n)))$ tends to $E(f(X))$ (or $F(f)$). If h is a function from $\mathbb{R}^k$ into $\mathbb{R}^p$, F_X (or F) a.s. continuous, the sequence $(h(X_n))$ converges in distribution to $h(X)$ (or $h(F)$).* This is the interpretation of Theorem 3.4.33.

Convergence in distribution cannot imply any other type of convergence, because it only deals with distributions. Let us take for example $\Omega = \{0,1\}$ with the uniform probability. The sequence identical to $1_{\{0\}}$ with distribution $b(\frac{1}{2})$ converges to $1_{\{1\}}$ in distribution,... , but in no other manner. On the other hand, convergence in probability of a sequence (X_n) of r.v.'s to X implies convergence in distribution by the following proof.

Let (X_n) and X be random vectors of dimension k. Assume that for all $\varepsilon > 0$, the sequence $(P(\|X_n - X\| > \varepsilon))$ tends to zero. Let $f \in C_K$, and let ω_f be its modulus of continuity. For all $\eta > 0$ and $\varepsilon = \omega_f(\eta)$,

$$
\begin{aligned}
&|E[f(X_m)] - E[f(X)]| \\
&\leqslant \int_{\{|X_m - X| < \varepsilon\}} |f(X_m) - f(X)| dP \\
&\quad + \int_{\{|X_m - X| \geqslant \varepsilon\}} |f(X_m) - f(X)| dP \\
&\leqslant \eta + \|f\| P(|X_m - X| \geqslant \varepsilon).
\end{aligned}
$$

Thus $(E[f(X_n)])$ converges to $E[f(X)]$ and $(X_n(P))$ tends vaguely to $X(P)$. Following Theorem 3.4.29, (X_n) converges in distribution to X.

Exercises 3.4.

E.1. Give the criteria of vague, weak, or narrow convergence of (F_n) in terms of the properties of the sequence (a_n) of positive numbers for

(a) $F_n = \delta_{a_n}$ (b) $F_n = N(a_n,1)$

(c) $F_n = N(0,a_n)$ (d) $F_n = N(a_n,a_n)$.

When the sequence (a_n) converges to a number a, state the possible limit.

E.2. *Stieltjes' integral.* Generalize Corollary 3.4.34 by considering the Stieltjes' sums associated with an arbitrary sequence of countable partitions of $\mathbb{R}$, the maximum width of which tends to 0.

E.3. *Operations on convergence in distribution.* Let (X_n) and (Y_n) be two sequences of random vectors on $(\Omega,\mathcal{A},P)$ which converge in distribution to X and Y, respectively.

(a) Show by an example that we can neither affirm that $(X_n + Y_n)$ tends to $X + Y$, nor that (X_nY_n) tends to XY.
(b) Prove that, if (Y_n) tends in distribution to a constant a, then (Y_n) converges in probability.
(c) Assume that (Y_n) converges in distribution to a. Prove that (X_n,Y_n) converges in distribution to (X,a). Deduce from this that the properties studied in (a) are then true.

E.4. We are given, on $\mathbb{R}$, a sequence (F_n) of distributions which converge to F, where F is continuous. Show that the sequence of distribution functions (F_n) converges uniformly to F.

E.5. On $\mathbb{R}^k$, assume that (F_n) converges vaguely to F and that the sequence $(\int|x|^r dF_n(x))$ is bounded, for one $r > 0$. Show that $(\int f\ dF_n)$ converges to $\int f\ dF$ for every real continuous function f, for which $\lim_{x\to\infty}|x|^{-r}f(x) = 0$.

E.6. Let F_n have density f_n with respect to Lebesgue measure. If (F_n) tends narrowly to F, F having density f, show by an example that it does not necessarily follow that $(f_n) \to f$ (a.s. or in L^1).

E.7. *Levy distance.* Let F and G be two probabilities on $\mathbb{R}$ and set

$$d(F,G) = \inf\{h > 0;\ F(\cdot - h) - h \leqslant G(\cdot) \leqslant F(\cdot + h) + h\}.$$

(a) Show that d is a distance;
(b) Show that "(F_n) tends narrowly to F" is equivalent to "$(d(F_n,F)) \to 0$."

Extension. Let $(X_t)_{t\in\overline{\mathbb{R}}}$ and X be r.v.'s with distributions $(F_t)_{t\in\overline{\mathbb{R}}}$ and F. We say that X converges in distribution to X if $t \to a$ when $\lim_{t\to a} d(F_t,F) = 0$. This signifies that, for every sequence (t_n) which tends to a,

$$(X_{t_n}) \xrightarrow{\mathcal{D}} X$$

(a property of metric spaces).

E.8. *Inversion of the Laplace transform.* (a) Let N_n have distribution $p(n\theta)$. Show that N_n/n converges in distribution to θ and $(N_n - n\theta)/\sqrt{n\theta}$ to $\mathcal{N}(0,1)$.

(b) For $\theta > 0$ and $x > 0$, consider the sequence

$$\left[e^{-n\theta} \sum_{0\leqslant k\leqslant[nx]} \frac{(n\theta)^k}{k!}\right]$$

($[nx]$ is the integer part of nx). Show that this converges to 0, 1/2, or 1, according to the relative positions of x and θ.

(c) Let F be a distribution on $[0,\infty[$ and ϕ its Laplace transform. Prove the "inversion formula": the sequence

$$\left[\sum_{k\leqslant nx} \frac{(-1)^k}{k!}\, n^k \phi^{(k)}(n)\right]$$

tends to $F(x)$ at each point x where F is continuous.

E.9. *Inversion of the Fourier transform.* (a) Let F be a bounded measure on $\mathbb{R}^k$, with Fourier transform ϕ_F, assumed integrable with respect to Lebesgue meausre. Calculate as a function of ϕ_F for $\sigma > 0$ and for K a compact set in $\mathbb{R}^k$ the integral

$$\left[(\sigma\sqrt{2\pi})^{-k}\right] \int 1_K(y)dy \left[\int \exp\left(-\frac{|x-y|^2}{2\sigma^2}\right) dF(x)\right].$$

Let σ tend to 0 and prove that F has a density

$$x \longmapsto (2\pi)^{-k} \int \exp(-i\langle x,u\rangle)\phi_F(u)du.$$

(b) Calculate the Fourier transform of the distribution on $\mathbb{R}$ with density $x \longmapsto \frac{1}{2} e^{-|x|}$. Show that the inversion formula applies in this case, and deduce that $x \longmapsto e^{-|x|}$ is the Fourier transform of the Cauchy distribution with density $x \longmapsto 1/\pi(1 + x^2)$.

Bibliographic Notes

The theories of measure and integration are given in very many works which may or may not use probabilistic vocabulary. Amongst the pure analysis works, let us mention Rudin (Chapters 1 - 4, and 9), and Halmos.

Other works fall into a probabilistic framework: Metivier and Billingsley. At a more advanced level, Chapters 1 and 2 of Neveu [1] give detailed material and numerous important exercises. For convergence in distribution more specifically, see Billingsley [1] (Chapter 1) and Parthasarathy.

Chapter 4
INDEPENDENCE: STATISTICS BASED ON THE OBSERVATION OF A SAMPLE

Objectives

Knowledge of a random phenomenon rests naturally on experiment, most often on the repetition of independent observations of this phenomenon. The frequency of an event gives an idea of its probability, which improves as the number of observations increases.

It is this fundamental idea of probabilistic intuition which we are analyzing here. A mathematical model of independence is based on the theory of product measures, which we shall rapidly cover. We then state the links between frequencies and probabilities and deal with some of their uses in statistics. These are very rough methods but apply to very general models.

4.1. Sequence of *n* Observations — Product Measure Spaces

4.1.1. Product of Measurable Spaces

Assume that, on $(\Omega,\mathcal{A})$, we observe $X_1, ..., X_n$, n measurable functions taking values in $(E_1,\mathcal{E}_1), ..., (E_n,\mathcal{E}_n)$. The total observation is $X = (X_1, ..., X_n)$ taking values in $E_1 \times ... \times E_n$. A rectangle in $E_1 \times ... \times E_n$ is a set $\Gamma = \Gamma_1 \times ... \times \Gamma_n$ where $\Gamma_1 \in \mathcal{E}_1, ..., \Gamma_n \in \mathcal{E}_n$. The set $\{X \in \Gamma\} = \bigcap_{i=1}^n \{X_i \in \Gamma_i\}$ is then an event in $\mathcal{A}$. From Exercise 3.1.6 the same applies to $\{X \in \Gamma\}$ if Γ is in the σ-algebra generated by the rectangles or product σ-algebra.

Definition 4.1.1. and Notations. For $(E_1,\mathcal{E}_1)$, ..., $(E_n,\mathcal{E}_n)$ n measurable spaces, the **product σ-algebra** $\mathcal{E}_1 \otimes ... \otimes \mathcal{E}_n$ is the σ-algebra of $E_1 \otimes ... \otimes E_n$ generated by the **rectangles** $\Gamma_1 \times ... \times \Gamma_n$ for $\Gamma_1 \in \mathcal{E}_1$, ..., $\Gamma_n \in \mathcal{E}_n$.

The measure space thus defined is the **product** of the n measurable spaces. Denote

$$(E_1 \times ... \times E_n, \mathcal{E}_1 \otimes ... \otimes \mathcal{E}_n) = \left[\prod_{i=1}^{n} E_i, \bigotimes_{i=1}^{n} \mathcal{E}_i\right]$$

$$(E \times ... \times E, \mathcal{E} \otimes ... \otimes \mathcal{E}) = (E,\mathcal{E})^n = (E^n,\mathcal{E}^{\otimes n}).$$

The set of rectangles is closed under finite intersections. From Tricks 3.1.13 and 3.1.11 we deduce the following properties.

4.1.2. Properties of Measurable Spaces. (a) *A function X from Ω into $E_1 \times ... \times E_n$, with components X_1, ..., X_n, taking values in E_1, ..., E_n is measurable from $(\Omega,\mathcal{A})$ into $\prod_{i=1}^n(E_i,\mathcal{E}_i)$ if, and only if, each component X_i is measurable from $(\Omega,\mathcal{A})$ into $(E_i,\mathcal{E}_i)$ $(i = 1, ..., n)$.*

(b) *For* r.v.'s f_1, ..., f_n *on $(E_1,\mathcal{E}_1)$, ..., $(E_n,\mathcal{E}_n)$ respectively, the function $(x_1, ..., x_n) \longmapsto \prod_{i=1}^n f_i(x_i)$ denoted by $\otimes_{i=1}^n f_i$ is an* r.v. *on $\prod_{i=1}^n(E_i,\mathcal{E}_i)$.*

(c) *Associativity Properties: For $j_0 = 0 < j_1 < ... < j_\ell = n$,*

$$\prod_{i=1}^{n} (E_i,\mathcal{E}_i) = \prod_{k=0}^{\ell-1} \left[\prod_{i=j_k+1}^{j_{k+1}} (E_i,\mathcal{E}_i)\right].$$

(d) $(\mathbb{R},\mathcal{B}_{\mathbb{R}})^k = (\mathbb{R}^k, \mathcal{B}_{\mathbb{R}^k})$ *(since the σ-algebras $\mathcal{B}_{\mathbb{R}^k}$ and $(\mathcal{B}_{\mathbb{R}^k})^{\otimes k}$ are generated by the product of k open intervals of $\mathbb{R}$).*

(c) *A probability on $\prod_{i=1}^n(E_i,\mathcal{E}_i)$ is characterized by its values on the rectangles. In particular, if $(\Omega,\mathcal{A})$ is equipped with a probability P, the distribution of $(X_1, ..., X_n)$, $F_X = F_{X_1,...,X_n}$ is characterized by its values on the rectangles, given by*

$$F_{X_1,...,X_N}(\Gamma_1 \times ... \times \Gamma_n) = P[X_1 \in \Gamma_1, ..., X_n \in \Gamma_n],$$

for $\Gamma_1 \in \mathcal{E}_1, \ldots, \Gamma_n \in \mathcal{E}_n$.

Convention. The σ-algebra used on E_i is often taken to be understood (the Borel σ-algebra, for example, for $E_i = \mathbb{R}^k$). If we speak about $\prod_{i=1}^n E_i$, it is understood that the product of these implied σ-algebras is used.

4.1.2. Product of σ-Finite Measures

Definition 4.1.3. and Notations. Let $(E_i,\mathcal{E}_i,\mu_i)_{1 \leqslant i \leqslant n}$ be n measurable spaces, the measures μ_i being σ-finite. There exists a unique measure on $\prod_{i=1}^n (E_i,\mathcal{E}_i)$, the **product measure** of $\mu_1, \ldots, \mu_n$ denoted by $\mu_1 \otimes \ldots \otimes \mu_n$, such that for all rectangles $\Gamma_1 \times \ldots \times \Gamma_n$ we have

$$\mu_1 \otimes \ldots \otimes \mu_n (\Gamma_1 \times \ldots \times \Gamma_n) = \mu_1(\Gamma_1) \ldots \mu_n(\Gamma_n).$$

The measure space thus constructed is the product of the n measure spaces. Denote

$$\mu \otimes \ldots \otimes \mu = \mu^{\otimes n}$$

$$(E_1 \times \ldots \times E_n, \mathcal{E}_1 \otimes \ldots \otimes \mathcal{E}_n, \mu_1 \otimes \ldots \otimes \mu_n) = \prod_{i=1}^{n} (E_i,\mathcal{E}_i,\mu_i)$$

$$(E^n,\mathcal{E}^{\otimes n},\mu^{\otimes n}) = (E,\mathcal{E},\mu)^n.$$

The uniqueness of the product measure follows from the above. We leave it to a course on measure theory to prove the existence of this product measure and the following theorem.

Theorem 4.1.4. (Fubini). (a) *For a positive* r.v. X *on* $(E_1 \times E_2, \mathcal{E}_1 \otimes \mathcal{E}_2)$ *the function* $x_1 \longmapsto \int X(x_1,x_2)d\mu_2(x_2)$ *is measurable on* $(E_1,\mathcal{E}_1)$ *and*

$$\int \left[\int X(x_1,x_2)d\mu_2(x_2) \right] d\mu_1(x_1) = \int X \, d\mu_1 \otimes d\mu_2.$$

From this the following useful tool is deduced.

Trick 4.1.5. *For* $1 \leqslant i \leqslant n$, *let* f_i *be* r.v.'s *on* $(E_i,\mathcal{E}_i)$ *and* $\otimes_{i=1}^n f_i$ *the function* $(x_1, \ldots, x_n) \longmapsto \prod_{i=1}^n f_i(x_i)$ *defined on* $\prod_{i=1}^n E_i$.

(a) *If the* r.v.'s f_i *are positive, we have*

$$\int \bigotimes_{i=1}^{n} f_i \, d \bigotimes_{i=1}^{n} \mu_i = \prod_{i=1}^{n} \int f_i d\mu_i.$$

If f_i is finite and if $f_i \cdot \mu_i$ is the measure having density f_i with respect to μ_i, then

$$\bigotimes_{i=1}^{n} (f_i \cdot \mu_i) = \left[\bigotimes_{i=1}^{n} f_i\right]\left[\bigotimes_{i=1}^{n} \mu_i\right].$$

(b) *The* r.v. $\otimes_{i=1}^{n} f_i$ *is* $\otimes_{i=1}^{n} \mu_i$ *- integrable if and only if each* f_i *is* μ_i*-integrable, and we then have*

$$\int \bigotimes_{i=1}^{n} f_i d \bigotimes_{i=1}^{n} \mu_i = \prod_{i=1}^{n} f_i d\mu_i.$$

Particular Case. Take for $j = 1, ..., n$, $k_j \geqslant 1$, and

$$(E_j, \mathcal{E}_j) = (\mathbb{R}^{k_j}, \mathcal{B}_{\mathbb{R}^{k_j}}).$$

Let $u_j \in \mathbb{R}^{k_j}$ for each j. Take for f_j the bounded function $x_j \longmapsto \exp i\langle u_j, x_j\rangle$ and for μ_j a bounded measure. We obtain

$$\int \exp i\left\{\sum_{j=1}^{n} \langle u_j, x_j\rangle\right\} d \bigotimes_{j=1}^{n} \mu_j(x_j)$$
$$= \prod_{j=1}^{n} \int \exp i\langle u_j, x_j\rangle d\mu_j.$$

The Fourier transform ϕ of a product of n bounded measures μ_j on $\mathbb{R}^{k_j}$, the Fourier transforms of which are ϕ_j for $j = 1, ..., n$, is $\phi = \phi_1 \otimes ... \otimes \phi_n$. The fourier transform characterizes the measure, thus *a measure in* $\mathbb{R}^{k_1+...+k_n}$ *is the product of bounded measures on* $\mathbb{R}^{k_j}$ $(j = 1, ..., n)$ *if and only if its Fourier transform is* $\otimes_{j=1}^{n} \phi_j$, ϕ_j *being the Fourier transform of* μ_j.

Exercises 4.1

E.1. *Integration by parts formula.* Let F and G be two increasing functions, continuous on the right, from $\mathbb{R}$ into $\mathbb{R}_+$, and let $-\infty < a < b < \infty$. Prove

$$F(b)G(b) - F(a)G(a)$$
$$= \int_{]a,b]} F(s_-)dG(s) + \int_{]a,b[} G(s)dF(s).$$

E.2. Let $(\Omega,\mathcal{A},\mu)$ be a measure space and X a positive r.v. on $(\Omega,\mathcal{A})$. Show that the sets $G = \{(\omega,X(\omega));\ \omega \in \Omega\}$ and $H = \{(\omega,X);\ \omega \in \Omega,\ 0 \leqslant x \leqslant X(\omega)\}$ are in $\mathcal{A} \otimes \mathcal{B}_{\mathbb{R}}$. Denoting by λ, Lebesgue measure on $\mathbb{R}$, show that

$$\int X\, d\mu = \int_0^\infty dx\ \mu[X \geqslant x] = (\mu \otimes \lambda)(H).$$

E.3. Consider the two r.v.'s X and Y on $(\Omega,\mathcal{A},P)$ with distributions F and G.

(a) Prove that if X (resp. X^2) is integrable, we have

$$E(X) = \int_0^\infty (1 - F(y))dy - \int_{-\infty}^0 F(y)dy$$

$$(\text{resp. } E(X^2) = 2\int_0^\infty x(1 - F(x) + F(-x))dx).$$

(b) Prove that for X and Y square integrable,

$$\Gamma(X,Y) = \iint \Gamma(1_{(X>s)},1_{(Y>t)})ds\ dt.$$

E.4. *Exchangeable distributions.* Let $\mathcal{S}$ be the set of permutations of $\{1,2, ..., n\}$. A distribution or a function defined on $\mathbb{R}^n$ is said to be exchangeable if it is invariant under each $\sigma \in \mathcal{S}$ by the function

$$(x_1, ..., x_n) \longmapsto (x_{\sigma(1)}, ..., x_{\sigma(n)}).$$

(a) Let F be a distribution on $\mathbb{R}^n$. Verify that the following properties are equivalent:

(i) F is exchangeable;
(ii) for all sequences $(B_1, ..., B_n)$ of n Borel sets of $\mathbb{R}$ and every $\sigma \in \mathcal{S}$,

$$F(B_1 \times ... \times B_n) = F(B_{\sigma(1)} \times ... \times B_{\sigma(n)});$$

(iii) the distribution function of F is exchangeable;
(iv) the Fourier transform of F is exchangeable.

(b) Let $(X_1, ..., X_n)$ be an exchangeable random vector of dimension n, i.e., the distribution of which is exchangeable. Show that the n marginal distributions are equal. Show that for $1 \leqslant i,j \leqslant n$, $X_i - X_j$ and $X_j - X_i$ have the same distribution. Are these properties sufficient for $(X_1, \ldots, X_n)$ to be exchangeable?

E.5. *Associated random variables.* Let $\mathbb{R}^n$ be given the partial order

$$(x_1, \ldots, x_n) \leqslant (y_1, \ldots, y_n) \Leftrightarrow x_1 \leqslant y_1, \ldots, x_n \leqslant y_n.$$

The notion of increasing, used below, is relative to this order.

(a) Let X be an r.v. and f and g two increasing functions from $\mathbb{R}$ into $\mathbb{R}$ such that $f(X)$ and $g(X)$ are square integrable. Show that $\Gamma(f(X),g(X)) \geqslant 0$.

(b) Let $X = (X_1, \ldots, X_n)$ be a random vector of dimension n. Its components are said to be associated (or X is associated) if for each pair (f,g) of increasing functions from $\mathbb{R}^n$ to $\mathbb{R}$ such that $f(X)$ and $g(X)$ are square integrable, the covariance $\Gamma[f(X),g(X)]$ is positive. Prove that if X is square-integrable, two components of X then have a positive covariance. Prove that in the definition it is sufficient to consider functions f and g taking values in $\{0,1\}$ (see Exercise 4.1.3). Let ϕ be an increasing function from $\mathbb{R}^n$ into $\mathbb{R}^k$. Show that $\phi(X)$ is associated when X is associated. If the associated components of X take their values in $\{0,1\}$, prove

$$P\left[\prod_{i=1}^{n} X_i = 1\right] \geqslant \prod_{i=1}^{n} P(X_i = 1)$$

$$P\left[\prod_{i=1}^{n} X_i = 0\right] \geqslant \prod_{i=1}^{n} P(X_i = 0).$$

(It should be noted that the functions which associate $\prod_{i=1}^{n} x_i$ or $1 - \prod_{i=1}^{n}(1 - x_i)$ to $(x_1, \ldots, x_n)$ are increasing from $\{0,1\}^n$ into $\{0,1\}$.) For an arbitrary associated X and $(x_1, \ldots, x_n) \in \mathbb{R}^n$, prove

$$P(X_1 > x_1, \ldots, X_n > x_n) \geqslant \prod_{i=1}^{n} P(X_i > x_i)$$

$$P(X_1 \leqslant x_1, \ldots, X_n \leqslant x_n) \geqslant \prod_{i=1}^{n} P(X_i \leqslant x_i).$$

4.2. Independence

4.2.1. Probability Conditional on an Event

Let us assume that a random phenomenon is described by a probability space $(\Omega, \mathcal{A}, P)$. Let $A \in \mathcal{A}$ be an event. If the supplementary information "the event A is realized" is given, the study of an event B is in fact that of $A \cap B$; the probability P must be replaced by the trace probability of P on A normalized in order that the total mass is 1. The events are independent when knowing that A has taken place does not change the probability of B. This leads to the following definition.

Definition 4.2.6. For $A \in \mathcal{A}$ with nonzero probability, the **conditional probability of** $B \in A$ given A is

$$P(B|A) = \frac{P(A \cap B)}{P(A)}.$$

The expectation of an r.v. X with respect to this probability is denoted by $E(X|A)$ and is called the **conditional expectation of** X given A. The **events** A and B are said to be **independent** when we have

$$P(B) = P(B|A), \quad \text{i.e.,} \quad P(A \cap B) = P(A)P(B).$$

Let an observation X be a measurable function from $(\Omega, \mathcal{A})$ into $(E, \mathcal{E})$, the **conditional distribution of** X given A is the probability which associates with $\Gamma \in \mathcal{E}$, $P[X \in \Gamma|A]$. This is the image probability of $P(\cdot|A)$ by X.

If the distribution F_X of X has a density f with respect to a measure μ on $(E, \mathcal{E})$, for $\Delta \in \mathcal{E}$, we have

$$P[X \in \Gamma|X \in \Delta] = \frac{P(X \in \Gamma \cap \Delta)}{P(X \in \Delta)}$$

$$= \frac{\int f(x) 1_{\Gamma \cap \Delta}(x) \mu(dx)}{\int f(x) 1_{\Delta}(x) \mu(dx)}.$$

The conditional distribution of X on $\{X \in \Delta\}$ has, with respect to μ, the density

$$x \longmapsto f(x|\Delta) = \frac{f(x)\mathbf{1}_\Delta(x)}{\int f(y)\mathbf{1}_\Delta(y)\mu(dy)}.$$

This function $f(\cdot|\Delta)$ is the **conditional density** of X given $\{X \in \Delta\}$.

4.2.2. Independent Observations

Let two observations X and Y be measurable functions from $(\Omega,\mathcal{A})$ into $(E,\mathcal{E})$ and $(E',\mathcal{E}')$, respectively, with distributions $F_X = X(P)$ and $F_Y = Y(P)$. They are independent if for all $\Gamma \in \mathcal{E}$ and $\Gamma' \in \mathcal{E}'$, the events $\{X \in \Gamma\}$ and $\{Y \in \Gamma'\}$ are independent. In other words, with $F_{X,Y}$ being the distribution of (X,Y),

$$P[X \in \Gamma, Y \in \Gamma'] = F_{X,Y}(\Gamma \times \Gamma') = P(X \in \Gamma)P(Y \in \Gamma')$$

$$= F_X(\Gamma)F_Y(\Gamma')$$

$$F_{X,Y} = F_X \otimes F_Y.$$

If $X_1, \ldots, X_n$ taking values in $(E_1,\mathcal{E}_1), \ldots, (E_n,\mathcal{E}_n)$ are observed in succession and if we assume that the ith observation is always independent of the preceding $X_1, \ldots, X_{i-1}$, then

$$F_{X_1,X_2} = F_{X_1} \otimes F_{X_2};$$

$$F_{X_1,X_2,X_3} = F_{X_1,X_2} \otimes F_{X_3} = F_{X_1} \otimes F_{X_2} \otimes F_{X_3} \ldots .$$

Definition 4.2.7. n observations $X_1, \ldots, X_n$ defined on the probability space $(\Omega,\mathcal{A},P)$ taking values in $(E_1,\mathcal{E}_1), \ldots, (E_n,\mathcal{E}_n)$ are **independent** when

$$F_{X_1, \ldots, X_n} = F_{X_1} \otimes \ldots \otimes F_{X_n}.$$

Note. *Pairwise independence of the* r.v.'s *is not sufficient* (Exercise 2.1.5).

Theorem 4.2.8. *Let $X_1, \ldots, X_n$ be n independent observations on $(\Omega,\mathcal{A},P)$, taking values in $(E_1,\mathcal{E}_1), \ldots, (E_n,\mathcal{E}_n)$, and for $1 \leqslant i \leqslant n$, let f_i be a measurable function on $(E_i,\mathcal{E}_i)$.*

(a) *The observations $f_i(X_i)$, $1 \leq i \leq n$, are independent.*
(b) *If f_i is a positive* r.v., *for $i = 1, \ldots, n$ we have*

$$E\left[\prod_{i=1}^{n} f_i(X_i)\right] = \prod_{i=1}^{n} E[f_i(X_i)].$$

(c) *If f_i is an* r.v. *for $i = 1, \ldots, n$, $\prod_{i=1}^{n} f_i(X_i)$ is P-integrable, if and only if each $f_i(X_i)$ is P-integrable and then relation* (b) *is again true.*

Proof. (a) Let f_i be measurable from $(E_i, \mathcal{E}_i)$ into $(E_i', \mathcal{E}_i')$ and let $\Gamma_i \in \mathcal{E}_i'$;

$$P\left[\bigcap_{i=1}^{n} (f_i(X_i) \in \Gamma_i)\right] = P\left[\bigcap_{i=1}^{n} (X_i \in f_i^{-1}(\Gamma_i))\right]$$

$$= \prod_{i=1}^{n} F_{X_i}(f_i^{-1}(\Gamma_i)) = \prod_{i=1}^{n} f_i(F_{X_i})(\Gamma_i)$$

$$= \prod_{i=1}^{n} F_{f_i(X_i)}(\Gamma_i).$$

Thus the distribution of $(f_1(X_1), \ldots, f_n(X_n))$ is $\otimes_{i=1}^{n} F_{f_i(X_i)}$: (b) and (c) follow from 4.1.5.

Corollary 4.2.9. (a) *Let X and Y be two independent integrable* r.v.'s; *then XY is integrable and the covariance*

$$\Gamma(X,Y) = E(XY) - E(X)E(Y) = E[(X - E(X))(Y - E(Y))]$$

is zero.

(b) *Let $X_1, \ldots, X_n$ be n integrable independent* r.v.'s; $\prod_{i=1}^{n} X_i$ *is integrable and*

$$E\left[\prod_{i=1}^{n} X_i\right] = \prod_{i=1}^{n} E(X_i).$$

(c) *Let $X_1, \ldots, X_n$ be n square integrable independent* r.v.'s. *The variance of $(X_1 + \ldots + X_n)$ is the sum of the variances of the X_i for* i $= 1, \ldots, n$.

Part (c) follows from the relations

$$\sigma^2(X_1 + \ldots + X_n) = E\left[\sum_{i=1}^{n} (X_i - E(X_i))^2\right]$$

$$= \sum_{i=1}^{n} \sigma^2(X_i) + \sum_{\substack{i \neq j \\ i,j=1,\ldots,n}} \Gamma(X_i, X_j)$$

$$= \sum_{i=1}^{n} \sigma^2(X_i).$$

Independent R.V.'s *Without Ties. Let* (X,Y) *be a pair of independent* r.v.'s *with distributions* F_X *and* F_Y; *if one of these two distributions is continuous, the probability* $P(X = Y)$ *is zero. The* r.v.'s *cannot be tied* a.s. Indeed, if F_Y is continuous, we write by virtue of Fubini's theorem.

$$P(X = Y) = \int dF_X(x) F_Y(\{x\}) = 0.$$

Using 4.1.5. again, the following criteria are deduced.

Independence Criteria 4.2.10. (a) *For each* i, μ_i *is a* σ*-finite measure on* $(E_i, \mathcal{E}_i)$ *and* f_i *is a positive* r.v. *on* $(E_i, \mathcal{E}_i)$. *If the distribution of* X_i *has a density* f_i *with respect to* μ_i, $X_1, \ldots, X_n$ *are independent if and only if the distribution of* $(X_1, \ldots, X_n)$ *has density* $\otimes_{i=1}^{n} f_i$ *with respect to* $\otimes_{i=1}^{n} \mu_i$.

(b) *Let* $(E_i, \mathcal{E}_i) = (\mathbb{R}^{k_i}, \mathcal{B}_{\mathbb{R}^{k_i}})$ *and* $\tilde{F}_{X_i}$ *the distribution function of* X_i. *Then* $X_1, \ldots, X_n$ *are independent if and only if the distribution function of* $(X_1, \ldots, X_n)$ *is* $\tilde{F}_{X_1} \otimes \ldots \otimes \tilde{F}_{X_n}$.

As in Chapter 3 we could confuse the notations F_{X_i} for the distribution functions and the distributions ... . The symbol $F_{X_1} \otimes \ldots \otimes F_{X_n}$ does not lead to confusion.

Under the same hypotheses, denoting by $\phi_{X_1}, \ldots, \phi_{X_n}$ the Fourier transforms of $F_{X_1}, \ldots, F_{X_n}$ (characteristic functions of $X_1, \ldots, X_n$): $X_1, \ldots, X_n$ *are independent if and only if the characteristic function of* $(X_1, \ldots, X_n)$ *is*

$$\phi_{X_1, \ldots, X_n} = \phi_{X_1} \otimes \ldots \otimes \phi_{X_n}$$

4.2.3. Families of Independent Observations

Definition 4.2.11. Let I be an arbitrary set. For each $i \in I$, consider X_i, measurable from $(\Omega, \mathcal{A})$ into $(E_i, \mathcal{E}_i)$. The family $(X_i)_{i \in I}$ is independent on $(\Omega, \mathcal{A}, P)$ if for all finite subsets J of I, $(X_j)_{j \in J}$ is an independent family. A sequence $(A_i)_{i \in I}$ of events is independent if and only if $(1_{A_i})_{i \in I}$ is independent.

Remark. Saying that $(X_i)_{i \in I}$ is independent is equivalent to saying that, for each family $(\Gamma_i)_{i \in I}$ of $\prod_{i=1}^n \mathcal{E}_i$, $(X_i \in \Gamma_i)_{i \in I}$ is independent, or again that the family $(\sigma(X_i))_{i \in I}$ is independent by the following definition.

Definition 4.2.12. A family $(\mathcal{G}_i)_{i \in I}$ of sub σ-algebras of $\mathcal{A}$ is independent if each family $(G_i)_{i \in I}$ of $\prod_{i \in I} \mathcal{G}_i$ is independent.

Proposition 4.2.13. *Let $\mathcal{C}$ and $\mathcal{D}$ be two families of subsets of $\mathcal{A}$ closed under finite intersections and such that, for all $C \in \mathcal{C}$ and all $D \in \mathcal{D}$, C and D are independent. Then, $\sigma(\mathcal{C})$ and $\sigma(\mathcal{D})$ are independent.*

Proof. Let us consider the set of pairs $(A,B) \in \mathcal{A} \times \mathcal{A}$ satisfying $P(A) \times P(B) = P(A \cap B)$. It contains $\mathcal{C} \neq \mathcal{D}$, thus $\sigma(\mathcal{C}) \times \mathcal{D}$, then $\sigma(\mathcal{C}) \times \sigma(\mathcal{D})$ by applying Trick 3.1.11 twice.

Corollary. (Associative Property). *Let I_1 and I_2 be two disjoint sets of I. If $(\mathcal{G}_i)_{i \in I}$ is independent, then the σ-algebras $\sigma(\mathcal{G}_i;\ i \in I_1)$ and $\sigma(\mathcal{G}_i;\ i \in I_2)$ are also independent.*

4.2.4. Sequences of Independent Observations

Let us assume that a sequence of experiments has been carried out. The nth experiment allows the observation of events in the σ-algebra $\mathcal{G}_n$ contained in $\mathcal{A}$. Then the past events constitute the σ-algebra $\mathcal{F}_n = \sigma(\mathcal{G}_p;\ p \leq n)$ and the σ-algebra of future events is $\mathcal{F}^n = \sigma(\mathcal{G}_p;\ p \geq n)$. To say that the σ-algebras $(\mathcal{G}_n)_{n \in \mathbf{N}}$ are independent is equivalent to saying that, for all n, $\mathcal{G}_n$ and $\mathcal{F}_{n-1}$ are independent or again that, for all n, $\mathcal{F}^n$ and $\mathcal{F}_{n-1}$ are independent: all this results from the associativity (Section 4.2.3). In particular, for a sequence of measurable functions, we can take $\mathcal{G}_n = \sigma(X_n)$. A first important asymptotic result is now established.

Definition 4.2.14. Let $(\mathcal{G}_n)$ be a sequence of sub σ-algebras of $\mathcal{A}$. The **tail σ-algebra** (or **asymptotic σ-algebra**) of this sequence is

$$\mathcal{L} = \bigcap_n \sigma(\mathcal{G}_p;\ p \geqslant n).$$

An **event** of $\mathcal{L}$ (resp. an r.v. on $(\Omega, \mathcal{L})$) is said to be a **tail** (or **asymptotic**) event: it depends only on the events following the nth experiment, for all n.

Examples. (1) Let $(B_n) \subset \mathcal{A}$ with $B_n \in \mathcal{G}_n$ for all n. $\overline{\lim}_{n\to\infty} B_n$ and $\underline{\lim}_{n\to\infty} B_n$ are tail events.

(2) Let (X_n) be a sequence of r.v.'s on $(\Omega,\mathcal{A})$. Take $\mathcal{G}_n = \sigma(X_n)$. Then $\{\omega;\ \lim_{n\to\infty} X_n(\omega)$ exists$\}$ and $\{\omega;\ \underline{\lim}_{n\to\infty}(1/(n+1))\Sigma_{k=0}^n X_k(\omega)$ exists$\}$ are tail events The r.v.;s $\overline{\lim}_{n\to\infty} X_n$, $\underline{\lim}_{n\to\infty} X_n$, $\overline{\lim}_{n\to\infty}(1/(n+1))\Sigma_{k=0}^n X_k$, and $\underline{\lim}_{n\to\infty}(1/(n+1))\Sigma_{k=0}^n X_k$ are tail events.

Theorem 4.2.15. (Kolmogorov 0–1 Law). *Let $(\mathcal{G}_n)_{n\in\mathbb{N}}$ be a sequence of independent sub σ-algebras of $\mathcal{A}$. The probability of every tail event is equal to* 0 *or* 1; *all tail* r.v.'s *are constant* (P–a.s.).

Proof. From the associativity property, $\mathcal{F}_{n-1}$ and $\mathcal{F}^n$, thus $\mathcal{F}_{n-1}$ and $\mathcal{L}$, are independent. Thus $\sigma\{\cup \mathcal{F}_n\} = \mathcal{F}_\infty$ and $\mathcal{L}$ are independent, i.e., $\mathcal{L}$ and $\mathcal{L}$ are independent. Let A and B be, in $\mathcal{L}$,

$$P(A \cap B) = P(A)P(B).$$

Take $B = A$: $P(A) = [P(A)]^2$; $P(A) = 0$ or 1.

Let X be a tail r.v.; X is (P-a.s.) equal to $\beta = \sup\{\alpha \in \mathbb{R};\ P(X \leqslant \alpha) = 0\}$. In the particular case of the tail event $\overline{\lim}_{n\to\infty} A_n$, for a sequence of independent events (A_n), we can characterize simply the cases where the probability equals 0 or 1 thanks to the following lemma.

Borel–Cantelli Lemma 4.2.16. *Let $(A_n) \subset \mathcal{A}$:*

(a) *For $\Sigma\, P(A_n) < \infty$, $P[\overline{\lim}_{n\to\infty} A_n] = 0$.*
(b) *If the sequence (A_n) is a sequence of independent events and if $\Sigma_n P(A_n)$ diverges,*

$$P[\overline{\lim_{n\to\infty}}\ A_n] = 1.$$

Proof. (a) has already been proved (Lemma 3.1.4). Let us prove (b):

$$1 - P[\overline{\lim_{n\to\infty}}\ A_n] = P[\underline{\lim_{n\to\infty}}\ A_n^c] = \lim_{n\to\infty} P\left[\bigcap_{p>n} A_p^c\right]$$

$$P\left[\bigcap_{p>n} A_p^c\right] = \lim_{N\to\infty} P\left[\bigcap_{n<p<N} A_p^c\right] = \lim_{N\to\infty} \prod_{n<p<N} P(A_p^c)$$

$$= \prod_{n<p} P(A_p^c) = \prod_{n<p} (1 - P(A_p));$$

$$\operatorname{Log} P\left[\bigcap_{p>n} A_p^c\right] = \sum_{p>n} \operatorname{Log}(1 - P(A_p))$$

$$\leqslant \sum_{p>n} (-P(A_p)) = -\infty.$$

Thus, $P[\bigcap_{p>n} A_p^c] = 0$ and

$$P\left[\overline{\lim_{n\to\infty}}\ A_n\right] = 1 - \lim_{n\to\infty} P\left[\bigcap_{p>n} A_p^c\right] = 1.$$

Exercises 4.2. *Distributions conditional on an event and failure rates* (Exercises 1 – 4).

E.1. Let X,Y be two independent r.v.'s, both having exponential distributions. Give the densities of the distributions of (X,Y), X, Y, and $Y - X$ conditional on the event $\{X \leqslant Y\}$.

E.2. Let T be a positive r.v. such that for all $t \in \mathbb{R}_+$, $P(T > t)$ is nonzero. Let us assume $P(T > s + t|T > t) = P(T > s)$. Find the form of the distribution of T. Explain why this result may be useful in modelling the time between two breakdowns of a machine, two arrivals of customers at a ticket office, or two catches by a fisherman.

E.3. *Failure rates.* A machine breaks down at a random time T, T being a positive r.v. from the distribution F having a continuous density f. If, at time t, the machine is still functioning, the risk of immediate breakdown is measured by the **failure rate**

$$\lim_{\substack{u \downarrow 0 \\ u>0}} \frac{1}{u} P[t < T \leqslant t + u \mid T > t].$$

(a) Calculate $h(t)$ as a function of f (when $P(T > t)$ is nonzero). What does h equal for an exponential distribution? We shall say that there is wear and tear when h increases and breaking in when h decreases. Prove, by setting $\Lambda(t) = \int_0^t h(s)ds$,

$$P(T > t) = e^{-\Lambda(t)}.$$

(b) Assume that T has a gamma distribution $\gamma(a,\lambda)$ (Exercise 3.1.13). Study whether there is wear and tear or breaking in according to the values of a. Prove that $\lambda = \lim_{t\to\infty} h(t)$; this is the asymptotic failure rate.

(c) Within this framework the *Weibull distribution* is often used; the distribution of a positive r.v. with a density, the failure rate of which is $t \longmapsto a\lambda^a t^{a-1}$. When is there wear and tear? or breaking in?

(d) For two components placed in parallel, it is natural to take for the time of breakdown $\sup(T_1,T_2)$ where T_1 and T_2 are independent r.v.'s with exponential distributions with parameters λ_1,λ_2. Show that, first of all, there is wear and tear then breaking in (wear and tear only for $\lambda_1 = \lambda_2$).

(e) Assume that there is wear and tear for arbitrary t. Show that this is equivalent to assuming for every pair (s,t):

$$\bar{F}(s + t) \geqslant \bar{F}(s)\bar{F}(t) \quad \text{with} \quad \bar{F} = 1 - F;$$

or again to assuming that Λ is convex.

Assume T is integrable with mean m. Calculate $E(\Lambda(t))$ and show

$$E(T \mid T > t) \leqslant t + m,$$

$$E[\inf(T,t)] \geqslant mF(t),$$

$$\Lambda(E(T)) \leqslant 1$$

$$P(T \geqslant t) \geqslant \exp\left[-\frac{t}{E(T)}\right] \quad \text{for} \quad 0 \leqslant t \leqslant E(T).$$

E.4. *Bivariate exponential distribution.* Let U_1, U_2, and U_{12} be three independent exponential r.v.'s with respective parameters λ_1, λ_2, and λ_{12}. Let $T_1 = \inf(U_1,U_{12})$ and $T_2 = \inf(U_2,U_{12})$. Calculate, for $(t_1,t_2) \in \mathbb{R}_+^2$, $P[T_1 > t_1, T_2 > t_2]$. Determine the marginal distributions of T_1 and T_2, their covariance, and the distribution of $T = \inf(T_1,T_2)$. Show that T is independent of the events $\{T_1 < T_2\}$, $\{T_1 > T_2\}$, and $\{T_1 = T_2\}$ as well as of the variable $|T_1 - T_2|$.

Calculate $P[T_1 > t_1, T_2 > t_2 \mid U_{12} \leq \inf(U_1,U_2)]$ and $P[T_1 > t_1, T_2 > t_2 \mid U_{12} > \inf(U_1,U_2)]$ and decompose the distribution of (T_1,T_2) into the sum of a measure concentrated on a set of Lebesgue measure zero and a measure having a density. Prove that a pair (T_1,T_2) follows the above distribution if and only if its marginal distributions are exponential and for all $(s_1,s_2,t) \in \mathbb{R}_+^3$,

$$P[T_1 > s_1 + t, T_2 > s_2 + t \mid T_1 > t, T_2 > t]$$

$$= P[T_1 > s_1, T_2 > s_2].$$

E.5. In Exercise 3.3.7, which pairs of r.v.'s (X,Y) are independent?

E.6. *The form of the support of the distributions of a pair of independent r.v.'s.* (a) Consider a pair of r.v.'s (X,Y), the distribution of which is uniform on a Borel subset A of $\mathbb{R}^2$. Under what conditions on A are X and Y independent?

(b) If μ is a measure on a topological space E together with its Borel σ-algebra $\mathcal{B}_E$, its support S_μ is the set of points, of which μ charges every neighborhood. Let μ' be another measure of this type and $S_{\mu'}$ its support. What is $S_{\mu \otimes \mu'}$?

E.7. Let X and Y be two independent r.v.'s with densities f and g with respect to Lebesgue measure. Prove that XY has density

$$x \longmapsto \int f(x/y)g(y) \frac{1}{|y|} \, dy.$$

E.8. Let X and Y be two independent $N(0,1)$ r.v.'s. Show that X/Y has a Cauchy distribution. Show that this is also the case for the distribution with density

$$x \longmapsto \frac{\sqrt{2}}{\pi} \frac{1}{1 + x^4}.$$

E.9. Let X and Y be two independent r.v.'s. Assume that X has a $\gamma(a + b,\lambda)$ distribution and that Y has a $\beta(a,b)$ distribution. Prove that XY and $X(1 - Y)$ are also independent and have gamma distributions (see Exercises 3.3.13 and 3.3.14).

E.10. *Independent events.* Let (A_n) be a sequence of events of $\mathcal{A}$. They are said to be independent if the r.v.'s (1_{A_n}) are independent. Prove that this is equivalent to saying that, for every finite subset J of $\mathbb{N}$, we have

$$P\left[\bigcap_{i \in J} A_i\right] = \prod_{i \in J} P(A_i).$$

E.11. Let $(X_i)_{1 \leq i \leq n}$ be n r.v.'s, the distributions of which are discrete. Prove that they are independent if, and only if, for every point $(x_i)_{1 \leq i \leq n}$ we have

$$P\left[\bigcap_{i=1}^{n} (X_i = x_i)\right] = \prod_{i=1}^{n} P(X_i = x_i).$$

E.12. Let X be a random variable such that X and $-X$ have the same distribution F. Let ε be a random variable independent of X such that

$$P(\varepsilon = 1) = p = 1 - P(\varepsilon = -1) \qquad (0 < p < 1)$$

(a) Calculate the distribution of εX.
(b) Under what conditions is the correlation between X and εX zero? (Assume $E(X^2) < \infty$.)
(c) Set $Y = 1_{(X>0)} - 1_{(X<0)}$. What are the distributions of the variables Y, YX, and the correlation between $|X|$ and Y?

E.13. *Bivariate Gaussian distribution.* Let U_1, U_2, and U_{12} be three independent centered Gaussian r.v.'s with nonzero variances v_1, v_2, and v_{12}. Let $X_1 = U_1 + U_{12}$ and $X_2 = U_2 + U_{12}$. What are the distributions of X_1 and of X_2, their means, their variances σ_1^2 and σ_2^2? What is the correlation coefficient ρ of (X_1,X_2)? Show that (X_1,X_2) has density

$$(x_1,x_2) \longmapsto \frac{1}{2\pi\sqrt{1 - \rho^2}\sigma_1\sigma_2} \exp\left[-\frac{1}{2(1-\rho^2)}\left(\frac{x_1^2}{\sigma_1^2} - 2\rho\frac{x_1x_2}{\sigma_1\sigma_2} + \frac{x_2^2}{\sigma_2^2}\right)\right].$$

Repeat the above for $X_2 = U_2 - U_{12}$. For $\sigma_1 = \sigma_2 = 1$ prove that

$$P[X_1 \geqslant 0, X_2 \geqslant 0] = P[X_1 \leqslant 0, X_2 \leqslant 0] = \frac{1}{4} + \frac{1}{2\pi}\text{Arc}\sin\rho$$

E.14. Let F be a distribution on $\mathbb{R}$ with characteristic function ϕ. Show

$$F(\{0\}) = \lim_{T\to\infty} \frac{1}{2T}\int_{-T}^{T} \phi(t)dt.$$

Show that, if (A_n) is a sequence of atoms of F, possibly finite or empty (Exercise 3.2.13),

$$\Sigma[F(A_n)]^2 = \lim_{T\to\infty} \frac{1}{2T}\int_{-T}^{T}|\phi(t)|^2dt.$$

E.15. Let (X_n) be a sequence of independent r.v.'s on (Ω,A,P) having the same distribution.

(a) Prove that there are two cases:

$$\text{Case 1}\quad \begin{cases} E(|X_1|) = +\infty \\ P\{\overline{\lim}\,(|X_n| \geqslant n)\} = 1 \end{cases}$$

$$\text{Case 2}\quad \begin{cases} E(|X_1|) < +\infty \\ P\{\overline{\lim}\,(|X_n| \geqslant n)\} = 0. \end{cases}$$

(It could be shown that $\Sigma_{n\geqslant 1}P\{|X_1| \geqslant n\} \leqslant \int|X_1|dP \leqslant \Sigma_{n\geqslant 0}P\{|X_1| \geqslant n\}$.)

(b) Show that if X_1 is not integrable on the set of ω, such that $((1/n)\Sigma_{k=1}^{n}X_k(\omega)$ converges, has measure zero.

E.16. *Maximum and Minimum.* Let (X_n) be a sequence of independent r.v.'s from the distribution F. What is the distribution function of $M_n = \sup_{p\leqslant n}X_p$? of $m_n = \inf_{p\leqslant n}X_p$? Show that the sequences (M_n) and (m_n) converges almost surely to the constants ess - sup $F = \sup(x;\ F(x\_) < 1\}$ and ess - inf $F = \inf\{x;\ F(x) > 0\}$ (first of all, study their convergence in distribution). Assuming F uniform on $[0,1]$, prove that (m_n) converges a.s., and that (nm_n) converges in distribution. Identify the limits.

E.17. *Distributions on $\mathbb{R}^2$ invariant under rotation.* Let $\mathcal{D}$ be the set of distributions on $\mathbb{R}^2$ invariant under rotations with center 0 (the image of $\nu \in \mathcal{D}$ under each of these rotations is ν).

(a) Show that the distribution ν is in $\mathcal{D}$ if and only if its Fourier transform is of the form $(u,v) \longmapsto \phi(u^2 + v^2)$, for a suitable function ϕ.

(b) Let (X,Y) be an r.v. taking values in $\mathbb{R}^2$, the distribution of which is in $\mathcal{D}$. Show that X and Y are independent if, and only if, X is distributed as $N(0,\sigma^2)$. Let $X = R \cos \theta$ and $Y = R \sin\theta$ with $R \geqslant 0$ and $0 \leqslant \theta < 2\pi$. Show that R and θ are independent if X and Y are independent. Then calculate the distribution of θ and that of R^2.

(c) Let μ be a probability on $\mathbb{R}_+$. Show that there exists a unique $\nu_\mu \in \mathcal{D}$, such that, for $r \in \mathbb{R}_+$, $\mu([0,r]) = \nu_\mu\{(x,y); x^2 + y^2 \leqslant r\}$. For $\mu = \delta_r$, set $\nu_\mu = \sigma_r$, the uniform distribution on the circumference of center 0, with radius r. Show that for every Borel set Γ of $\mathbb{R}^2$, we have $\nu_\mu(\Gamma) = \int \mu(dr)\sigma_r(\Gamma)$ ("decomposition of the measure ν_μ").

(d) Show that the function $\mu \longmapsto \nu_\mu$ is a bijection from the set $\mathbf{P}(\mathbb{R}_+)$ of probabilities on $\mathbb{R}_+$ into $\mathcal{D}$.

(e) Let γ be the function $\mathbb{R}^2 \to \mathbb{R}$, $(x,y) \longmapsto x$. Show that for $\mu \in \mathbf{P}(\mathbb{R}_+)$, with $\mu(0) = 0$, the image of ν_μ by γ, $\gamma(\nu_\mu)$ has density

$$x \longmapsto \frac{1}{\pi} \int_{|x|}^{\infty} \frac{1}{(r^2 - x^2)^2}\, d\mu(r).$$

E.18. To every $\theta \in [0,2\pi[$, associate the point $M(\theta) = (\cos\theta, \sin\theta)$ of the Euclidean plane. Let X and Y be two independent r.v.'s uniformly distributed on $[0,2\pi]$. What is the distribution of the distance from $M(X)$ to $M(Y)$?

4.3. Distribution of the Sum of Independent Random Vectors

If $X_1, \ldots, X_n$ are the successive wins obtained in a game on the first, ..., nth round, we are interested in the total winnings $S_n = X_1 + \ldots + X_n$, hence in its distribution. Assume then that $X_1, \ldots, X_n$ take values in $\mathbb{R}^k$. The distribution of S_n is the image measure of $F_{X_1,\ldots,X_n}$ by the function $(x_1, \ldots, x_n) \longmapsto$

$x_1 + \dots + x_n$ of $(\mathbb{R}^k)^n$ into $\mathbb{R}^k$. When $F_{X_1,\dots,X_n}$ equals $F_{X_1} \otimes \dots \otimes F_{X_n}$, this is the convolution of the measures $F_{X_1}, \dots, F_{X_n}$.

Definition 4.3.17 and Notation. The **convolution** (or **convolution product**) of the σ-finite measures $\mu_1, \dots, \mu_n$ on $\mathbb{R}^k$ is the measure on $\mathbb{R}^k$ given by the image of $\mu_1 \otimes \dots \otimes \mu_n$ under $(x_1, \dots, x_n) \longmapsto (x_1 + \dots + x_n)$. Denote this by $\mu_1 * \dots * \mu_n$ or $*_{i=1}^n \mu_i$; $\mu * \dots * \mu = \mu^{*n}$.

Theorem 4.3.18. *If $X_1, \dots, X_n$ are n independent observations taking values in $\mathbb{R}^k$, then*

$$F_{X_1+\dots+X_n} = F_{X_1} * \dots * F_{X_n}.$$

The following properties of the convolution are easily deduced from the properties of image measures (Proposition 3.3.20). These properties are easily interpreted when convolution is introduced as the distribution of the sum of independent random vectors.

Properties 4.3.19. (a) *The* r.v. *f on $\mathbb{R}^k$ is $\mu_1 * \dots * \mu_n$ integrable if, and only if, $(x_1, \dots, x_n) \longmapsto f(x_1 + \dots + x_n)$ is $\mu_1 \otimes \dots \otimes \mu_n$ integrable. Then*

$$\int f(z)d\mu_1 * \dots * \mu_n(z)$$
$$= \int f(x_1 + \dots + x_n)d\mu_1 \otimes \dots \otimes \mu_n(x_1, \dots, x_n).$$

(b) *The convolution product is commutative and associative.*
(c) *Dirac measure at* 0 *is the identity element for convolution.*

In view of the importance of the convolution of measures, it is essential to know how to calculate them.

Calculation Rules 4.3.20. (a) *Measures concentrated on* $\mathbb{Z}$. $(p_n)_{n\in\mathbb{Z}}$ and $(q_n)_{n\in\mathbb{Z}}$ being two positive sequences

$$\left[\sum_{n\in\mathbb{Z}} p_n \delta_n\right] * \left[\sum_{n\in\mathbb{Z}} q_n \delta_n\right] = \sum_{n\in\mathbb{Z}} \left[\sum_{m\in\mathbb{Z}} q_{n-m} p_m\right] \delta_n.$$

For measures concentrated on $\mathbb{N}$ we obtain

$$\left[\sum_{n\in\mathbb{N}} p_n\delta_n\right] * \left[\sum_{n\in\mathbb{N}} q_n\delta_n\right] = \sum_{n\in\mathbb{N}}\left\{\sum_{m=0}^{n} p_{n-m}q_m\right\}\delta_n.$$

If these two distributions are the distributions of independent r.v.'s X_1 and X_2, these relations are interpreted as has been done in Section 2.1.2. In this case it has been seen that *the generating function of the sum of two* r.v.'s *concentrated on* $\mathbb{Z}$ *is the product of their generating functions.*

(b) *Measures having a density with respect to Lebesgue measure* λ *on* $\mathbb{R}^k$. *Let* f *be a positive* r.v. *finite on* $\mathbb{R}^k$ *and* ν *a* σ*-finite measure on* $\mathbb{R}^k$. *Then* $(f\cdot\lambda) * \nu$ *has a density with respect to* λ, *which is the function* $x \longmapsto \int f(x-y)d\nu(y)$. In particular, letting g be a positive, finite r.v. on $\mathbb{R}^k$, we have

$$(f\cdot\lambda) * (g\cdot\nu) = (f*g)\cdot\lambda,$$

where $f * g$ is the **convolution of the functions** f and g defined by

$$(f*g)(x) = \int f(x-y)g(y)dy = \int g(x-y)f(y)dy.$$

When f and g are zero outside $\mathbb{R}_+$, $f(x-y)g(y)$ is zero for y outside $[0,x]$ and

$$(f*g)(x) = \int_0^x f(x-y)g(y)dy.$$

(c) Let $\nu_1, \ldots, \nu_n$ be n bounded measures on $\mathbb{R}^k$. *The Fourier transform of* $\nu_1 * \ldots * \nu_n$ *is the product of the Fourier transforms of* $\nu_1, \ldots, \nu_n$. (The characteristic function of the sum of n independent random vectors is the product of their characteristic functions.)

Note. Do not confuse this with the criterion of Independence 4.2.10: here we have a property which is not a sufficient condition for independence.

(d) *If* $\nu_1, \ldots, \nu_n$ *are bounded measures concentrated on* $\mathbb{R}_+$, *then the Laplace transform of* $\nu_1 * \ldots * \nu_n$ *is the product of their Laplace transforms.*

(e) Let F be a distribution $\mathbb{R}$, $I_F = \{t;\ \int e^{tx}dF(x) < \infty\}$. On I_F the Laplace transform $\phi_{F^{*n}}$ of F^{*n} is $(\phi_F)^n$; $\psi_{F^{*n}} = \text{Log}\,\phi_{F^{*n}} =$

$n\,\text{Log}\,\phi_F$. *The Cramer transform of* F^{*n} *thus satisfies, for* $a \in \mathbb{R}$, *the relation*

$$h_{F^{*n}}(na) = \sup_{u \in I_F} (u\, na - \psi_{F^{*n}}(u)) = nh_F(a).$$

Exercises 4.3. *Some calculations*

E.1. Let $a < b$ and $c < d$. Calculate: $\mathcal{U}_{[a,b]} * \delta_c$; $\mathcal{U}_{[a,b]} * \mathcal{U}_{[c,d]}$.

E.2. Let m_1, m_2 be in $\mathbb{R}$ and σ_1, σ_2 be in $]0,+\infty[$. Calculate $\mathcal{N}(m_1,\sigma_1^2) * \mathcal{N}(m_2,\sigma_2^2)$.

The Gamma distribution (Exercise 3.3.13) *and* χ^2 *distribution* (Exercises 3 – 5).

E.3. (a) Give the Laplace transform of $\gamma(a,\lambda)$; calculate its moment of order k. Determine $\gamma(a,\lambda) * (\gamma(b,\lambda)$.

(b) Let X be an r.v. with normal distribution $\mathcal{N}(0,1)$. Prove that X^2 has distribution $\gamma(1/2,\ 1/2)$. For $X_1, \ldots, X_n$ independent with distribution $\mathcal{N}(0,1)$, the distribution of $X_1^2 + \ldots + X_n^2$ is designated by $\chi^2(n)$. Determine its Laplace transform, density, mean, and variance.

E.4. (a) Let X and Y be two finite independent r.v.'s and F the distribution function of X. Assume that Y has a density g with respect to Lebesgue measure λ, and that g is zero on $]-\infty,0[$. Show that the distribution function of X/Y equals $\int_0^\infty F(ty)g(y)dy$ at each point $t \in \mathbb{R}$.

(b) Let $X, Y_1, \ldots, Y_n$ be $n + 1$ independent and normal r.v.'s. The **Student distribution on** n **degrees of freedom** is the distribution of $\sqrt{n}X/(Y_1^2 + \ldots + Y_n^2)^{1/2}$. Verify that its density is

$$x \longmapsto \frac{\Gamma[(n+1)/2]}{\Gamma[n/2]\sqrt{n\pi}} \left(1 + \frac{x^2}{n}\right)^{-(n+1)/2}.$$

Show that it is centered and that its variance equals $n/(n - 2)$ for $n > 2$. What is the limit of this distribution as n tends to ∞?

(c) Let X_1 and X_2 be two independent r.v.'s with distributions $\gamma(a,\lambda)$ and $\gamma(b,\lambda)$. What is the distribution of X_1/X_2? Let n and m be integers ≥ 1. We call the **Fisher–Snedecor distribution on** n **and** m **degrees of freedom** the distribution of $(X/n)/(Y/m)$, for X and Y independent, X with distribution $\chi^2(n)$ and Y with distribution $\chi^2(m)$. Show that its density is

$$x \longmapsto \frac{\Gamma((n+m)/2)n^{n/2}m^{m/2}}{\Gamma(n/2)\Gamma(m/2)} \frac{x^{n/2-1}}{(m+nx)^{(n+m)/2}}$$

and that, for $m > 4$, its mean is $m/(m-2)$ and its variance

$$\frac{2m^2(m+n-2)}{n(m-4)(m-2)^2}.$$

E.5. *The Noncentral Gamma distribution.* (a) Set, for $a > 0$, $c > 0$, $\lambda \geqslant 0$,

$$\gamma(a,c,\lambda) = \sum_{m=0}^{\infty} \frac{e^{-c}c^m}{m!} \gamma(a+m,\lambda).$$

Calculate the Laplace transform of $\gamma(a,c,\lambda)$. Show $\gamma(a,c,\lambda) * \gamma(b,c',\lambda) = \gamma(a+b, c+c', \lambda)$.

(b) Let $(X_i)_{1 \leqslant i \leqslant n}$ be n independent r.v.'s with respective distributions $N(m_i,1)$. Show that $\Sigma_{i=1}^n X_i^2$ has distribution

$$\gamma\left(\frac{n}{2}, \frac{m_1^2 + \dots + m_n^2}{2}, \frac{1}{2}\right),$$

the **noncentral χ^2 distribution on n degrees of freedom** with parameter $\Lambda\|m\|^2 = m_1^2 + \dots + m_n^2$; $\chi^2(n,\|m\|^2)$. The **noncentral Fisher Distribution on n and m degrees of freedom** with parameter γ^2 is the distribution of $(X/n)/(Y/m)$, where X and Y are independent, X with distribution $\chi^2(n,\gamma^2)$ and Y with distribution $\chi^2(m)$. Determine its density, and its mean for $m > 2$.

E.6. *A sample of exponential variables.* Let (X_n) be a sequence of independent r.v.'s having the exponential distribution with parameter λ. Let $T_n = X_1 + \dots + X_n$.

(a) Calculate the density of the distribution of T_n (the Erlang distribution with parameter λ and of order n). Verify that $2\lambda S_n$ has distribution $\chi^2(2n)$ (cf. Exercise 4.3.3).

(b) Let $N_t = \Sigma_{n \geqslant 1} 1_{[T_n \leqslant t]}$. Show that N_t has a Poisson distribution with parameter λt. Show that if t tends to $+\infty$, (N_t/t) tends to λ a.s. and $((N_t - \lambda t)/\sqrt{\lambda t})$ tends to $N(0,1)$ in distribution (cf. Exercise 3.4.7). Show that for all $\xi > 0$ there exists a constant $k(\xi)$, strictly positive, such that

$$P\left[\left|\frac{N_t}{t} - \lambda\right| \geqslant \xi\right] \leqslant 2e^{-tk(\xi)};$$

the convergence of N_t/t to λ is exponential (cf. Exercise 2.1.3). Calculate the density of $(T_1, \dots, T_n)$ conditional on $(N_t = n)$

and, for $0 < u < t$, the distribution of N_u conditional on $(N_t = n)$ and the distribution of (N_u, N_{t-u}).

E.7. Let μ and ν be two bounded measures in $(\mathbb{R}, \mathcal{B}_{\mathbb{R}})$. Show that $\mu * \nu$ equals δ_0 if, and only if, there exists a constant k and an $a \in \mathbb{R}$ such that $\mu = k\delta_a$, $\nu = (1/k)\delta_{-a}$.

4.4. A Sample from a Distribution and Estimation of This Distribution

4.4.1. An n-Sample from a Distribution

Let F be a distribution on $(E, \mathcal{E})$. In this section we are trying to clarify the idea of frequency: a sequence of independent observations from an unknown distribution F will allow us to estimate the characteristics of F.

Definition 4.4.21. **An n-sample from a distribution** F is a sequence $(X_1, \ldots, X_n)$ of n independent observations from the distribution F.

It is not necessarily useful to give details of the probability space $(\Omega, \mathcal{A}, P)$ on which these observations are defined. The *canonical representation* of the n-sample can be taken: $(\Omega, \mathcal{A}, P) = (E, \mathcal{E}, F)^n$ and X_i is the ith coordinate of E^n. When n-samples from several distributions defined on $(E, \mathcal{E})$ are to be studied, we shall take $(\Omega, \mathcal{A}) = (E, \mathcal{E})^n$ and shall denote P_F for $F^{\otimes n}$ and E_F the expectation associated with P_F. Finally, X will designate an arbitrary observation amongst $X_1, \ldots, X_n$.

4.4.2. Sample Mean of an n-sample from a Distribution F on $\mathbb{R}$

Assume that F has a mean $(\int |x| dF(x) < \infty)$. Let $m = E(X) = \int x \, dF(x)$. The intuitive frequency idea leads to considering the **sample** (or **empirical**) **mean** $\overline{X}_n = (X_1 + \ldots + X_n)/n$ and comparing it with m: m is estimated by $\overline{X}_n$. We will see in what sense $\overline{X}_n$ is a "good" estimator for m (cf. Section 2.3). When n is clear, $\overline{X}$ is used in place of $\overline{X}_n$.

Assume that $\int x^2 dF(x) = E(X^2)$ is finite, and denote by σ^2 the variance of F.

(a) $E_F(\overline{X}_n) = m$: the estimator $\overline{X}_n$ is "unbiased";

(b) $E_F(\overline{X}_n - m)^2 = \sigma^2(\overline{X}_n) = \sigma^2/n$. The "quadratic error" made by replacing m by $\overline{X}_n$ tends to 0 as n tends to $+\infty$. It follows from this, by Tchebychev's inequality, that for all $a > 0$

$$P_F[|\overline{X}_n - m| \geqslant a] \leqslant \frac{\sigma^2}{na^2}.$$

This is the "weak law of large numbers." The sequence $(\overline{X}_n)$ tends to m in probability.

(c) Assume that $I_F = \{t;\ E_F(e^{tX}) < \infty\}$ is a neighborhood of 0. Let h_F be the Cramer transform of F. The Cramer transform of F^{*n} equals $nh_F(a)$ at the point $na \in \mathbb{R}$ (Rule 4.3.20). Thus, following Proposition 3.3.25, for $\varepsilon > 0$,

$$P_F[X_1 + \dots + X_n \geqslant n(m + \varepsilon)] = P_F(\overline{X}_n \geqslant m + \varepsilon) \leqslant e^{-nh_F(m+\varepsilon)}$$

$$P_F(\overline{X}_n \leqslant m - \varepsilon) \leqslant e^{-nh_F(m-\varepsilon)}.$$

The probability that $|\overline{X}_n - m|$ exceeds ε decreases exponentially quickly. From this we deduce, by the proof of Theorem 2.5.10, the "strong law of large numbers": $(\overline{X}_n)$ converges a.s. to m. However, to obtain this result it is sufficient to hypothesize "X is integrable": this will be seen in Volume 2. (See Exercises 4.4.2 for the case where X is square integrable.) The following theorem has been obtained.

Theorem 4.4.22. *If $(X_1, \dots, X_n)$ is an n-sample from an integrable distribution F, with mean m, its sample mean $\overline{X}_n$ satisfies the following properties.*

(a) *It is an "unbiased" estimator of m: $E_F(\overline{X}_n) = m$.*

(b) *If F has variance σ^2, its variance is σ^2/n. The "weak law of large numbers" holds: $\overline{X}_n$ converges in probability to m, the sequence of estimators $(\overline{X}_n)$ of m is "weakly consistent." "In probability" can in fact be replaced by "a.s."; in other words "weak" can be replaced by "strong" without assuming σ^2 to be finite.*

(c) *If $\{t\colon \int e^{tx} dF(x) < \infty\}$ is a neighborhood of 0, the sequence of estimators $(\overline{X}_n)$ is exponentially consistent. If h_F is the Crâmer transform of F, then for all $\varepsilon > 0$,*

$$P_F[\overline{X}_n - m \geqslant \varepsilon] \leqslant e^{-nh_F(m+\varepsilon)};$$

$$P_F[\overline{X}_n - m \leqslant -\varepsilon] \leqslant e^{-nh_F(m-\varepsilon)}.$$

(d) An alternative way of stating the error made in replacing m by X_n, results in the following essential theorem.

4.4.23. Central Limit Theorem. *If $\overline{X}_n$ is the sample mean of an n-sample from a distribution with mean and variance σ^2, the sequence $\{\sqrt{n}(\overline{X} - m)\}$ converges in distribution to $N(0,\sigma^2)$ (in other words, $\{(1/\sqrt{n})(X_1 + \dots + X_n - nm)\}$ converges in distribution to $N(0,\sigma^2)$).*

Proof. Since the distribution F is square integrable, its Fourier transform is twice differentiable (Exercise 3.3.9). Assume m is zero, then $\phi'(0)$ is zero and $\phi''(0)$ equals σ^2. The characteristic function of $\sqrt{n}\,\overline{X}_n$ equals, for all $t \in \mathbb{R}$,

$$\psi_n(t) = E[\exp(it\sqrt{n}\,\overline{X}_n)]$$

$$= E\left[\exp i\,\frac{t}{\sqrt{n}}(X_1 + \dots + X_n)\right] = \left[\phi\left(\frac{t}{\sqrt{n}}\right)\right]^n.$$

But

$$\phi\left(\frac{t}{\sqrt{n}}\right) = 1 - \frac{t^2}{2n}\,\sigma^2 + o\left(\frac{1}{n}\right)$$

$$\lim_{n\to\infty} \operatorname{Log} \psi_n(t) = \lim_{n\to\infty} n \operatorname{Log}\left[1 - \frac{t^2}{2n}\,\sigma^2 + o\left(\frac{1}{n}\right)\right]$$

$$= -\frac{t^2\sigma^2}{2}$$

$$\lim_{n\to\infty} \psi_n(t) = \exp\left[-\frac{t^2\sigma^2}{2}\right].$$

Thus, following Levy's Theorem 3.4.30, $\sqrt{n}\,\overline{X}_n$ converges to $N(0,\sigma^2)$. The theorem follows from this by considering in the general case the centered r.v.'s $(X_n - m)$.

Note. *Properties* (c) *and* (d) *measure asymptotically the deviation of $\overline{X}_n$ and m from two distinct points of view. In* (c), *the interval $]-\infty,\overline{X}_n + \varepsilon]$ or $[\overline{X}_n - \varepsilon,+\infty[$ is given and it is shown that the probability of being incorrect in stating that m is in this interval decreases exponentially quickly. In* (d) *a probability of* *error α is fixed, and we try to determine "confidence intervals of level close to $1 - \alpha$" in which we can state that m lies, with an error probability close to α (for large n).*

This is what we have done in Section 2.4.5. For $\alpha < 1/2$, ϕ_α can again be defined by

$$\frac{1}{\sqrt{2\pi}} \int_{\phi_\alpha}^{\infty} \exp\left[-\frac{x^2}{2}\right] dx = \alpha.$$

For large n we again obtain confidence intervals of levels close to $1 - \alpha$:

$$]-\infty, \bar{X}_n + \frac{\sigma}{\sqrt{n}}\phi_\alpha]; \quad [\bar{X}_n - \frac{\sigma}{\sqrt{n}}\phi_\alpha, \infty[\quad \text{(one-sided intervals)}$$

$$[\bar{X}_n - \frac{\sigma}{\sqrt{n}}\phi_{\alpha/2}, \bar{X}_n + \frac{\sigma}{\sqrt{n}}\phi_{\alpha/2}] \quad \text{(two-sided interval).}$$

Remember the usual values: $\phi_{0.025} \simeq 2$, $\phi_{0.001} \simeq 3$.

An Example of an Application. The Monte Carlo Method for the calculation of integrals. Let $(X_1, \ldots, X_n)$ be an n-sample from a uniform distribution $A \in \mathcal{B}_{\mathbb{R}^k}$ with volume 1 and let ϕ be measurable from A into $\mathbb{R}$, integrable with respect to Lebesgue measure. Then, for large n, with an error risk close to 0.05 it can be stated that

$$\left|\frac{1}{n}\sum_{i=1}^{n} \phi(X_i) - \int_A \phi(x)dx\right| \leqslant \frac{2K}{\sqrt{n}},$$

where K is the standard deviation of $\phi(X_1)$.

We know, with the help of a computer, how to obtain an approximation to an n-sample from this uniform distribution on A (see Exercise 4.5.6). This gives, neglecting the error risk of 0.05, a procedure for calculating multiple integrals which converges as $1/\sqrt{n}$. It has the advantage over deterministic numerical methods of calculation of not making any regularity assumptions for ϕ, and of converging as quickly for multiple integrals as for simple integrals.

4.4.3. Empirical Estimators

n observations $X_1(\omega), \ldots, X_n(\omega)$ are made taking values in $(E,\mathcal{E})$. If we have no idea about their distributions no particular outcome will be favored. The distribution of $X^{(n)} = (X_1, \ldots, X_n)$ will be estimated by the empirical distribution, uniform on $(X_1(\omega), \ldots, X_n(\omega))$,

$$\bar{F}_n(\omega, \cdot) = \frac{1}{n} \sum_{i=1}^{n} \delta_{X_i}(\omega).$$

Definition 4.4.24. The **empirical distribution** $\bar{F}_n$ of n observations taking values in $(E,\mathcal{E})$ is defined, for each $\omega \in \Omega$, by

$$\bar{F}_n(\omega, \cdot) = \frac{1}{n} \sum_{i=1}^{n} \delta_{X_i}(\omega).$$

For $\Gamma \in \mathcal{E}$, $\omega \longmapsto \bar{F}_n(\omega,\Gamma)$ is an r.v. on $(\Omega,\mathcal{A})$; and for fixed ω, $\Gamma \longmapsto \bar{F}_n(\omega,\Gamma)$ is a probability. We shall say that $\bar{F}_n$ is a **random measure.**

Consider a sequence (X_n) of independent r.v.'s from the distribution F. Let f be an r.v. on $(E,\mathcal{E})$ integrable with respect to F. The law of large numbers is then written as $(\bar{F}_n(\omega,f))$ tends to $F(f)$ for almost all ω. The r.v. $\omega \longmapsto \bar{F}_n(\omega,f)$ will be designated by $\bar{F}_n(f)$.

For $E = \mathbb{R}^k$, to each $u \in \mathbb{R}^k$ corresponds a set of measure zero N_u, such that for $\omega \notin N_u$, $(\int \exp i < u, x > \bar{F}_n(\omega,dx))$ tends to $\int \exp i<u,x>dF(x)$. Take $N = \bigcup_{u \in \mathbb{Q}^k} N_u$: for $\omega \in N$, the sequence $(\bar{F}_n(\omega, \cdot))$ converges narrowly to F (the convergence of the Fourier transforms on $\mathbb{Q}^k$ is clearly sufficient in Theorem 3.4.30). For $k = 1$, the distribution function of $\bar{F}_n(\omega, \cdot)$ will be designated by $x \longmapsto \bar{F}_n(\omega,x)$; and for $x \in \mathbb{R}$, the r.v. $\omega \longmapsto \bar{F}_n(\omega,x)$ (resp. $\bar{F}_n(\omega,x-)$) by $\bar{F}_n(x)$ (resp. $\bar{F}_n(x-)$).

Glivenko–Cantelli Theorem 4.4.25. (a) *For a sequence (X_n) of independent random vectors taking values in $\mathbb{R}^k$, from the distribution F, the empirical distribution functions $(\bar{F}_n(\omega, \cdot))$ converge narrowly to F for almost all ω.*

(b) *For $k = 1$, the convergence of the distribution functions is* a.s. *uniform: for almost all ω, $(\sup_x|\bar{F}_n(\omega,x) - F(x)|)$ tends to zero.*

Proof of (b). Let N be an integer and, for $0 \leqslant j \leqslant N$,

$$x_{j,N} = \inf\{x;\ F(x) \geqslant j/N\};$$

certain of the $N + 1$ points may be confounded. We then have

$$F(x_{j,N}) \geqslant \frac{j}{N} \geqslant F(x_{j,N}-);$$

from which

$$F(x_{j,N}) + \frac{1}{N} \geqslant F(x_{j+1,N}-).$$

Let $x \in [x_{j,N}; x_{j+1,N}[$:

$$\bar{F}_n(x_{j,N}) - F(x_{j+1,N}-) \leqslant \bar{F}_n(x) - F(x) \leqslant \bar{F}_n(x_{j+1,N}-) - F(x_{j,N});$$

$$\bar{F}_n(x_{j,N}) - F(x_{j,N}) - \frac{1}{N} \leqslant \bar{F}_n(x) - F(x) \leqslant \bar{F}_n(x_{j,N}) - F(x_{j,N}) + \frac{1}{N};$$

$$\sup_x |\bar{F}_n(x) - F(x)| \leqslant \frac{1}{N} + \sup_{0 \leqslant j \leqslant N} |\bar{F}_n(x_{j,N}) - F(x_{j,N})|.$$

Let $\Gamma \subset \Omega$ be a set of measure zero such that, for $\omega \notin \Gamma$, the sequences $(\bar{F}_n(\omega, x_{j,N}))$ converge to $F(x_{j,N})$ for all $N \geqslant 1$, $j = 0, \ldots, N$. For $\omega \notin \Gamma$, the above inequality gives

$$\overline{\lim_n} \sup_x |\bar{F}_n(\omega, x) - F(x)| \leqslant \frac{1}{N}.$$

This being true for all N, the theorem is established.

Definition 4.4.26. Empirical Estimator. Let an n-sample be from a distribution F on $(E, \underline{E})$ with empirical distribution function $\bar{F}_n$. Let $\mathcal{F}_n$ be the set of uniform probabilities concentrated on a set of n points of E:

$$\mathcal{F}_n = \left\{\frac{1}{n} \sum_{i=1}^{n} \delta_{X_i}\ ;\ (X_i)_{1 \leqslant i \leqslant n} \in E^n\right\}.$$

For a functional T defined on a set of probabilities containing F and $\mathcal{F}_n$, the **empirical estimator** of $T(F)$ will be $T(\bar{F}_n)$.

In the examples which follow, assume again that $(E,\mathcal{E}) = (\mathbb{R},\mathcal{B}_{\mathbb{R}})$.

Example 1. Method of Moments. Let p be an integer $\geqslant 1$ and $m_p(F) = \int x^p dF(x)$ a function defined on $\Pi_p = \{F: \int |x|^p dF(x) < \infty\}$. $m_p(F)$ is then estimated by $m_p(\overline{F}_n) = (1/n)\Sigma_{j=1}^n X_j^p$. Similarly, for $T(F) = f(m_1(F), ..., m_p(F))$, T being defined on Π_p by an r.v. f defined on $\mathbb{R}$, the estimator of $T(F)$ obtained by the method of moments if $f(m_1(\overline{F}_n), ..., m_p(\overline{F}_n))$.

Particular Case. (a) $T(F) = \sigma^2(F) = m_2(F) - (m_1(F))^2$ leads to the estimator of $\sigma^2(F)$, called the empirical variance:

$$\overline{\sigma}_n^2 = \frac{1}{n}\sum_{j=1}^n X_j^2 - \left[\frac{1}{n}\sum_{j=1}^n X_j\right]^2 = \frac{1}{n}\sum_{j=1}^n (X_j - \overline{X}_n)^2.$$

This is not an unbiased estimator; its mean is $[(n-1)/n]\sigma^2(F)$. The *unbiased estimator of* $\sigma^2(F)$

$$\overline{S}_n^2 = \frac{1}{n-1}\sum_{j=1}^n (X_j - \overline{X}_n)^2$$

is most often called the **sample variance.**

(b) Assume that F is a mixture of Poisson distributions with respective parameters λ_1 and λ_2; for $0 \leqslant a \leqslant 1$ we have

$$F(k) = a\,\frac{\lambda_1^k}{k!}\,e^{-\lambda_1} + (1-a)\,\frac{\lambda_2^k}{k!}\,e^{-\lambda_2}.$$

For X from the distribution F, we calculate

$$T_1(F) = E(X) = a\lambda_1 + (1-a)\lambda_2$$

$$T_2(F) = E[X(X-1)] = a\lambda_1^2 + (1-a)\lambda_2^2$$

$$T_3(F) = E[X(X-1)(X-2)] = a\lambda_1^3 + (1-a)\lambda_2^3,$$

from which the calculation of (a,λ_1,λ_2) as a function of $T_1(F)$, $T_2(F)$, $T_3(F)$ and their estimators, obtained by replacing $T_i(F)$ by $T_i(\overline{F}_n)$, $i = 1,2,3$, follows.

(c) Consider an n-sample $(X_i,Y_i)_{1\leqslant i\leqslant n}$ from a distribution F on $\mathbb{R}^2$; the r.v.'s X and Y being square integrable, the **empirical covariance** $\overline{\Gamma}_n$, and the **empirical correlation** $\overline{\rho}_n$ are

$$\overline{\Gamma}_n = \overline{X_n Y_n} - \overline{X}_n\,\overline{Y}_n = \frac{1}{n}\sum_{i=1}^{n}(X_i - \overline{X}_n)(Y_i - \overline{Y}_n),$$
$$\overline{\rho}_n = \frac{\overline{\Gamma}_n}{\overline{\sigma}_n(X)\overline{\sigma}_n(Y)}.$$

If $(X_i,Y_i)_{1\leqslant i\leqslant n}$ are the outcomes of an experiment, $\overline{\Gamma}_n$ and $\overline{\rho}_n$ are the quantities calculated and studied in Section 1.2.

The regression line was then introduced:

$$y = \overline{Y}_n + \frac{\overline{\Gamma}_n}{\overline{\sigma}_n^2(X)}(x - \overline{X}_n)$$

which we will call here the **empirical regression line** of Y with respect to X, or the **empirical orthogonal regression line** with equation $(y - \overline{Y})\sin\overline{\theta}_n + (x - \overline{X})\cos\overline{\theta}_n = 0$, with

$$\tan 2\overline{\theta}_n = \frac{2\overline{\Gamma}_n}{\overline{\sigma}_n^2(X) - \overline{\sigma}_n^2(Y)}.$$

These regression lines correspond to the method of moments estimator for the regression line of Y on X or the orthogonal regression line (Exercise 3.2.9).

Example 2. Estimation of the Bounds of F. Let m and M be the essential lower or upper bounds of F:

$$m = \sup\{a;\ F(a) = 0\},\quad M = \inf\{a;\ F(a) = 1\}.$$

Their respective empirical estimators are

$$m_n = \inf(X_i;\ 1 \leqslant i \leqslant n),\quad M_n = \sup(X_i;\ 1 \leqslant i \leqslant n).$$

Example 3. Quantiles and Order Statistics

Definition 4.4.26. Let $\alpha \in]0,1[$. The function F increases from 0 to 1, hence there exists an x such that $F(x^-) \leqslant \alpha \leqslant F(x)$. The function F may be constant on an interval I_α (possibly reduced to x) containing x. I_α is called the **interval of quantiles** of order α, each $x \in I_\alpha$ is a **quantile** of order α. It is sometimes convenient to take for the quantile, the middle of the interval of quantiles, denoted by q_α. If F charges q_α, we can have $F(q_\alpha-) < \alpha < F(q_\alpha)$. A quantile of order 1/2 is a **median**. We will often denote by F_α^- a quantile of order α, and by F^+_α a quantile of order $1 - \alpha$. This is what we have done in Section 2.4.3 for the binomial distribution.

Definition 4.6.27. Order Statistic. Consider the mapping of $\mathbb{R}^n$ into $\mathbb{R}_n = \{(x_1, ..., x_n); x_1 \leqslant ... \leqslant x_n\}$ which associates with $x = (x_1, ..., x_n)$ the vector $\tilde{x}$ obtained by arranging the components of x in increasing order. The variable $\tilde{X} = (X_{(1)}, ..., X_{(n)})$ is the **order statistic** of $X = (X_1, ..., X_n)$; $X_{(i)}$ is the ith order statistic.

The sample estimates $\bar{q}_\alpha$ and q_α can be determined with the help of these order statistics.

– For the sample median $\bar{q}_{1/2}$, we obtain $X_{(n+1)}$ for $n = 2m + 1$ and $([X_{(m)} + X_{(m+1)}]/2)$ for $n = 2m$;
– When α is not a multiple of $1/n$, we have

$$\bar{q}_\alpha = \frac{X_{([n\alpha])} + X_{([n\alpha]+1)}}{2} .$$

4.4.4. Order and Rank Statistics

We have just associated, with $x = (x_1, ..., x_n)$ in $\mathbb{R}^n$, the ordered sequence $\tilde{x} = (\tilde{x}_1, ..., \tilde{x}_n)$. Let $r_i = \Sigma_{j=1}^n \mathbf{1}_{(x_i \geqslant x_j)}$ be the rank of x_i in the sequence x, and $r = (r_1, ..., r_n)$. To x corresponds a unique pair $(\tilde{x},r)$ and $x_i = \tilde{x}_{r_i}$; when there are ties, $x_i = x_j$, then $r_i = r_j$. If all the components of x are distinct, r is an element of $\mathcal{S}_n$, the set of permutations of $\{1, ..., n\}$.

Definition 4.4.28. Let $(X_1, ..., X_n) = X^{(n)}$ be a random vector in $\mathbb{R}^k$. Its **rank vector** $R = (R_1, ..., R_n)$ is defined by $R_i = \Sigma_{j=1}^n \mathbf{1}_{(X_i \geqslant X_j)}$, for $i = 1, ..., n$. For an n-sample from a continuous distribution on $\mathbb{R}$, R takes values in $\mathcal{S}_n$ a.s. and there are a.s. no ties (this has been seen in Section 4.2.3).

Theorem 4.4.29. *Let $X^{(n)}$ be an n-sample from a continuous distribution F on $\mathbb{R}$.*

(1) *Its order statistic* $\widetilde{X}^{(n)}$ *and its rank vector* R *are independent.*
(2) R *has a uniform distribution on* $\mathfrak{S}_n$.
(3) *The distribution of* $\widetilde{X}^{(n)}$ *is the trace of* $n!F^{\otimes n}$ *on* $\widetilde{\mathbb{R}}_n = \{(x_1, ..., x_n); x_1 \leqslant ... \leqslant x_n\}$.

Proof. Let $\sigma \in \mathfrak{S}_n$; $X_\sigma^{(n)} = (X_{\sigma(1)}, ..., X_{\sigma(n)})$ is also an n-sample from F. (2) is obtained by writing

$$\begin{aligned} P[R = \sigma] &= P[X_{\sigma^{-1}(1)} < ... < X_{\sigma^{-1}(n)}] \\ &= P[X_1 < ... < X_n] = \frac{1}{n!}. \end{aligned}$$

For a Borel set A in $\mathbb{R}$ contained in $\widetilde{\mathbb{R}}_n$, we have

$$P[\widetilde{X}^{(n)} \in A, R = \sigma] = P[X^{(n)}_{\sigma^{-1}} \in A] = F^{\otimes n}(A)$$

$$P[\widetilde{X}^n \in A] = \sum_{\sigma \in \mathfrak{S}_n} P[\widetilde{X}^n \in A, R = \sigma] = n!F^{\otimes n}(A).$$

From which (1) and (3) follow.

Exercises 4.4.

E.1. Let $(X_n)_{n \geqslant 1}$ be a sequence of independent r.v.'s such that each one has a Cauchy distribution with parameter $\alpha > 0$. Let $S_n = \Sigma_{p=1}^n X_p$. Give the distribution of

$$\left(\frac{S_{n+p}}{n+p} - \frac{S_n}{n}\right).$$

Show that (S_n/n) converges in distribution but not in probability. Why does the weak law of large numbers not apply?

E.2. *Law of large numbers for identically distributed square integrable, uncorrelated r.v.'s.* Let (X_n) be a sequence of identically distributed, (i.e., having the same distribution) square integrable r.v.'s, with mean m and variance σ^2. Assume that the r.v.'s are pairwise uncorrelated. For $n \geqslant 1$, let $S_n = X_1 + ... + X_n$.

(a) What is the variance of S_n/n? Show that S_n/n tends to m in quadratic mean.

(b) Take $m = 0$. Write down Tchebychev's inequality for S_{n^2}/n^2: prove that the sequence (S_{n^2}/n^2) tends a.s. to zero.

Let

$$Y_n = \sup_{n^2+1 \leq k < (n+1)^2} [S_k - S_{n^2}];$$

prove that (Y_n/n^2) tends a.s. to zero.

(c) Prove that (S_n/n) tends a.s. to m.

E.3. *Rounding errors in a computer program.* Assume that in a computer program each operation is rounded to the Jth decimal place. n operations are carried out and rounding occurs at each step. It is hypothesized that the rounding errors add together, are independent, and uniform on $[-\frac{1}{2}10^{-J}, \frac{1}{2}10^{-J}]$. For large n, give to within 0.05 an upper confidence bound for the absolute value of the error in the final result. Knowing n (large), what J can be chosen to guarantee a result accurate to within 10^{-2} with a probability of the order of 0.95?

E.4. *Code* (following on from Exercise 1.3.15). Let $p = (p_1, ..., p_k)$ be a probability on $\{1, ..., k\}$, and $H = -\Sigma_{i=1}^k p_i \log_2 p_i$ its entropy. For integer n and $\varepsilon > 0$, define the following subset of $\{1, ..., k\}^n$:

$$\Delta_{n,\varepsilon} = \left\{(x_1, ..., x_n); \frac{1}{n}\sum_{j=1}^{n} \log_2 \frac{1}{p_{x_j}} < H + \varepsilon\right\}.$$

Demonstrate that $\Delta_{n,\varepsilon}$ has at most $2^{n(H+\varepsilon)}$ elements. For $(X_1, ..., X_n)$ an n-sample from p, show that $P((X_1, ..., X_n) \notin \Delta_{n,\varepsilon})$ decreases exponentially in n. Compare 2^{nH} and k^n; what significance has this if we are interested in words of length n from the alphabet $(A_1, ..., A_k)$ of Exercise 1.3.15e)?

E.5. Consider an n-sample $(X_1, ..., X_n)$ where the X_i have a **log-normal** distribution, i.e., Log X_i has an $N(m,\sigma^2)$ distribution. Estimate m and σ^2 by the method of moments.

E.6. For an n-sample from a distribution F with variance σ^2, let $\overline{\sigma^2}$ be the empirical estimator of σ^2. Determine an unbiased estimator S^2 proportional to $\overline{\sigma^2}$. Assume that F has a moment α_4 of order 4. Calculate the correlation between S^2

and $\overline{X}$ and show that it tends to zero if n tends to $+\infty$.

E.7. (a) Let $(X_1, ..., X_n)$ be an n-sample from a Gaussian distribution $N(m,\sigma^2)$. For $2 \leqslant i \leqslant n$, set $Y_i = X_i - X_1$. Show that the empirical mean $\overline{X}$ is independent of the random vector $(Y_i)_{2\leqslant i\leqslant n}$. What is the distribution of $(Y_i)_{2\leqslant i\leqslant n}$? Show that the empirical variance S^2 is a function of $(Y_i)_{2\leqslant i\leqslant n}$. Deduce from this the independence of $\overline{X}$ and S^2.

(b) We try to show that property (a) characterizes the Gaussian distribution. To this end consider two independent r.v.'s X_1 and X_2 both having the characteristic function ϕ. Assume that $X_1 - \overline{X} = (1/2)(X_1 - X_2)$ and $\overline{X}$ are independent. Prove that for $(u,v) \in \mathbb{R}^2$:

$$\phi(u + v)\phi(v - u) = |\phi(u)|^2(\phi(v))^2.$$

Deduce from this that (X_1,X_2) is a 2-sample from a Gaussian distribution.

E.8. Let $\mathcal{U}_{a,b}$ be a uniform distribution on $[a,b]$; an n-sample $(X_1, ..., X_n)$ from $\mathcal{U}_{a,b}$ is observed in order to estimate the mean $(a+b)/2$ of this distribution.

(a) What is the quadratic error $E(\overline{X} - (a+b)/2))^2$ of the sample mean.

(b) Let $T = (1/2)[\sup_{1\leqslant i\leqslant n}(X_i) - \inf_{1\leqslant i\leqslant n}(X_i)]$. Calculate $E(T)$ and $E[T - (a+b)/2)]^2$. Is the estimator $\overline{\text{X}}$ admissible in the sense of Definition 2.3.7?

E.9. Let $R = (R_1, ..., R_n)$ be the rank vector of an n-sample from a continuous distribution on $\mathbb{R}$. By using Exercise 1.3.9, calculate, for $1 \leqslant i,j \leqslant n$, the mean and variance of R_i, and the covariance of R_i and R_j. What is the distribution of R_j conditional on $\{R_i = r\}$?

E.10. *Distributions of the order statistics* (Exercise 1.3.9). Let $(X_1, ..., X_n)$ be an n-sample from the distribution F on $\mathbb{R}$.

(a) What is the distribution function of $X_{(r)}$ for $1 \leqslant r \leqslant n$ (in particular for $X_{(1)} = \inf X_i$, $X_{(n)} = \sup X_i$)?

(b) Let $I_1, \ldots, I_k$ be k disjoint Borel sets the union of which is $\mathbb{R}$. Denote by Y_i the r.v. equal to $\Sigma_{j=1}^{k} j 1_{(X_i \in I_j)}$. Prove that the r.v.'s $(Y_i)_{1 \leqslant i \leqslant n}$ form an n-sample from a distribution. Which distribution? Calculate $P[X_{(r-1)} \leqslant x, x < X_{(r)} \leqslant x + h, x + h < X_{(r+1)}]$.

(c) Assume that F has a density f with respect to Lebesgue measure. Prove that $X_{(r)}$ has density f_r with

$$f_r(x) = \frac{n!}{(r-1)!(n-r)!} F^{r-1}(x)(1 - F(x))^{n-r} f(x).$$

Similarly we could use (b) to calculate the density of the pair $(X_{(r)}, X_{(s)})$, for $r < s$, and the density of $(X_{(r_1)}, \ldots, X_{(r_k)})$ for $1 \leqslant r_1 < r_2 \ldots < r_k \leqslant n$. What is the density of $(X_{(1)}, \ldots, X_{(n)})$?

(d) Assume that F has an exponential distribution; $X_1, \ldots, X_n$ are the failure times for n identical components started at the same time, $X_{(r)}$ is the time of the rth observed failure. Show that the r.v.'s $(X_{(r)} - X_{(r-1)})_{1 \leqslant r \leqslant n}$ are independent. What is the distribution of $X_{(r)} - X_{(r-1)}$? Interpret this.

(e) In the framework of (c) what is the density of $(X_{(s)} - X_{(r)}, X_{(r)})$? Then that of $X_{(s)} - X_{(r)}$? What is the distribution function of the dispersion $X_{(n)} - X_{(1)}$?

(f) Repeat calculations (c) and (e) for a uniform distribution on $[0,1]$. Prove that the r.v.'s

$$\frac{X_{(1)}}{X_{(2)}}, \left[\frac{X_{(2)}}{X_{(3)}}\right]^2, \left[\frac{X_{(3)}}{X_{(4)}}\right]^3, \ldots, \left[\frac{X_{(n-1)}}{X_{(n)}}\right]^{n-1}, [X_{(n)}]^n$$

are independent and identically distributed. (Which distribution?)

E.11. *Renewal process.* This is by definition the sequence of successive instants $(T_n)_{n \geqslant 1}$ at which events occur (breakdowns, equipment replacements, ...). Assume that T_1 has distribution G and is independent of the sequence $(\tau_n = T_{n+1} - T_n)_{n \geqslant 1}$ of independent r.v.'s from the same distribution F; F and G are concentrated on $]0,\infty]$. For $t \geqslant 0$, define $F(t) = P(\tau_1 \leqslant t)$ and $N_t = \Sigma_{n \geqslant 1} 1_{(T_n \leqslant t)}$ the number of events preceding t: $(N_t)_{t \geqslant 0}$ is the **counting process** associated with the renewal process.

(a) Assume $F = G$. Prove that, for all $t > 0$, an integer k can be found satisfying $F^{*k}(t) < 1$. Then verify $P[T_{nk} \leqslant t] \leqslant (F^{*k}(t))^n$ and $E(N_t) < \infty$. Let $\nu = \inf\{n; \tau_n = \infty\}$. Verify'

$$\lim_{t\to\infty} E(N_t) + 1 = E(\nu) = \frac{1}{P(\tau_1 = \infty)} \leqslant \infty.$$

(b) Assume that $F(\infty) = \sup_{t<\infty} F(t) < 1$ and $F = G$. Let $V = \sup\{T_n; T_n < \infty\}$. Prove

$$E(V) = \frac{1}{1 - F(\infty)} \int_0^\infty [F(\infty) - F(t)]dt.$$

A pedestrian sees cars coming in a one way street. The intervals between the random arrival times of a car in front of him are exponential with parameter $\lambda > 0$. Let (T_n) be the sequence of these passage times, $T_0 = 0$; the pedestrian crosses at time $S = \inf\{T_n; T_{n+1} - T_n \geqslant 1\}$; calculate $E(S)$.

(c) Assume that F has a finite mean m and that $F = G$. Show

$$\lim_{n\to\infty} \frac{T_n}{n} = m, \qquad \lim_{t\to\infty} \frac{N_t}{t} = \frac{1}{m} \quad \text{(a.s.)}.$$

If it is also assumed that F has a variance σ^2, show that $(\sqrt{n}/\sigma)(T_n/n - m)$ converges in distribution to $N(0,1)$ and that $\sqrt{t}((N_t/t) - (1/m))$ converges in distribution to $N(0,\sigma^2/m^3)$ (cf. Exercise 3.4.7).

(d) Still assuming that F has a finite mean m, let ϕ be its Laplace transform. Prove that for $u > 0$,

$$\int_0^\infty e^{-ut} \frac{1}{m} F([t,\infty[)dt = \frac{1 - \phi(u)}{mu}.$$

Assume $t \longmapsto E(N_t)$ is linear: the measure $G * \sum_{n\geqslant 0} F^{*n}$ is proportional to Lebesgue measure on $[0,\infty[$. By identifying their Laplace transforms, prove that G has, with respect to Lebesgue measure, the density $t \longmapsto (1/m)F([t,\infty[)$. Prove the converse.

(e) Let $t > 0$, $n \geqslant 1$ and $T_n(t) + t = T_{N_t+n}$ the time of the nth event following t. Prove that for arbitrary G, $(T_{n+1}(t) - T_n(t))_{n\geqslant 1}$ is a sequence of independent variables with distribution F and that this sequence is independent of $T_1(t)$. What is the counting process of $(T_n(t))_{n\geqslant 1}$? Prove that, if T_1 has a distribution with density $u \longmapsto (1/m)F([u,\infty[)$ with respect to Lebesgue measure, the same applies to $T_1(t)$.

E.12. *Poisson Process.* This is the counting process associated with a renewal process defined in Exercise 4.4.11, for which F and G are exponential distributions with parameter λ.

(a) Prove that, for all t, the sequence $\{T_1(t), (T_{n+1}(t) - T_n(t))_{n\geqslant 1}\}$ is a sequence of independent exponential r.v.'s with parameter λ. Recall the idea of Exercise 4.2.2. The exponential time intervals correspond to a phenomenon without wear and tear or without memory (if observations start at time t, it is the same thing as starting at time 0).

(b) Show that the distribution of $(T_1, ..., T_n)$ conditional on $(N_t = n)$ is the same as that for the order statistic of an n-sample from the uniform distribution on $[0,t]$ (Exercise 4.4.10).

E.13. *Gambler's ruin.* For a sequence $(X_n)_{n\geqslant 1}$ of independent identically distributed r.v.'s not identically 0 and a and b two numbers > 0, consider

$$S_n = X_1 + ... + X_n, \quad \nu_b = \inf\{n; S_n \geqslant b\},$$

$$\nu_a = \inf\{n; S_n \leqslant -a\}, \qquad \nu_{ab} = \inf(\nu_a,\nu_b).$$

(a) Adapt the proof of Lemma 2.5.12, and prove that ν_{ab} has moments of all orders.

(b) Assume X is integrable with mean m. A **stopping time relative to the sequence** (X_n), ν, is an r.v. taking values in $\mathbb{N} \cup \{\infty\}$ satisfying, for arbitrary n, $\{\nu = n\} \in \sigma(X_1, ..., X_n)$. Prove that if ν is a stopping time, $\inf(\nu,n)$ is a stopping time and that ν_a, ν_b and ν_{ab} are stopping times. Prove *Wald's theorem*: for ν integrable $E(S_\nu) = mE(\nu)$. (Adapt Exercise 2.5.4 and take first of all X positive and $\nu \leqslant n$.) In Exercise 4.4.11(c), prove $E(T_{N_t}) = mE(N_t)$.

(c) Assume $m \geqslant 0$; prove that ν_a is not integrable. Let $I = \inf_{n\geqslant 1}S_n$, $M = \sup_{n\geqslant 1}S_n$. Show that for $m < 0$, $P(I > -a)$ is zero and ν_a is finite a.s.

(d) For $m > 0$ and $X \leqslant K < \infty$, show, by approximating ν_b by $\inf(\nu_b,n)$ for $n \geqslant 0$,

$$mE(\nu_b) \leqslant K + b.$$

However, without assuming $X \leqslant K$, we can find a K such that $Y = \inf(X,K)$ has a strictly positive mean. Prove, for such a K,

$$E(\nu_b) \leqslant \frac{K+b}{E(Y)} .$$

Consequence of (c) *and* (d): *for* $m > 0$,

$$P(M = \infty) = P(I > -\infty) = 1;$$

ν_b *is integrable and*

$$\lim_{b\to\infty} \frac{E(\nu_b)}{b} = \frac{1}{m};$$

ν_a *is not integrable.*

(e) Assume M is finite a.s.; let G be its distribution. Prove

$$\begin{cases} G(x) = E[G(x - X)1_{(X<x)}] \\ P(M \leqslant 0) > 0 \end{cases} .$$

Show $P(\nu_b = \infty) > 0$. Show that there exists a $c > 0$ such that for all x,

$$\Sigma\, P(x \leqslant S_n < x + c) \leqslant 1/P(M > -c) < \infty.$$

Deduce from this, $S_n \xrightarrow{\text{a.s.}} -\infty$.

(f) *For* $m = 0$, verify

$$P(M = \infty) = P(I = -\infty) = 1.$$

The two times ν_a *and* ν_b *are finite* a.s., *but not integrable.*

4.5. Nonparametric Tests

4.5.1. Testing a New Drug

A team of doctors wants to check the effectiveness of a new drug B, of which it is certain that there are no harmful side-effects. If the disease to be treated is widespread, i.e., if the population of patients is large, we can find a fairly homogeneous subpopulation (age, previous medical history,...)

and can choose at random m individuals from this subpopulation to whom the traditional drug A will be administered and n individuals who will take B. The recovery time (if the disease is easily cured) or the survival time is measured, and A and B are compared. The choice of m and n will depend on the experimental possibilities. The patients treated by A and B, respectively, may be considered as forming an m-sample $X^{(m)}$ and an n-sample $Y^{(n)}$ from the respective distributions F_0 and F_1: this is what we shall do in Section 4.5.2.

If, on the other hand, the disease is rare, we shall only have at our disposal a small number of patients and we shall not reduce such a restricted population by taking samples. It is the statistician who, by allocating drugs at random to the patients, will create an interpretable situation. This will be studied in Section 4.5.3 by an example. This is an important distinction.

4.5.2. Hypothesis Tests H_0: "$F_0 = F_1$" Against H_1: "$F_0 < F_1$"

(a) In the first setup described above, we thus observe $X^{(m)}$, an m-sample from F_0, and $Y^{(n)}$, an n-sample from F_1; these two samples are independent of each other. The hypothesis H_0 "the drugs A and B are equivalent" comes down to "$F_0 = F_1$." Hypothesis H_1 "drug B is preferable" comes down to "$F_0 < F_1$," for an order relation on the distributions F_0 and F_1, signifying that in a sense to be specified, X has a tendency to be smaller than Y. For example, we could define $F_0 < F_1$ by the fact that the distribution function of F_0 is larger than that of F_1 (cf. 3.3.6).

The experimenter observes $Z = (X^{(m)}, Y^{(n)})$ from the distribution $F_0^{\otimes(n+m)}$ under H_0. The idea is to set aside H_0 when a suspect configuration of Z is found. The statistician is thus responsible for defining a region D, the rejection region, in $\mathbb{R}^{m+n}$ and for advising the experimenter: "see if Z is in D, and if it is, accept the hypothesis H_1" ..., i.e., accept the use of drug B. We are concerned with testing H_0 against H_1. How is this zone D chosen? We only want to change the usual practices if it is worth the trouble. A small number $\alpha > 0$ is therefore fixed and we want to choose D such that, under the hypothesis H_0, the probability of rejecting H_0 is less than α for arbitrary F_0.

In any case, the framework defined here is very general, and the choice of the rejection region is rather arbitrary. Two different choices can lead to different conclusions. We shall be guided in the choice of D by a qualitative examination of the hypothesis H_1. Even if we only want to make incorrect changes with probability less than α, we would like also to have the greatest possible chances to make changes if H_1 is correct! We would like to maximize the **power**, the probability of accepting H_1 if it is true. In general this probability depends on the pairs (F_0,F_1) and cannot simultaneously be maximized for all pairs. Another difficulty is that F_0 is unknown. The calculations will be simplified if we can find a **distribution free** statistic under H_0, i.e., a random variable $\phi(Z)$ the distribution of which (i.e., the image under ϕ of $F_0^{\otimes(n+m)}$) does not depend on F_0. We could then *look for a rejection region of the form* $\{\phi(Z) \in C\}$. Denote by $P_{H_0}(\phi(Z) \in C)$ the probability of rejection, independent of F_0 under H_0. It is frequently the case that the region $\{\phi(Z) \in C\}$ is of the form $\{\phi(Z) \geqslant t\}$; t is called the **threshold** of the test; the level is $P_{H_0}(\phi(Z) \geqslant t) = \alpha_t$. If Z is observed, we reject H_0 at all levels less than $\alpha_{\phi(Z)}$. This number $\alpha_{\phi(Z)}$ is called the **significance level** of the hypothesis when Z is observed.

A test is not a proof, and it is important to associate with the result of an experiment, its significance level.

(b) *Wilcoxon's Test.* Assume F_0 and F_1 are continuous; $N = n + m$. Let $R = (S_1, ..., S_m, T_1, ..., T_n)$ be the rank vector of Z and $W_S = S_1 + ... + S_m$, $W_T = T_1 + ... + T_n$ the sum of the ranks of X and Y. Since, under H_0, the distribution of R is uniform on $\mathfrak{S}_N$, the distributions of W_S and W_T are defined under H_0: these distribution-free statistics are tabulated. Their *Wilcoxon distribution* has been studied in Exercise 1.3.9(c). If H_0 is tested against H_1, notice that, for $F_1 > F_0$, it must be expected that W_T takes large values. We are thus led to take for the rejection region $D_\alpha = (W_T > t_\alpha)$, with t_α determined by $t_\alpha = \inf(t; P_{H_0}(W_T > t) \leqslant \alpha)$. (The distribution of W_T being discrete, we cannot in general find a t_α, such that $P_{H_0}(W_T > t_\alpha) = \alpha$.)

Note on the Tables of Wilcoxon's Distribution. Under H_0, for $\sigma \in \mathfrak{S}_N$, R and $\sigma \circ R$ have the same distribution, uniform on

$\mathcal{S}_N$. Taking $\sigma(i) = N - i + 1$, for $1 \leqslant i \leqslant N$, we obtain

$$\sigma(R) = (N - S_1 + 1, \ldots, N - S_m + 1,$$
$$N - T_1 + 1, \ldots, N - T_n + 1),$$

$$P_{H_0}[W_S = k] = P_{H_0}[W_S = m(N+1) - k].$$

The distribution of W_S is symmetric about $m(N+1)/2$. However $W_S + W_T$ equals $N(N+1)/2$; thus

$$W_T - \frac{n(N+1)}{2} = \frac{m(N+1)}{2} - W_S$$

has the same distribution as $W_S - m(N+1)/2$. By adding $nm/2$, we obtain $W_{X,Y} = W_S - m(m+1)/2$ and $W_{Y,X} = W_T - n(n+1)/2$ have the same distribution. This is why the tables give the distribution of $W_{X,Y}$ for $m \leqslant n$ only. This is sufficient by changing possibly the roles of X and Y.

(c) **The Kolmogorov–Smirnov Test.** Let us assume again that F_0 is continuous. Let $\bar{F}_X$ and $\bar{F}_Y$ be the empirical distribution functions of $X^{(m)}$ and $Y^{(n)}$, and $D_{m,n} = \sup_x[\bar{F}_X(x) - \bar{F}_Y(x)]$:

$$D_{m,n} = \sup[\bar{F}_X(x) - \bar{F}_Y(x);\ x \in \{X_1, \ldots, X_m, Y_1, \ldots, Y_n\}]$$
$$= \sup\left[\frac{i}{m} - \frac{j}{n};\ 1 \leqslant i \leqslant m,\ 0 \leqslant j \leqslant n,\ Y_{(j)} \leqslant X_{(i)} < Y_{(j+1)}\right]$$

(denoting $Y_{(n+1)} = \infty$, $Y_{(0)} = 0$). Set $U_i = F_0(X_i)$, $V_j = F_0(Y_j)$. Under H_0, $(U_1, \ldots, U_m, V_1, \ldots, V_n)$ form an N-sample from a uniform distribution on $[0,1]$ (Exercise 3.3.5) and since F_0 is increasing, $Y_{(j)} \leqslant X_{(i)}$ is equivalent a.s. to $V_{(j)} \leqslant U_{(i)}$. From this the statistic

$$D_{m,n} = \sup\left[\frac{i}{m} - \frac{j}{n};\ 1 \leqslant i \leqslant m,\ 0 \leqslant j \leqslant n,\ V_{(j)} \leqslant U_{(i)} < V_{(j+1)}\right]$$

is distribution free under H_0.

If H_1 is true, since the values of X are rather less than those of Y, we expect to observe large values of $D_{m,n}$. We thus take the largest possible $d_{m,n,\alpha}$ such that $P_{H_0}[D_{m,n} \geq d_{m,n,\alpha}] \leq \alpha$, and choose for the rejection region $\{D_{m,n} \geq d_{m,n,\alpha}\}$. We cannot, without restricting H_1, say which of the tests of Wilcoxon (b) or of Kolmogorov–Smirnov (c) is better, i.e., which will make us choose H_1 with the greatest probability if it is true. The two tests can give different conclusions. The r.v.'s $\sqrt{mn/(m+n)}\,D_{mn}$ tend in distribution to a distribution D (this will be seen in Volume II). Accepting this, it gives for large n and m approximations to $d_{m,n,\alpha}$ (tabulated).

4.5.3. Another Problem Leading to Nonparametric Tests

(a) *The sign test.* In order to try out a new fertilizer B to improve the yield of wheat, we have at our disposal n fields (n small, of the order of 10) at an experimental station, each of the fields being divided into two plots. Allocate at random with probability 1/2 (with the help of a coin or some other means) fertilizer B to one plot on a field and the normal fertilizer A to the other. For field i, let (X_i,Y_i) be the respective yields of the plots which have received A and B. Let us stress an important problem in agronomics. The group of fields at the disposal of the experimenter is very limited. It is unreasonable to assume that $(X_i,Y_i) = Z_i$ is, for $1 \leq i \leq n$, an n-sample from a distribution on $\mathbb{R}^2$. This would amount to saying that the n fields are of the same type and are independent.

We then test H_0: "B and A give the same yield" against H_1: "B improves the yield." Consider then, under H_0, the probability (due solely to the random draw) of having $(X_i < Y_i)$. Since the fertilizers have been allocated with probability 1/2 to each of the plots, $P(X_i > Y_i) = P(X_i < Y_i)$. Assume that $P(X_i = Y_i)$ is zero. Then the r.v.'s $U_i = 1_{(X_i<Y_i)}$ are Bernoulli r.v.'s. Due to the fact that the n random draws are independent, the U_i are independent, thus $S = U_1 + \dots + U_n$ has a distribution $b(n,1/2)$. Under the hypothesis H_1, we shall observe more often $(X_i < Y_i)$ and S will be larger. From which we obtain the rejection region of the **sign test**: s_α is defined as the smallest integer s such that

$$\sum_{k=s+1}^{n} \frac{1}{2^k} \binom{n}{k} \leqslant \alpha,$$

and the rejection region is taken as $D_\alpha = \{S > s_\alpha\}$. Note that, if H_0 is false and $p = P(X_i < Y_i) > 1/2$, the probability of D_α can still be calculated. Since p is unknown, it is estimated by $\overline{p} = S_n/n$ and the error of wrongly accepting H_0 is estimated by

$$1 - \sum_{k=s_\alpha+1}^{n} \binom{n}{k} \overline{p}^k (1 - \overline{p})^{n-k}.$$

The probability of D_α, as a function of p, is the power of the test.

(b) *Sign and Rank test.* Here is a situation which generalizes that of Section 4.5.2 when $m = n$ and that of 4.5.3. n pairs of independent r.v.'s $(X_i,Y_i)_{1 \leqslant i \leqslant n}$ are observed; for each i, the pair (X_i,Y_i) has distribution F_i. We test the hypothesis H_0: "for all i, the distribution F_i is exchangeable, i.e., (X_i,Y_i) and (Y_i,X_i) have the same distribution." Then (Exercise 4.1.4) the r.v.'s $D_i = X_i - Y_i$ and $-D_i$ have the same distribution.

If it is assumed, as we shall do here, that the distributions of the D_i are continuous, the sign test can be modified by taking into account the sequence of the observed values $|D_i|$. Consider the vector $(|D_i|)_{1 \leqslant i \leqslant n}$, $R = (R_i)_{1 \leqslant i \leqslant n}$ its rank vector and the increasing sequence $(S_1, \dots, S_\nu)$ of ranks of the ν r.v.'s $|D_i|$ such that D_i is positive (ν is random). Let $V_S = S_1 + \dots + S_\nu$; a large V_S signifies that X_i is frequently clearly less than Y_i (whereas the "clearly" is omitted in the sign test).

Assume that the hypothesis H_0 is true; $\varepsilon_i = \mathbf{1}_{(D_i>0)} - \mathbf{1}_{(D_i<0)}$ and $|D_i|$, i fixed, are independent, since, for $x > 0$,

$$P(\varepsilon_i = 1, |D_i| > x) = \frac{1}{2} P(|D_i|) > x)$$
$$= P(\varepsilon_i = 1) P(|D_i| > x).$$

Thus $\varepsilon = (\varepsilon_i)_{1 \leqslant i \leqslant n}$ and $|D| = (|D_i|)_{1 \leqslant i \leqslant n}$ are independent. R is a function only of $|D|$ and is independent of ε. Let m be an integer, $1 \leqslant m \leqslant n$, and $I = \{i_1, \dots, i_m\}$ a subset of $\{1, \dots, m\}$. We have

$$P\left[\bigcap_{i \in I, j \in I^c} (\varepsilon_i = 1, \varepsilon_j = -1)\right] = \frac{1}{2^n}.$$

For $I \subset \{1, ..., n\}$, denote $R_I = \{R_i; i \in I\}$. Let $I(\varepsilon) = \{i; 1 \leqslant i \leqslant n, \varepsilon_i = 1\}$. Let $v \in \mathbb{N}$,

$$I_v = \{I; I = \{i_1, ..., i_m\} \subset \{1, ..., n\}, i_1 + ... + i_m = v\}$$

and let $|I_v|$ be the number of elements of I_v. Then

$$P[V_S = v] = P[R_{I(\varepsilon)} \in I_v] = \sum_{I \in I_v} P[I(\varepsilon) = I] = 2^{-n}|I_v|.$$

Thus under H_0 the statistic V_S does not depend on the exchangeable distributions $F_1, ..., F_n$, on condition that we assume that the distribution of the r.v.'s $|D_1|, ..., |D_n|$ are continuous. To test H_0 against H_1 "the r.v.'s X_i tend to be smaller than Y_i, for $1 \leqslant i \leqslant n$," we use the fact that V_S tends to be larger under H_1 than under H_0. A **sign and rank test** of level α is obtained by taking the rejection region $\{V_S > t_\alpha\}$ with

$$P_{H_0}(V_S > t_\alpha) \leqslant \alpha < P_{H_0}(V_S > t_\alpha - 1).$$

General Notes. (a) *The size and level.* The tests tackled here of level α have rejection regions of the form $\{T > t_\alpha\}$, T a distribution free statistic under H_0 taking integer values and $t_\alpha \in \mathbb{N}$: $P(T > t_\alpha) \leqslant \alpha < P(T > t_\alpha - 1)$. In general $P(T > t_\alpha)$, the **size** of the test, is $< \alpha$. If the size is less than α, the following test can be proposed. For $T > t_\alpha$ reject H_0, for $T < t_\alpha$ accept it. For $T = t_\alpha$ make a random choice with the help of a roulette, independent of the problem, which gives the value 1 with probability p and 0 with probability $1 - p$, and reject the hypothesis if the roulette gives a 1. The size, the probability of rejecting H_0 if it is true, is then

$$P_{H_0}(T > t_\alpha) + pP_{H_0}(T = t_\alpha):$$

p can be chosen such that the size is α.

(b) *Ties.* We have, in all of this section, excluded the possibility of ties. For the sign test, there is no problem; it is sufficient to remove the observed pairs of ties. We shall not tackle the problem for the other tests here. In practice they are very often used for discontinuous distributions in spite of ties.

Exercises 4.5

E.1. *Wilcoxon's statistic.* With the notations of Section 4.5.2 show

$$W_{X,Y} = \sum_{i=1}^{m} \sum_{j=1}^{n} 1_{[X_i > Y_j]} \cdot$$

Show that $W_{X,Y} - mn/2$ and $mn/2 - W_{X,Y}$ have the same distribution.

E.2. *The Kolmogorov test of fit for a continuous distribution* F. (a) Let $(X_1, ..., X_n)$ be an n-sample from a continuous distribution F and $\underline{F}_n$ its empirical distribution. Let $D_n = \sup_x |F(x) - \underline{F}_n(x)|$. Prove that the distribution of D_n does not depend on F: it is tabulated.

(b) Deduce from (a) a natural test of fit for F. Having observed $(X_1, ..., X_n)$, test H_0 "this is an n-sample from F" against H_1 "H_0 is false."

E.3. *Spearman's test.* Ten *rose wine tasters* are asked to choose from amongst five wines the one which they prefer. In fact, it is the same wine but served in different colored glasses [from the darkest color, 1, to the lightest color, 5 (clear glass)]. The hypothesis we want to test is H_0: "the color of the glass does not affect the choice." The following table of results is obtained.

	Rank of the wine chosen as best in a glass of color				
	1	2	3	4	5
	dark				→ clear
Number of wine tasters choosing this rank	1	0	3	2	4

Construct a test of H_0 based on the Spearman coefficient. From what level is H_0 acceptable? (See Exercises 1.2.5 and 1.3.11(d).)

E.4. (*continuation of E.2*). Let X be a uniformly distributed variable on the set $\{0,1,2,3,4,5,6\}$ and Y a variable with uniform density on the interval $[0,1]$. Consider a sample of size 1000 of the random variable Z, which grouped in classes of the form $]m/2, (m+1)/2]$ with $0 < m < 12$ or $[0,1/2]$, is distributed in the following way.

Class	$0 \leqslant \cdot \leqslant 0.5$	$0.5 < \cdot \leqslant 1$	$1 < \cdot \leqslant 1.5$	$1.5 < \cdot \leqslant 2$	$2 < \cdot \leqslant 2.5$	$2.5 < \cdot \leqslant 3$	$3 < \cdot \leqslant 3.5$	$3.5 < \cdot \leqslant 4$	$4 < \cdot \leqslant 4.5$	$4.5 < \cdot \leqslant 5$	$5 < \cdot \leqslant 5.5$	$5.5 < \cdot \leqslant 6$
Size	197	220	112	115	71	94	61	45	36	24	9	16

At the 0.05 level, can we accept that the variable Z has the same distribution as the variable XY? Use Exercise 2, even though the distribution of XY is not continuous.

E.5. *Median test.* (a) Consider an n-sample $(X_1, \ldots, X_n)$ from a continuous distribution F_0 on $\mathbb{R}$ with median m_0. What is the distribution of

$$\sum_{i=1}^{n} \mathbf{1}_{[X_i \leqslant m_0]} = S_n?$$

(b) We want to test H_0 "the hypothesis made in (a) is correct" against H_1 "this is an n-sample from a continuous distribution F with median $m > m_0$": propose a natural test of level α. Likewise by replacing in H_1 "$m > m_0$," by "$m < m_0$," then by "$m \neq m_0$." (Use a method similar to Section 2.4.4.) Calculate, for $m \neq m_0$, the power of the test, the probability of accepting H_1.

E.6. *Random numbers.* (a) Let (X_n) be a sequence of independent uniform r.v.'s on $\{0,1, \ldots, 9\}$. What is the distribution of $0 \cdot X_1 X_2 \ldots X_n$? What is the distribution of $0 \cdot X_1 X_2 \ldots X_n \ldots$? Starting from a large sample from the uniform distribution on $\{0,1, \ldots, 9\}$ how can we construct a sample from the uniform distribution on

$$\left\{ \frac{i}{10^k} ; 0 \leqslant i < 10^k \right\}?$$

For large k show that this distribution is approximated by $\mathcal{U}_{[0,1]}$. Starting from "random numbers" on {0,1, ..., 9}, we can, up to 10^{-k}, construct "random numbers" on [0,1].

(b) Starting from an n-sample $(X_1, \ldots, X_n)$ from the distribution $\mathcal{U}_{[0,1]}$ how can we construct an n-sample from a distribution F (Exercise 3.3.5)?

(c) Let A be a Borel set of $[0,1]^k$ charged by Lebesgue measure and $(X_1, \ldots, X_k) = X$ a k-sample from $\mathcal{U}_{[0,1]}$. What is the distribution of X conditional on $(X \in A)$? Starting with a large sample from $\mathcal{U}_{[0,1]}$, how can a sample from the uniform distribution on A be constructed? Look up some "random numbers" tables (or those produced by a computer subprogram). Think up and carry out tests to see if they can be considered as samples from a uniform distribution on {0, ..., 9} (Median test, Exercise 4.5.5; Kolmogorov's test, Exercise 4.5.2).

E.7. *Comparison of k samples.* Let $(Y_{j1}, \ldots, Y_{jn_j})$ for $1 \leqslant j \leqslant k$ be k independent samples; the jth being an n_j-sample from F_j. We want to test by a nonparametric method the hypothesis

$$H_0: \text{"}F_1 = F_2 = \ldots = F_k\text{"}.$$

To this end, denote by $R_{j\ell}$ the rank of $Y_{j\ell}$ in the ordered set of all the $(Y_{ji})_{\substack{j=1,\ldots,k \\ i=1,\ldots,n_j}}$. Set

$$R_{j.} = \frac{1}{n_j} \sum_{\ell=1}^{n_j} R_{j\ell}, \quad R_{..} = \frac{1}{k} \sum_{j=1}^{k} R_{j.}\,.$$

Let

$$T = \frac{\sum_{j=1}^{k} (R_{j.} - R_{..})^2 n_j}{\sum_{j=1}^{k} \sum_{\ell=1}^{n_j} (R_{j\ell} - R_{j.})^2}.$$

Show that T is distribution free under H_0. Think up a suitable test using T, of H_0 against an alternative of the form "$F_j(\cdot) = F(\cdot - \tau_j)$," where F is a given distribution and $\tau_j \in \mathbb{R}$, $j = 1, \ldots, n$.

E.8. *Analysis of variance by ranks.* k cheeses $F_1, \ldots, F_j, \ldots, F_k$ are tasted by n tasters $G_1, \ldots, G_i, \ldots, G_n$. Each taster ranks, in increasing order of preference, the k cheeses. Of course, in general the n different rankings are not identical. We are thus led to identifying a taster with a test. To see if the cheeses are not identical and if there is a significant "order", the following model is proposed. The rank R_{ij} given by the ith taster to the jth cheese, is the rank of X_{ij} amongst the $(X_{i\ell})_{\ell=1,\dots,k}$, where X_{ij} is the mark given by taster i to cheese j; $X_{ij} = \tau_j + e_{ij}$, $(\tau_j)_{j=1,\dots,k}$ is a sequence of positive numbers and (e_{ij}) is a family of independent, centered variables from continuous distributions. For fixed i, the $(e_{ij})_{j=1,\dots,k}$ have the same distribution.

(a) Show that with this model, each taster ranks the k cheeses without ties. How can the τ_j and e_{ij} be interpreted? What is the relationship between the τ_j if a cheese is intrinsically better than all the others? or if all the cheeses are equivalent?

(b) We want to test the hypothesis H_0 "the cheeses are equivalent" against H_1 "they are not equivalent."
Let $R_{.j} = (1/n)\Sigma_{i=1}^{n} R_{ij}$, $R_{..} = (1/k)\Sigma_{j=1}^{k} R_{.j}$. Calculate $R_{..}$. The *variation between treatments* is to be used $S = \Sigma_{j=1}^{n}(R_{.j} - R_{..})^2$. Show that S is distribution free under H_0. Deduce from this a test of H_0 against H_1. Calculate, for $k = 2$, $n = 3$, the table for S (the distribution under H_0).

(c) Let us test H_0 against H_1': "$\tau_1 < \tau_2 < \dots < \tau_k$." Show that $L = \Sigma_{j=1}^{n} a(j)R_{.j}$, where $a(j)$ is a sequence of numbers ≥ 0, is distribution free under H_0. Construct a natural test with the help of L. Indicate why it is necessary to choose an increasing sequence $a(j)$ with not too rapid growth in order to construct a powerful enough test.

E.9. *Sign and Rank test.* We wish to study the influence of group life in a young children's nursery, in particular the stimulation of the children. We can assess the children on a certain scale between 0 and 100. The values obtained are not sufficiently certain to be treated numerically. More precisely, a child having a rating of 60 does not have a perception twice that of a child with a rating of 30, but on the other hand the difference between 60 and 40 is certainly higher than between 40 and 30. The differences are not numerically exact, however they have *sufficient meaning in order that we can arrange them as a function of their value.* A study will therefore be carried out on twins, one being in the

Rating for the twin

Pair	At the nursery	At home
a	82	63
b	69	42
c	73	74
d	43	37
e	58	51
f	56	43
g	76	80
h	65	62

nursery, the other staying at home, the choice being made by a random draw. Test by the sign and rank method H_0 "the perceptions at the nursery and at home are not different."

E.10. Let F and G be two continuous distributions on $\mathbb{R}$.

(a) Let (X,X') and (Y,Y') be two independent 2-samples from F and G, respectively. Consider the events

$$A = \{\sup(X,X') < \inf(Y,Y')\}$$

and

$$B = \{\sup(Y,Y') < \inf(X,X')\}.$$

Prove $P(A) = \int(1 - G)^2 d(F^2)$, and calculate $p = P(A) + P(B)$ as a function of $\Delta = \int(F - G)^2 d(F + G)$. Show that Δ is zero and p is a minimum for $F = G$ (and only in this case).

(b) Consider n independent observations $((X_i,X_i'),(Y_i,Y_i'))_{1\leqslant i\leqslant n}$, each with the distribution of the pair $((X,X'),(Y,Y'))$ in (a) and associate with the ith observation the events A_i and B_i as in (a). Let $V_i = 1_{(A_i \cup B_i)}$; construct with the help of $S = \Sigma_{i=1}^n V_i$ a natural test of H_0 "$F = G$" against H_1 "$F \neq G$."

Bibliographic Notes

We find in the works referred to in [3] expositions on independence. Let us cite Neveu ([1], Chapter 3) and also Loève and Chow-Teicher. Limit theorems are detailed in Feller, in particular Chapter 15, Volume 2, and in Loève. An older but very complete account is Gnedenko and Kolmogorov. On historic grounds, we cite P. Levy [2], a decisive work in the development of the theory.

Nonparametric tests receive an elementary treatment in Siegel, Mosteller and Rourke; Lehmann [2] is a very clear account, easy to read on rank tests. It can be complemented on the theoretical level by Hajek–Sidak.

Random numbers are studied in Hammersley, where several applications to approximate calculations will be found. The notion of associated r.v.'s is studied in Barlow and Proschan. In Volume 2 we shall come back to renewal and Poisson processes; simple accounts appear in Neveu [2], and Ross.

Chapter 5
GAUSSIAN SAMPLES, REGRESSION, AND ANALYSIS OF VARIANCE

Objectives

Gaussian distributions play a fundamental role in probability because of the central limit theorems. Many variables can be considered, at least in the first approximation, to be Gaussian. Measurement errors in physics, yield from a plot in agriculture, the amplitude of a signal picked up on a noisy line, etc., fall into this class, the variations of which, sometimes very large, are due to a large number of factors, none of which predominates.

In this section we study the transmitted signal or the average yield. We are dealing with estimation and testing problems starting from a noisy observation or the yield from several plots, or the problem of the influence of various factors.

5.1. Gaussian Samples

5.1.1. Distributions Associated with a Gaussian n-Sample

If X is an r.v. with distribution $\mathcal{N}(0,1)$, the distribution of X^2 has Laplace transform ϕ_{X^2} defined by

$$\phi_{X^2}(t) = E(e^{-tX^2}) = \frac{1}{\sqrt{2\pi}}\int \exp\left[-(2t+1)\frac{x^2}{2}\right]dx = \frac{1}{\sqrt{2t+1}}.$$

χ^2 Distribution. If $X = (X_1, \ldots, X_n)$ is an $N(0,1)$-sample, the distribution of $X_1^2 + \ldots + X_n^2$ has Laplace transform $t \mapsto (2t+1)^{-n/2}$. This distribution is said to be chi-squared with n degrees of freedom and denoted by $\chi^2(n)$. It is a $\gamma(n/2, 1/2)$ distribution with density

$$x \mapsto \frac{1}{\Gamma\left(\frac{n}{2}\right)} \left(\frac{1}{2}\right)^{n/2} e^{-x/2} x^{n/2-1} 1_{(x>0)}$$

(Exercises 4.3.3 and 3.3.13); its mean is n and its variance $2n$; it is tabulated. If a change of scale is made $x \mapsto \sigma x$, $X_1^2 + \ldots + X_n^2$ is multiplied by σ^2; its distribution is denoted by $\sigma^2\chi^2(n)$. (If an r.v. Z has distribution F let us agree to denote by αF the distribution of αZ, for $\alpha \in \mathbb{R}$.)

Fisher–Snedecor Distribution. Let Y_1, Y_2 be two independent r.v.'s, Y_1 with distribution $\chi^2(n_1)$, Y_2 with distribution $\chi^2(n_2)$. The distribution $(Y_1/n_1)/(Y_2/n_2)$, denoted by $F(n_1,n_2)$ is called Fisher's distribution with (n_1,n_2) degrees of freedom (Exercise 4.3.4).

Student's Distribution. Let X with distribution (0,1), and Y, with distribution $\chi^2(n)$, be independent. The distribution of $\sqrt{n}X/\sqrt{Y}$, denoted by $t(n)$, is called Student's distribution with n degrees of freedom. Its density resembles an overdispersed Gaussian density

$$t(n) \xrightarrow{\mathcal{D}} N(0,1) \quad \text{if } n \to \infty \quad \text{(Exercise 4.3.4).}$$

Quantiles. Let $\alpha \in]0,1[$. If G is a distribution on $\mathbb{R}$, its distribution function continuous and strictly increasing, its quantile g_α of order $1 - \alpha$ is defined by $G([g_\alpha, \infty[) = \alpha$ (Definition 4.4.26). In particular, $X^2_{n,\alpha}$, $F_{n_1,n_2,\alpha}$, $t_{n,\alpha}$ designate

quantiles of order $1 - \alpha$ for the distributions we have just introduced and ϕ_α that of $N(0,1)$.

5.1.2. Linear Transformation of a Gaussian n-Sample

Let $X = (X_1, \ldots, X_n)$ be a random vector of dimension n. It is an n-sample from $N(0,\sigma^2)$ if and only if its characteristic function ϕ_X is defined, for all $t = (t_1, \ldots, t_n) \in \mathbb{R}^n$, by

$$\phi_X(t) = E[\exp i<t,X>] = \prod_{j=1}^{n} E[\exp i(t_j X_j)]$$
$$= \prod_{j=1}^{n} \exp\left[-\frac{\sigma^2}{2}t_j^2\right] = \exp\left[-\frac{\sigma^2}{2}\|t\|^2\right].$$

Applying this to λt for $\lambda \in \mathbb{R}$, it can be seen that $<t,X>$ has distribution $N(0,\sigma^2\|t\|^2)$.

Let $(e_1, ..., e_n)$ be an orthonormal basis of $\mathbb{R}^n$. The random vector $\{<X,e_1>, ..., <X,e_n>\}$ is yet again an $N(0,\sigma^2)$ n-sample. In fact, its characteristic function equals, for $t = (t_1, ..., t_n) \in \mathbb{R}^n$:

$$E\left[\exp i \sum_{j=1}^{n} t_j<X,e_j>\right] = E\left[\exp i<X, \sum_{j=1}^{n} t_j e_j>\right]$$
$$= \exp\left[-\frac{\sigma^2}{2}\left\|\sum_{j=1}^{n} t_j e_j\right\|^2\right]$$
$$= \exp\left[-\frac{\sigma^2}{2}\|t\|^2\right].$$

This is equivalent to saying that the vector transform of X by an orthogonal matrix is an $N(0,\sigma^2)$ n-sample.

Let $\oplus_{j=1}^{p} E_j$ be a decomposition of $\mathbb{R}^n$ into p pairwise orthogonal subspaces, of dimension r_j $(1 \leqslant j \leqslant p)$; an orthonormal basis of $\mathbb{R}^n$ can be chosen such that each E_j is generated by r_j vectors $(e_{j_1}, ..., e_{jr_j})$ of this basis. X_{E_j}, the projection of X on E_j is $\Sigma_{k=1}^{r_j}<X, e_{jk}>e_{jk}$, and

$$\|X_{E_j}\|^2 = \sum_{k=1}^{r_j} <X, e_{jk}>^2$$

has distribution $\sigma^2\chi^2(r_j)$.

Cochran's Theorem 5.1.1. *Let $X = (X_1, ..., X_n)$ be an $N(0,\sigma^2)$ n-sample.*

(a) *The components of X in each orthonormal basis of $\mathbb{R}^n$ form an $N(0,\sigma^2)$ n-sample; the transform of X by an orthogonal transformation is an $N(0,\sigma^2)$ n-sample.*
(b) *Let $E_1 \oplus ... \oplus E_p$ be a decomposition of $\mathbb{R}^n$ into p pairwise orthogonal subspaces of dimension $r_1, ..., r_p$. The random vectors $X_{E_1}, ..., X_{E_p}$, the projections of X on $E_1, ..., E_p$, are independent; the* r.v.'s $\|X_{E_1}\|^2, ..., \|X_{E_p}\|^2$ *are independent*

with distributions $\sigma^2\chi^2(r_1), ..., \sigma^2\chi^2(r_p)$.

5.1.3. Fundamental Statistics in a Gaussian Sample

Let us take for E the line generated by $\mathbf{1}_n = (1, ..., 1)$ or by the normal vector $v = (1/\sqrt{n})\mathbf{1}_n$. For $x = (x_1, ..., x_n)$ and $\bar{x} = (1/n)\Sigma_{i=1}^n x_i$ we have $<x,v> = (1/\sqrt{n})\Sigma_{i=1}^n x_i = \sqrt{n}\,\bar{x}$,

$$\|x_E^\perp\|^2 = \|x\|^2 - n(\bar{x})^2 = \sum_{i=1}^n x_i^2 - n(\bar{x})^2 = \sum_{i=1}^n (x_i - \bar{x})^2$$

Thus, if $(X_1, ..., X_n)$ is an $N(0,1)$ n-sample, $\sqrt{n}\,\bar{X}$ and $\Sigma_{i=1}^n(X_i - \bar{X})^2$ are independent r.v.'s, $\sqrt{n}\,\bar{X}$ having distribution $N(0,1)$ and $\Sigma_{i=1}^n(X_i - \bar{X})^2$ having distribution $\chi^2(n-1)$.

Let X have distribution $N(m,\sigma^2)$, then $(X - m)/\sigma$ has a standard normal distribution. The key to the study of an $N(m,\sigma^2)$ n-sample is

Proposition 5.1.2. *Let* $X = (X_1, ..., X_n)$ *be an n-sample from an* $N(m,\sigma^2)$ *distribution. Associate with it the sample mean* $\bar{X} = (1/n)\Sigma_{i=1}^n X_i$ *and the sample variances* $S^2 = [1/(n-1)]\Sigma_{i=1}^n(X_i - \bar{X})^2$. *Then* $\bar{X}$ *and* S^2 *are independent and (denoting* ~ *for "is distributed as"):*

$$\frac{\sqrt{n}(\bar{X} - m)}{\sigma} \sim N(0,1), \quad \frac{1}{\sigma^2}\sum_{i=1}^n (X_i - m)^2 \sim \chi^2(n),$$

$$\frac{1}{\sigma^2}\sum_{i=1}^n (X_i - \bar{X})^2 \sim \chi^2(n-1), \quad \frac{\sqrt{n}(\bar{X} - m)}{S} \sim t(n-1).$$

Note. *The empirical estimator of the variance is* $[(n-1)/n]S^2$, *but as in Section 4.4 the unbiased estimator* S^2 *of the variance is called "the empirical variance." Note also the relation*

$$\sum_{i=1}^n (X_i - m)^2 = \sum_{i=1}^n (X_i - \bar{X})^2 + n(\bar{X} - m)^2.$$

5.1.4. Noncentrality

Let $X = (X_1, ..., X_n)$ be a random vector. Assume that the components of X are again independent, but the distribution of X_i is $N(m_i,\sigma^2)$, σ^2 being the same for all the X_j. Setting $m = (m_1, ..., m_n)$ we can write $X = m + Y$, where Y is an n-sample from $N(0,\sigma^2)$. The vector Y is the **white noise** in signal theory,

m representing the emitted signal; it is often the measurement error, the true value being m.

As a function of m, the distribution of $\|X\|^2$ depends only on $\|m\|$. In fact, let $X' = m' + Y$, with $\|m'\| = \|m\|$. There exists an orthogonal transformation a such that $a(m) = m'$, and thus such that $a(X) = m' + a(Y)$ and X' have the same distribution: $\|a(X)\| = \|X\|$ and $\|X'\|$ have the same distribution.

Definition 5.1.3. We call the **noncentral chi-squared distributions on n degrees of freedom** with parameter $\lambda \geq 0$ the distribution of $\|X\|^2 = \Sigma_{i=1}^n X_i^2$ when the X_i are independent $N(m_i,1)$ r.v.'s ($i = 1, ..., n$) and $\lambda = \Sigma_{i=1}^n m_i^2$. This distribution is denoted by $\chi'^2(n,\lambda)$ (λ is the **noncentrality parameter**). For $\lambda = 0$ we obtain the $\chi^2(n)$ distribution.

Consider again a decomposition $\oplus_{j=1}^p E_j$ of $\mathbb{R}^n$ into orthogonal subspaces E_j with orthonormal basis $(e_{j1}, ..., e_{jr_j})$.

The projection of X on E_j is

$$X_{E_j} = m_{E_j} + \sum_{k=1}^{r_j} e_{jk} \langle X, e_{jk} \rangle$$

and $\|X_{E_j}\|^2$ has distribution $\chi'^2(r_j, \|m_{E_j}\|^2)$, from which:

Corollary 5.1.4. (Cochran). *Let $(X_1, ..., X_n)$ be independent Gaussian* r.v.'s, *with distribution $N(m_1,\sigma^2) \dots N(m_n,\sigma^2)$, and $E_1 \oplus \dots \oplus E_p$ a decomposition of $\mathbb{R}^n$ into p orthogonal subspaces of dimensions $r_1, ..., r_p$. The random vectors $X_{E_1}, ..., X_{E_p}$, the projections of X onto $E_1, ..., E_p$, are independent; the* r.v.'s *$\|X_{E_1}\|^2, ..., \|X_{E_p}\|^2$ are independent with distributions $\sigma^2\chi'^2(r_1, \|m_{E_1}\|^2), ..., \sigma^2\chi'^2(r_p, \|m_{E_p}\|^2)$.*

Properties of the $\chi'^2(n,\lambda)$ Distribution. Take $m_1 = ... = m_{n-1} = 0$ and $m_n = \sqrt{\lambda}$. The distribution of $\|X\|^2$ is $\chi'^2(n,\lambda)$. For $X = m + Y$, we have $\|X\|^2 = \|Y\|^2 + 2Y_n\lambda + \lambda^2$; $\chi'^2(n,\lambda)$ has mean $n + \lambda$. The appearance of the graphs of the densities is as pictured on the following page.

The **noncentral Fisher Distribution** with parameter $\lambda > 0$, $F'(n_1,n_2,\lambda)$ is the distribution of $(Y_1/n_1)/(Y_2/n_2)$, if Y_1 and Y_2 are two independent r.v.'s, Y_1 distributed as $\chi'^2(n_1,\lambda)$ and Y_2

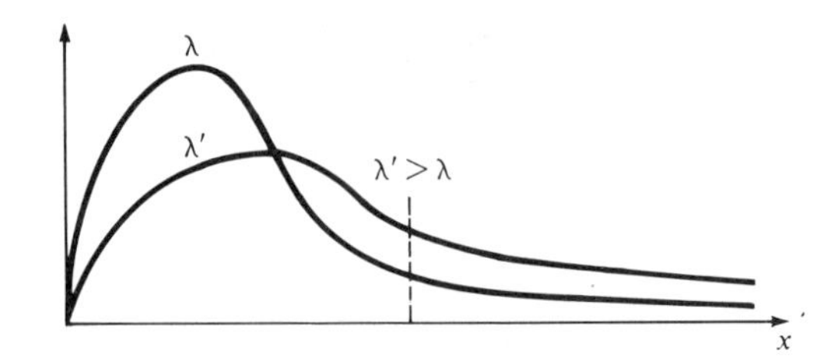

distributed as $\chi^2(n_2)$. The **noncentral Student distribution** with parameter $a \in \mathbb{R}$, $t'(n,a)$, is the distribution of $\sqrt{n}X/\sqrt{Y}$, if X and Y are independent, X distributed as $N(a,1)$, and Y distributed as $\chi^2(n)$ (Exercise 4.3.5). In actual practice, it is important to note that, *for every fixed* t, *the functions* $\lambda \longmapsto \chi'^2(n,\lambda)$ $([t,\infty[)$, $\lambda \longmapsto F'(n_1,n_2,\lambda)$ $([t,\infty[)$, *and* $a \longmapsto t'(n,a)$ $([t,\infty[)$, *are increasing* (Exercise 5.1.1).

5.1.5. Estimation and Tests for an $N(m,\sigma^2)$ *n-Sample*

In this section we are concerned with estimation or testing problems, analogous to those which were studied in Chapters 2 and 4. Estimating or testing m is studying the signal strength or the mean value of a production. Questions asked about the parameter of an n-sample of a Bernoulli distribution will be easily expressed for m. Statistics on σ^2 are concerned with the noise intensity or variations in production.

(a) **Estimation**. Assume first of all that σ^2 is known, m unknown. The sample mean $\bar{X}$ is an unbiased estimator of m. It is exponentially consistent (Theorem 4.4.22) and we shall see (Section 7.1.7) that, amongst unbiased estimators, it is the one with minimum variance. The distribution of $\sqrt{n}(\bar{X} - m)/\sigma$ is $N(0,1)$; for $0 < \alpha < 1$

$$\left[\bar{X} - \frac{\sigma\phi_{\alpha/2}}{\sqrt{n}}, \bar{X} + \frac{\sigma\phi_{\alpha/2}}{\sqrt{n}}\right]$$

is a two-sided confidence interval (Section 2.4.5) of level $(1 - \alpha)$, and $]-\infty, \bar{X} + \sigma\phi_\alpha/\sqrt{n}[$ is a one-sided confidence interval of the same level. Unlike Section 4.4.2, these results are valid for arbitrary n, and not only asymptotically.

When σ^2 is unknown, m can again be estimated by $\bar{X}$; $\sqrt{n}(\bar{X}-m)/S$ has a Student distribution $t(n - 1)$. A two-sided confidence interval of level $1 - \alpha$ for m is

$$\left[\bar{X} - t_{n-1,\alpha/2}\frac{S}{\sqrt{n}},\quad \bar{X} + t_{n-1,\alpha/2}\frac{S}{\sqrt{n}}\right].$$

In order to estimate σ^2, *m being known*, the empirical estimator $\bar{\sigma}^2 = (1/n)\Sigma_{i=1}^n(X_i - m)^2$ can be chosen, which is unbiased with distribution $(\sigma^2/n)\chi^2(n)$. *If m is unknown,* S^2 *is an unbiased estimator of* σ^2. We will return to these estimates later (Section 7.1 and Exercise 7.1.1).

It is important to know if the observations are not too dispersed around their mean value. The upper confidence bounds for σ^2 of level $1 - \alpha$ are used,

$$\frac{n}{\chi^2_{n,1-\alpha}}\,\bar{\sigma}^2$$

if m is known, or

$$\frac{n - 1}{\chi^2_{n-1,1-\alpha}}\,S^2$$

if m is unknown. We can claim that σ^2 is majorized by these numbers with a probability of error equal to α.

The preceding estimators are empirical estimators (up to the factor $n/(n - 1)$ for the sample variance). The principle of *maximum likelihood* could also be used, assuming, for example, (m,σ^2) unknown. The density of the *n-sample* is

$$f_{m,\sigma^2}(x_1, \ldots, x_n) = \frac{1}{(2\pi)^{n/2}\sigma^n}\exp\left[-\frac{1}{2\sigma^2}\sum_{i=1}^n (x_i - m)^2\right]$$

$$\phi(m,\sigma^2) = \text{Log } f_{m,\sigma^2}(x_1, \ldots, x_n)$$

$$= -\frac{n}{2}\text{ Log } 2\pi - \frac{n}{2}\text{ Log } \sigma^2 - \frac{1}{2\sigma^2}\sum_{i=1}^n (x_i - m)^2.$$

If $X_1, \ldots, X_n$ are the observed values, m and σ^2 are estimated by the pair $(\hat{m},\hat{\sigma}^2)$ which maximizes ϕ. In view of the form of ϕ, these estimators are obtained by writing

$$\frac{\partial\phi}{\partial m}(\hat{m},\hat{\sigma}^2) = 0, \qquad \frac{\partial\phi}{\partial\sigma^2}(\hat{m},\hat{\sigma}^2) = 0.$$

The first equation gives $\hat{m} = \overline{X}$, and the second is written

$$-\frac{n}{\hat{\sigma}^2} + \frac{1}{(\hat{\sigma}^2)^2}\sum_{i=1}^{n}(X_i - \hat{m})^2 = 0$$

$$\hat{\sigma}^2 = \frac{1}{n}\sum_{i=1}^{n}(X_i - \hat{m})^2 = \frac{n-1}{n}S^2.$$

The estimator $\hat{\sigma}^2$ obtained is thus biased.

If we want to estimate m by minimizing the function $t \longmapsto \sum_{i=1}^{n}(X_i - t)^2$, we again find $t = \overline{X}$ and the minimum value obtained is $n\overline{\sigma}^2$. This *estimation principle* is known as **least squares.** The *theory of errors* is based on this: the best estimate of the value of m starting from the measurements $m + Y_i$, where $(Y_i)_{1 \leq i \leq n}$ is an n-sample from an $N(0,\sigma^2)$ distribution (σ^2 unknown), is that which makes the average value of the amplitude Y^2 of the error Y the smallest possible, thus that which minimizes $(1/n)\sum_{i=1}^{n}Y_i^2$.

(b) **Tests.** When the two parameters are unknown and the hypothesis under test deals with only one of the two, the other is said to be a **nuisance parameter** for the testing problem. *Starting from the construction of confidence intervals, tests can be constructed on m and σ^2.* Here are a few examples, the reader can think up some others.

Student's Test of H_0: "$m \leq m_0$" against H_1: "$m > m_0$" of Level α. *This is the test for which we accept H_0 for*

$$\overline{X} \in \left]-\infty, t_{n-1,\alpha}\frac{S}{\sqrt{n}} + m_0\right].$$

Recall how we passed from confidence regions to tests. If m_0 is the mean,

$$m_0 \in \left[\overline{X} - t_{n-1,\alpha}\frac{S}{\sqrt{n}}, \infty\right[$$

with probability $1 - \alpha$, since $\overline{X}$ is in $]-\infty, t_{n-1,\alpha} S/\sqrt{n} + m_0]$ with probability $1 - \alpha$. If the parameter equals m we have $\overline{X} \in]-\infty, t_{n-1,\alpha} S/\sqrt{n} + m]$ with probability $1 - \alpha$; for $m \leq m_0$, $\overline{X} \in]-\infty, t_{n-1,\alpha} S/\sqrt{n} + m_0]$ with probability greater than $1 - \alpha$; and for $m > m_0$, $\overline{X} \in]-\infty, t_{n-1,\alpha} S/\sqrt{n} + m_0]$ with a smaller probability.

For the test defined above, the probability of deciding H_1 calculated for the value m of the mean is an increasing function β of m, the **power** of the test. Its **level** (maximum risk of error for $m \leqslant m_0$) is α. The probability that this test leads to a wrong decision H_1 is $1 - \beta(m)$. Since m is unknown, it is necessary to estimate $\beta(m)$; $\beta(\bar{X})$ is used as an estimate of the power and this can be calculated with the help of the noncentral Student distributions t' (tabulated).

To test H_0': "$m = m_0$" against H_1': "$m \neq m_0$", we use the *acceptance region of "$m = m_0$"*

$$\bar{X} \in \left[m_0 - t_{n-1,\alpha/2}\frac{S}{\sqrt{n}},\ m_0 + t_{n-1,\alpha/2}\frac{S}{\sqrt{n}}\right]$$

and an error is made with probability α by wrongly rejecting "$m = m_0$." The probability of accepting H_0 is a decreasing function of $|m - m_0|$.

The Fisher–Snedecor Test of Level α of "$\sigma^2 < \sigma_0^2$" Against "$\sigma^2 > \sigma_0^2$". *This is the test which accepts "$\sigma^2 < \sigma_0^2$" for*

$$S^2 \leqslant \frac{1}{n-1}\chi^2_{n-1,\alpha}\,\sigma_0^2.$$

The probability of rejecting H_0 is an increasing function of the variance σ^2 which can be calculated with the help of the noncentral Fisher distributions F' (tabulated).

5.1.6. Estimation of the Mean m of a Random Vector $X = m + Y$ under a Linear Constraint, Y being an n-Sample from an $N(0,\sigma^2)$ Distribution

The parameters m and $\sigma^2 > 0$ are unknown, but m is assumed to be in a subspace E of dimension k of $\mathbb{R}^n$. Let $(e_1, \ldots, e_k)$ be an orthonormal basis of E. The projection X_E of X on E is $m + \sum_{i=1}^k e_i\langle Y,e_i\rangle$, and $X - X_E$ coincides with $Y - Y_E$, the projection of Y on $E^\perp$. We deduce from Theorem 5.1.1,

Proposition 5.1.5. *Let E be a vector subspace of dimension k in $\mathbb{R}^n$, $m \in E$, Y an n-sample from $N(0,\sigma^2)$ with $\sigma^2 > 0$ and $X = m + Y$.*

(a) *X_E is a random vector with mean m (an unbiased estimator of m) and $\|X_E - m\|^2$ is distributed as $\sigma^2\chi^2(k)$.*
(b) *$X - X_E$ and X_E are independent. The 'residual variation' $\|X - X_E\|^2$ is distributed as $\sigma^2\chi^2(n - k)$ and the "residual variance" $[1/(n - k)]\|X - X_E\|^2$ is an unbiased estimator of σ^2.*

Notes. (a) *Let* $v \in \mathbb{R}^n$: $\langle v, X_E \rangle$ *is an unbiased estimator of* $\langle v, m \rangle$. *We will show later (Section 7.1 and Exercise 7.1.2) that this estimator of* $\langle v, m \rangle$ *and the residual variance for estimating* σ^2 *are the minimum variance unbiased estimators.*

(b) *The estimator* X_E *of* m *would also have been obtained by the method of maximum likelihood. However, the maximum likelihood estimator of* σ^2 *is biased (Exercise 5.1.11).*

A **confidence sphere** *for* m *can be given, by noting that*

$$\frac{\frac{1}{k}\|X_E - m\|^2}{\frac{1}{n-k}\|X - X_E\|^2}$$

has distribution $F(k, n-k)$. m *lies in* $B(X_E, r_\alpha)$, *the ball with center* X_E *and radius*

$$r_\alpha = \left[\|X - X_E\|^2 \frac{k}{n-k} F_{k,n-k,\alpha}\right]^{1/2}$$

with probability $1 - \alpha$.

Example. Let k independent samples from $N(m_i, \sigma^2)$ be of respective sizes n_i, $1 \leq i \leq k$; denote the ith by $X^{(i)} = (X_{i1}, \ldots, X_{in_i})$; σ^2 is unknown. For $n = n_1 + \ldots + n_k$, $X = (X^{(1)}, \ldots, X^{(k)})$ is a Gaussian vector of $\mathbb{R}^n$ of the form $X = m + Y$, Y an n-sample from $N(0, \sigma^2)$ and $m = (m_1 \mathbf{1}_{n_1}, \ldots, m_k \mathbf{1}_{n_k})$, where $\mathbf{1}_p$ is a sequence of p elements equal to 1. E is the subspace of $\mathbb{R}^n$ of dimension k, composed of vectors of the form

$$(\lambda_1 \mathbf{1}_{n_1}, \ldots, \lambda_k \mathbf{1}_{n_k}).$$

In what follows, the index of summation in an average is replaced by a dot: thus $X_{i.} = (1/n_i)\Sigma_{j=1}^{n_i} X_{ij}$ is the sample mean of $X^{(i)}$. We have

$$X_E = (X_{1.} \mathbf{1}_{n_1}, \ldots, X_{k.} \mathbf{1}_{n_k})$$

and

$$\|X - X_E\|^2 = \sum_{j=1}^{n_1} (X_{1j} - X_{1.})^2 + \ldots + \sum_{j=1}^{n_k} (X_{kj} - X_{k.})^2.$$

The **total variation** (*empirical inertia of the cloud of points* X_{ij}) is

$$\|X - X_{..}\|^2 = \|X - X_E\|^2 + \|X_E - X_{..}\|^2.$$

It is the sum of the variation between the samples $\|X_E - X_{..}\|^2$, the **between group variation**, and the variation within the groups $\|X - X_E\|^2$, the **within groups variation** (the **residual variation** of Proposition 5.1.5).

5.1.7. Analysis of Variance

Let us compare the means of the k Gaussian, independent samples of the preceding example; the variances are the same for all the observations.

First of all let us study the case of two independent samples from the above, $(X_1, ..., X_p)$, a p-sample from $N(m,\sigma^2)$, and $(Y_1, ..., Y_q)$, a q-sample from $N(m',\sigma^2)$. $\bar{X} - \bar{Y}$ has distribution

$$N\left[m - m', \sigma^2\left(\frac{1}{p} + \frac{1}{q}\right)\right]$$

and $\Sigma_{i=1}^p(X_i - \bar{X})^2 + \Sigma_{j=1}^q(Y_j - \bar{Y})^2$ has distribution $\sigma^2\chi^2(p + q - 2)$ and is independent of $\bar{X} - \bar{Y}$. Thus the r.v.

$$Z_{p,q} = \frac{(\bar{X} - \bar{Y} - (m - m'))/\sqrt{1/p + 1/q}}{\sqrt{\sum_{i=1}^p (X_i - \bar{X})^2 + \sum_{j=1}^q (Y_j - \bar{Y})^2}\Big/\sqrt{p + q - 2}}$$

has distribution $t(p + q - 2)$.

It is not difficult to deduce from this confidence intervals for $(m - m')$ and reasonable tests of comparison between m and m'. Set

$$T_{p,q} = \frac{(\bar{X} - \bar{Y})/\sqrt{1/p + 1/q}}{\sqrt{\sum_{i=1}^p (X_i - \bar{X})^2 + \sum_{j=1}^q (Y_j - \bar{Y})^2}\Big/\sqrt{p + q - 2}}$$

its distribution is

$$t'\left[p + q - 2, \frac{m - m'}{\sqrt{1/p + 1/q}}\right]$$

and the probability that T_p exeeds a given constant (**threshold**) increases as a function of $m - m'$. To test "$m \leq$

m'" against "$m > m'$," we are thus led to decide to accept "$m \leq m'$" on $\{T_{p,q} \leq t_{p+q-2,\alpha}\}$. This is a test of level α (its error risk, for $m \leq m'$ is less than α); and its power, the probability of rejecting "$m \leq m'$" is an increasing function of $m - m'$.

We could also have tested the equality of the means "$m = m'$" (or "$(m1_p, m'1_q)$ is in the one-dimensional subspace generated by 1_n"), against "$m \neq m'$". If we decide to accept "$m = m'$" on $\{-t_{p+q-2,\alpha/2} \leq T_{p,q} \leq t_{p+q-2,\alpha/2}\}$, we have a test of level α (the probability of making the wrong decision "$m \neq m'$" is α).

Let us generalize this last test. Let $X = (X_1, ..., X_n)$ be a random vector of the form $X = m + Y$, Y an n-sample from $N(0,\sigma^2)$, where m is in E, a subspace of dimension k of $\mathbb{R}^n$. Let $H \subset E$ be a subspace of dimension r; we want to test "$m \in H$" against "$m \in E \backslash H$." If we try to estimate m under the constraint $m \in H$; the estimator is X_H, whereas under the constraint $m \in E$ the estimator is X_E: $\|X_E - X_H\|^2$ thus represents the distance between the two estimators. From Cochran's theorem, the r.v.

$$Z = \frac{\|X_E - X_H\|^2/(k-r)}{\|X - X_E\|^2/(n-k)}$$

follows the distribution

$$F'\left(k-r, n-k, \frac{\|m - m_H\|^2}{\sigma^2}\right).$$

Given a threshold t, the probability that Z exceeds t is an increasing function of $\|m - m_H\|$. Let $\alpha \in]0,1[$. *We decide to reject the hypothesis "$m \in H$" on $\{Z > F_{k-r,n-k,\alpha}\}$. "$m \in H$" is then wrongly rejected with probability α; this is a test of level α. The power, the probability of rejecting "$m \in H$", is an increasing function of $\|m - m_H\|$* which can be calculated, if m and σ^2 are known (or estimated if not) with the help of the tables of the noncentral Fisher distribution.

Notes. (a) *As before, the total variation $\|X - X_H\|^2$ is the sum of the variation between groups $\|X_E - X_H\|^2$ and the residual variation $\|X - X_E\|^2$.*

(b) *The space H is invariant under the following transformations on $\mathbb{R}^n$:*

(1) *translations* $x \longmapsto x + c \quad (c \in H)$,

(2) *dilatations* $x \longmapsto \sigma x \quad (\sigma > 0)$,
(3) *orthogonal transformations leaving H and E invariant.*

The same applies to Z; this invariance of the test statistic, linked to the problem, is a property often asked for (we will come back to this in Section 8.4).

Example. For the samples $X^{(i)} = (X_{i1}, \ldots, X_{in_i})$ from the distribution $N(m_i,\sigma^2)$, $i = 1, \ldots, k$, studied in Section 5.1.6, we carry out the test of homogeneity "$m_1 = \ldots = m_k$" against "it is false." Here H is generated by $\mathbf{1}_n$ and is of dimension 1, E is generated by $(\mathbf{1}_{n_1}, \ldots, \mathbf{1}_{n_k})$. The random variable

$$Z = \frac{\sum_i n_i(X_{i.} - X_{..})^2/(k-1)}{\sum_{i,j} (X_{ij} - X_{i.})^2/(n-k)}$$

has distribution $F(k-1, n-k, \|m_E - m_H\|^2)$. Having chosen $\alpha \in]0,1[$, the hypothesis of equality of the means is rejected on $\{Z > F_{k-1,n-k,\alpha}\}$. It is a test of level α and the power, the probability of rejecting the equality of the means, is an increasing function of $\Sigma_{i=1}^k n_i(m_i - \overline{m})^2$, with $\overline{m} = (1/n)\Sigma_{i=1}^k n_i m_i$ ($\overline{m}\mathbf{1}_n$ is the projection of m on H).

Exercises 5.1.

E.1. Show that the functions $\lambda \longmapsto \chi'^2(n,\lambda)$, $\lambda \longmapsto F'(n_1,n_2,\lambda)$ and $a \longmapsto t'(n,a)$ are increasing for the order of the distributions defined in Exercise 3.3.6.

E.2. A bullet is fired at a target and hits it at point P. Assume that the horizontal and vertical distances of P from the center O of the target are independent normal variables. What is the distribution of OP?

E.3. The standard deviation of the content of a component in a drug is 8 milligrams. A new production process aims at lowering this standard deviation. For ten measurements of content on units produced by the new process, we obtain in mg;

725 722 727 718 723 731 719 724 726 726.

Assume that the measurements are independent Gaussian r.v.'s. Carry out a test to show if the aim of the research is achieved (state the chosen level).

E.4. The weights in grams of 1000 jars of jam coming out in succession from a packing machine were the following (the results are given in classes of width 2, the origin of the first being 2000 and the endpoint of the last 2024):

Class	1	2	3	4	5	6	7	8	9	10	11	12
number of jars	9	21	58	131	204	213	185	110	50	16	3	0

(a) By accepting that the weights of the jars follow a Gaussian distribution, estimate its mean and standard deviation (in each class, take as value the middle of the class).

(b) By accepting that the standard deviation of the machine does not vary with time (equal to that estimated in (a)) and that adjustment only affects the mean, what value must be chosen if we want the probability that a jar weighs less than 2000 g (contravention of the weights and measures regulations) to be less than 10^{-4}?

(c) When the machine has been adjusted, in the course of production, 8 jars are simultaneously weighed to control the adjustment of the machine. In which cases will we decide to modify the adjustment? (Give a surveillance limit and a control limit analogous to Section 2.4.2.)

E.5. The following table represents measurements made in three forests on the height of certain trees. In the ith forest, n_i trees are measured with heights $X_{i1}, ..., X_{in_i}$. Give, stating the hypotheses you make about the model, a test of level 0.05, in order to decide if the height of the trees is the same in the three forests. Give a confidence interval of level 0.99 for the mean height in each of the three forests.

	Forest 1	Forest 2	Forest 3
	23.4	22.5	18.9
	24.4	22.9	21.1
	24.6	23.7	21.2
	24.9	24.0	22.1
	25.0	24.4	22.5
	26.2	24.5	23.5
	26.3	25.3	24.5
	26.8	26.0	24.6
	26.8	26.2	26.2
	26.9	26.4	26.7
	27.0	26.7	
	27.6	26.9	
	27.7	27.4	
		28.5	
n_i	13	14	10
$\sum_{i=1}^{n_i} X_{ij}$	337.6	355.4	231.3
$\sum_{j=1}^{n_i} X_{ij}^2$	8789.36	9062.96	5403.51

E.6. *Test of an interaction.* The following table describes the yields obtained, in well defined conditions of culture, climate, ... , with three types T of land and four varieties V of corn.

The problems which may be asked are of different types.

(a) *The effect of different factors.* Does the variation in the quality of the land influence the yield? Do the different varieties of corn give on average equivalent yields.

(b) *The simultaneous effect of the factors.* Do the varieties of corn adapt in the same way to the different types of land? Denote by X_{ijk} the kth measurement made for variety V_i, $i = 1, ..., k$ and land T_j, $j = 1,2,3$. Take as a model

$$X_{ijk} = \mu + \alpha_i + \beta_j + \gamma_{ij} + \varepsilon_{ijk},$$

where the ε_{ijk} are independent with distribution $N(0,\sigma^2)$; μ, α_i, β_j, γ_{ij} ($i = 1, ..., 4$, $j = 1, ..., 3$) and σ^2 are unknown parameters. Assume $\Sigma_i\alpha_i = 0$, $\Sigma_j\beta_j = 0$, $\Sigma_i\gamma_{ij} = 0$, $\Sigma_j\gamma_{ij} = 0$ (verify that this

	V_1	V_2	V_3	V_4
T_1	37 34 40 39	39 40 41	37 38 39 36 55	35 39 37
T_2	32 35 31 32	40 43 41 39	45 43 42 39 40	39 37 36
T_3	39 37 36	37 39 40 41	41 39 40	38 39 37 36

can always be done without loss of generality). Estimate these parameters. Use this model to test the different hypotheses suggested above. Interpret the γ_{ij} as coefficients (called interaction coefficients) allowing quantification of problem (b).

E.7. *The Latin Square design.* This design is used when we want to study a treatment with 3 factors, each factor having the same number of levels and assumed to be additive. Consider the following model, where the ε_{ijk} are independently distributed as $N(0,\sigma^2)$:

$$Y_{ijk} = \mu + \alpha_i + \beta_j + \gamma_k + \varepsilon_{ijk},$$

the parameters satisfying

$$\sum_i \alpha_i = \sum_j \beta_j = \sum_k \gamma_k = 0 \quad (i = 1,2, ..., m;\ j = 1,2, ..., m;$$
$$k = 1,2, ..., m).$$

Latin Square: This is a square $m \times m$ table containing integers chosen between 1 and m, such that each of the m numbers $\{1, ..., m\}$ appears once and only once in each row and each column. We are going to deal with the following example ($m = 5$):

j \ i	1	2	3	4	5
1	4	5	3	1	2
2	5	1	4	2	3
3	3	4	2	5	1
4	2	3	1	4	5
5	1	2	5	3	4

Design

5	16	13	8	7
12	4	11	10	5
12	13	10	20	5
3	8	5	10	14
7	7	15	11	9

Results

The table on the left represents the **design** of an experiment. We have 5 breeds of cows, 5 levels of the fat content in milk (variable within each breed), and 5 storage temperatures. We decide to experiment on breed i and level j, at the temperature given by the index which appears as the (i,j)th entry in the table. It can be verified that there is no repetition of a same **treatment** (breed, level, temperature). The table of results gives Y_{ijk}, the time for complete fermentation of the milk; thus $Y_{114} = 5$. Show that, to test "the breed has no effect", it is suffiient to test "$\Sigma_{i=1}^{5}\alpha_i^2 = 0$." Give a suitable test of level 0.05 and apply it to the above results.

E.8. *Random effects models.* A characteristic X is observed in the descendants of k bulls: on a descendant in the pth generation from the ith bull kept in environment j, X_{ij}^{p} is observed, $1 \leqslant j \leqslant r$, $1 \leqslant p \leqslant n_i$. Let

$$X_{i.}^{p} = \frac{1}{r}\sum_{j=1}^{r} X_{ij}^{p}, \qquad n = \sum_{i=1}^{k} n_i$$

and

$$\overline{X}_i = \frac{1}{n_i}\sum_{p=1}^{n_i} X_{i.}^{p}\,.$$

Assume that X_{ij}^{p} has a random part E_{ij}^{p} due to the environment and a random part A_i^{p} due to heredity. Thus set $X_{ij}^{p} = m_i + A_i^{p} + E_{ij}^{p}$. The vector $m = (m_i)_{1\leqslant i\leqslant k}$ is constant, but unknown, A_i^{p} has a distribution $N(0,\sigma_a^2)$, and E_{ij}^{p} has a distribution $N(0,\sigma_e^2)$ with the variances σ_a^2 and σ_e^2 unknown. All the variables (A_i^{p}) and (E_{ij}^{p}) are independent.

We want to test the influence of heredity, hence, to test "$\sigma_a^2 = 0$" against "$\sigma_a^2 > 0$".

(a) Set

$$V_I = \sum_{i=1}^{k} \sum_{p=1}^{n_i} \sum_{j=1}^{r} (X_{ij}^p - X_{i.}^p)^2.$$

Show that V_I is distributed as $\sigma_e^2\chi^2(n(r-1))$.

(b) Let

$$V_E = \sum_{i=1}^{k} \sum_{p=1}^{n_i} (X_{i.}^p - \overline{X}_i)^2.$$

Show that V_E is independent of V_I and is distributed as $(\sigma_a^2 + \sigma_e^2/r)\chi^2(n-k)$.

(c) Deduce from this a test of "$\sigma_a^2 = 0$" against "$\sigma_a^2 \neq 0$" of level α, using the statistic V_E/V_I and express its power with the help of Fisher's distribution.

E.9. *Using Student's test for large, non-Gaussian samples.* Let $(X_1, \ldots, X_n)$ be an n-sample from a distribution with mean m and variance σ^2, $\overline{X}_n$ and S_n^2 its sample mean and variance. Show that (S_n^2) tends to σ^2 in probability and that $(\sqrt{n}(\overline{X}_m - m)/S_n)$ tends in distribution to $N(0,1)$. Study the sequence $(\sqrt{n}(\overline{X}_n - m')/S_n)$ for $m' \neq m$. Justify the use of Student's test (Section 5.1.5) for testing $m \leqslant m_0$ against $m > m_0$ for large samples.

E.10. *Maximum likelihood estimator.* Refer back to the situation in Section 5.1.6.

(a) Assume that E is the set of vectors in $\mathbb{R}^n$ the last $n - k$ coordinates of which are zero. Give the density of X and a maximum likelihood estimator of (m,σ^2).

(b) Deduce from this a maximum likelihood estimator within the framework of Section 5.1.7.

5.2. Gaussian Random Vectors

5.2.1. Definition

We have seen that, if $X = (X_1, \ldots, X_n)$ is a Gaussian n-sample, every linear combination $\langle t,X\rangle = \Sigma_{j=1}^n t_j X_j$ of the components is also Gaussian. We are led to the following definition.

Definition 5.2.6. A random vector X taking values in $\mathbb{R}^n$ is said to be **Gaussian** if every linear combination of its components is Gaussian.

Let X be a centered Gaussian vector. For $t \in \mathbb{R}^n$, the r.v. $<t,X>$ is centered, Gaussian and

$$E[\exp i<t,X>] = \exp\left[-\frac{1}{2}E(<t,X>^2)\right].$$

If γ is the bilinear form $(t,t') \longmapsto E[<t,X><t',X>]$, the associated quadratic form $t \longmapsto E(<t,X>^2)$ is positive. In the canonical basis, the matrix $\Gamma = \{E(X_iX_j)\}_{1\leqslant i,j\leqslant n}$ is associated with γ. It is a symmetric matrix and positive (with positive eigenvalues). We have $E(<t,X>^2) = <t,\Gamma t>$, and thus

$$E[\exp i<t,X>] = \exp\left[-\frac{1}{2}<t,\Gamma t>\right].$$

The distribution of X is characterized by its covariance Γ.

If X has mean m and covariance Γ, $X - m = Y$ is Gaussian and centered with covariance Γ. We thus have

$$E[\exp i<t,X>] = \exp\left[i<t,m> - \frac{1}{2}<t,\Gamma t>\right].$$

The distribution of X is characterized by the pair (m,Γ).

Definition 5.2.7. Denote by $N(m,\Gamma)$ the distribution of a Gaussian random vector X with mean m and covariance Γ. The characteristic function of X is

$$t \longmapsto \exp\left[i<t,m> - \frac{1}{2}<t,\Gamma t>\right].$$

We shall see later that if Γ is a symmetric and positive matrix and m is an arbitrary vector, we can find a Gaussian vector X with distribution $N_n(m,\Gamma)$; the case $\Gamma = 0$ corresponds to Dirac measure on m.

Let $X = (X_j)_{1\leqslant j\leqslant n}$ be distributed as $N_n(m,\Gamma)$ with $m = (m_j)_{1\leqslant j\leqslant n}$. For $1 \leqslant j \leqslant m$, the distribution of X_j has Fourier transform

$$E(\exp it_jX_j) = \exp\left[it_jm_j - \frac{\sigma_j^2\, t_j^2}{2}\right].$$

The diagonal elements of Γ are $\Gamma_{jj} = \sigma_j^2$ $(1 \leqslant j \leqslant n)$. The distribution of X_j is thus $N(m_j,\sigma_j^2)$. The components $(X_j)_{1\leqslant j\leqslant n}$

are independent if, and only if, $N_n(m,\Gamma)$ is the product $\otimes_{j=1}^{n} N(m_j,\sigma_j^2)$. In other words, if Γ is a diagonal matrix, or if the components $(X_j)_{1\leq j\leq n}$ are pairwise uncorrelated. This necessary condition for independence, which in general is not sufficient, suffices in the case of a Gaussian random vector.

Let a be a linear mapping $\mathbb{R}^n \to \mathbb{R}^k$ and ${}^t a$ its transpose from $\mathbb{R}^k$ into $\mathbb{R}^n$;

$$<y,a(x)> = <{}^t a(y),x> \quad \text{if } y \in \mathbb{R}^k,\ x \in \mathbb{R}^n.$$

Let us find the distribution $a(X)$. Let A be the matrix representing a with respect to the canonical basis of $\mathbb{R}^n$. We have, for $y \in \mathbb{R}^k$

$$\begin{aligned} E[\exp i<y,\ AX>] &= E[\exp i<{}^t Ay,\ X>] \\ &= \exp\left[-\frac{1}{2}<{}^t Ay,\ \Gamma^t Ay>\right] \\ &= \exp\left[-\frac{1}{2}<y,\ A\Gamma^t Ay>\right]. \end{aligned}$$

Thus $a(X)$ has distribution $N_k(0,\ A\Gamma^t A)$. Let us show that every Gaussian vector X is of the form $a(Z)$, where Z is a k-sample from $N(0,1)$ and a a linear mapping $\mathbb{R}^k \to \mathbb{R}^n$. The proof will show at the same time that to every $n \times n$ symmetric, positive matrix Γ we can associate a Gaussian matrix X with distribution $N_n(0,\Gamma)$.

Let Γ be a positive symmetric matrix of rank r, $(f_1, ..., f_r, f_{r+1}, ..., f_n)$ an orthonormal basis of eigenvectors of Γ and $(\lambda_1^2, ..., \lambda_r^2, 0, ..., 0)$ the associated eigenvalues.

Let U be an orthogonal matrix the column vectors of which are these vectors. The diagonal matrix $\Delta = {}^t U\Gamma U$ is the covariance matrix of the vector $Y = (Y_1, ..., Y_n)$ with $Y_i = <X,f_i>$, $(1 \leq i \leq n)$: $\sigma^2(Y_i) = \lambda_i^2$, for $1 \leq i \leq r$ and for $i > r$, $Y_i = 0$ a.s. Thus X takes a.s. its values in the subspace generated by $f_1, ..., f_r$. Consider the following two mappings:

$$a\colon (x_1, ..., x_r) \longmapsto \sum_{i=1}^{r} x_i\lambda_i f_i\ , \quad \mathbb{R}^r \to \mathbb{R}^n$$

and

$$b\colon x \longmapsto \left(\frac{1}{\lambda_i} <x,\ f_i>\right)_{1\leq i\leq r}, \quad \mathbb{R}^n \to \mathbb{R}^r.$$

Then $Z = b(X)$ is an r-sample from $N(0,1)$ and $X = a(Z)$. For every vector Z', an r-sample from $N(0,1)$, $a(Z')$ has distribution $N_n(0,\Gamma)$, from which the existence follows of a Gaussian vector with given covariance Γ.

Theorem 5.2.8. (a) *To every nonzero, symmetric and positive matrix* Γ, *and to every* $m \in \mathbb{R}^n$, *a Gaussian random vector can be associated with distribution* $N_n(m,\Gamma)$.

(b) *If* X *is a Gaussian vector with distribution* $N_n(m,\Gamma)$ *and if* a *is a linear transformation from* $\mathbb{R}^n$ *into* $\mathbb{R}^k$ *and* A *the associated matrix relative to the canonical basis,* $a(X)$ *has distribution* $N_k(a(m), A\Gamma {}^tA)$. *The components of* X *are independent if, and only if,* Γ *is diagonal.*

(c) *Every* $N_n(0,\Gamma)$ *Gaussian vector may be written as* $X = a(Z)$, *where* Z *is an r-sample from* $N(0,1)$, r *the rank of* Γ *and* a *a linear mapping from* $\mathbb{R}^r$ *into* $\mathbb{R}^n$.

5.2.2. Density of the Distribution $N_n(m,\Gamma)$

If the rank of Γ is $r < n$ and if X is a Gaussian vector with distribution $N_n(m,\Gamma)$, X is a.s. in a subspace of dimension r, of Lebesgue measure zero. Thus the distribution $N_n(m,\Gamma)$ has not got a density. For $r = n$, we apply the change of variables formula (Section 3.3.4). Let Z be an n-sample from $N(0,1)$; its density is f_Z, with

$$f_Z(x) = (2\pi)^{-n/2}\exp\left[-\frac{1}{2}\|x\|^2\right].$$

The vector $a(Z)$ has the distribution $N_n(0,\Gamma)$ (with the preceding notations). Let b be the inverse of a. The matrix associated with a is UD, denoting by D the diagonal matrix of which the diagonals is $\lambda_1, \ldots, \lambda_n$. Its determinant is $\lambda_1 \ldots \lambda_n$, the inverse of that of b. However the determinant $|\Gamma|$ of Γ is equal to that of ${}^tU\Gamma U$, thus to $(\lambda_1 \ldots \lambda_n)^2$. The density of $a(Z)$ is $x \longmapsto (1/\sqrt{|\Gamma|})f_Z(b(x))$. We write

$$\|b(x)\|^2 = {}^t(b(x))b(x) = {}^t(D^{-1}U^{-1}x)(D^{-1}U^{-1}x)$$

$$= {}^txU(D^{-1})^2\, {}^tUx = {}^tx\Gamma^{-1}x.$$

Let X be distributed as $N_n(m,\Gamma)$: $X = m + Y$, for Y distributed as $N_n(0,\Gamma)$, thus the density of $N_n(m,\Gamma)$ is

$$x \longmapsto \frac{1}{(2\pi)^{n/2}\sqrt{|\Gamma|}} \exp\left[-\frac{1}{2}\,{}^t(x-m)\Gamma^{-1}(x-m)\right].$$

Theorem 5.2.9. *When Γ is invertible, the distribution $\mathcal{N}_n(m,\Gamma)$ has the density*

$$x \longmapsto \frac{1}{(2\pi)^{n/2}\sqrt{|\Gamma|}} \exp\left[-\frac{1}{2}\,{}^t(x-m)\Gamma^{-1}(x-m)\right].$$

Exercises 5.2.

E.1. Consider a vector X from the distribution $\mathcal{N}_3(0,\Gamma)$ on $\mathbb{R}^3$, with

$$\Gamma = \begin{bmatrix} 3 & -1 & 0 \\ -1 & 3 & 0 \\ 0 & 0 & 2 \end{bmatrix}.$$

Find a vector of the form $a(X)$, where a is a linear operator from $\mathbb{R}^3$ into itself, the components of which are independent.

E.2. Let $(X_1, \ldots, X_n)$ be a centered Gaussian random vector. Prove

$$E(X_1X_2X_3X_4) = E(X_1X_2)E(X_3X_4) + E(X_1X_3)E(X_2X_4) + E(X_1X_4)E(X_2X_3).$$

E.3. Let X be a centered Gaussian random vector with covariance matrix Γ.

(a) Assume Γ is invertible. Show that ${}^tX\Gamma^{-1}X$ follows a $\chi^2(n)$ distribution.
(b) Show that we can always find a diagonal matrix D and an $n \times r$ matrix U such that ${}^tXUD{}^tUX$ follows a $\chi^2(r)$ distribution, where r is the rank of Γ, and $\Gamma UD{}^tU\Gamma = \Gamma$.

E.4. Let X be a random vector with distribution $\mathcal{N}_n(m,\Gamma)$, A and B two $n \times n$ matrices, $Y = AX$, $Z = BX$. Show that Y and Z are independent if, and only if, $A\Gamma{}^tB$ is zero.

E.5. Let (X_1,Y_1), (X_2,Y_2), ..., (X_n,Y_n) be an n-sample from

$$\mathcal{N}_2\left[(m_1,m_2), \begin{bmatrix} \sigma_1^2 & \rho\sigma_1\sigma_2 \\ \rho\sigma_1\sigma_2 & \sigma_2^2 \end{bmatrix}\right] \quad \text{with} \quad \sigma_1 > 0,\ \sigma_2 > 0,\ |\rho| < 1.$$

(a) Assume that m_2, σ_1, σ_2, and ρ are known and that m_1 is unknown. Determine the maximum likelihood estimator (MLE) of m_1. Is this estimator unbiased? What is its variance?

(b) Assume m_1 and m_2 are known, σ_1, σ_2, and ρ unknown. What are the MLE's of σ_1, σ_2, ρ?

(c) Assume that m_1, m_2, σ_1, σ_2, and ρ are unknown. What are the MLE's of m_1, m_2, σ_1, σ_2, ρ?

E.6. *Non-Gaussian pairs of Gaussian r.v.'s.* Look for a pair of r.v.'s (X,Y) of this type with a density of the form

$$(x,y) \longmapsto \frac{c^2}{2\pi} \exp\left(-\frac{x^2+y^2}{2}\right) - \lambda(x)\lambda(y)$$

for C a constant and a real function λ to be determined. Show that λ can be chosen such that X and Y are uncorrelated.

E.7. *Autoregressive Model (AR1).* Let (ε_n) be a sequence of independent r.v.'s with distribution $\mathcal{N}(0,\sigma^2)$ and let θ be a real parameter. Consider a sequence (X_n) defined by

$$X_1 = \varepsilon_1, \quad X_n = \theta X_{n-1} + \varepsilon_n.$$

What is the distribution of X_n? For what values of θ does it converge? What then is the limit distribution? Calculate the covariance of X_n and $X_{n-k} (k < n)$. What is the distribution of $(X_1, \ldots, X_n)$? Calculate its density. Determine the maximum likelihood estimator of (θ,σ^2) based on the observation of $(X_1, \ldots, X_n)$.

E.8. *Moving average.* With the data of Exercise 7, consider a sequence (Y_n) defined by

$$Y_1 = \varepsilon_1, \quad Y_n = \theta\varepsilon_{n-1} + \varepsilon_n.$$

What is the density of $(Y_1, \ldots, Y_n)$? Its covariance matrix? Its distribution? Determine the maximum likelihood estimator of (θ,σ^2) based on the observation (Y_1,Y_2).

E.9. *The use of Student distributions for statistics concerned with a Gaussian vector which is not a sample.* Let $(X_1, \ldots, X_n)$ be a Gaussian vector of dimension n. For $1 \leqslant i \leqslant n$ the mean of X_i is m and

$$\Gamma(X_i,X_j) = \begin{cases} \rho & \text{for } |i - j| = 1 \\ \sigma^2\rho & \text{for } |i - j| = 0 \\ 0 & \text{for } |i - j| > 1. \end{cases}$$

Let $\overline{X}_n$ and S_n^2 be the sample mean and variance. Calculate the variance of $\overline{X}_n$ and the mean of S_n^2. Study the asymptotic behavior of the sequences $(\sqrt{n}(\overline{X}_n - m))$ and (S_n^2); show that the sequence $(\sqrt{n}(\overline{X}_n - m)/S_n)$ converges in distribution to a distribution which is to be determined. By assuming ρ known, study the validity of Student's test of Section 5.1.5 for testing $(m \leqslant m_0)$ against $(m > m_0)$ when n is large. What form have the confidence intervals?

E.10. (a) Let (X_1,X_2) be a centered, Gaussian pair, with covariance matrix

$$\begin{bmatrix} 1 & \rho \\ \rho & 1 \end{bmatrix}.$$

Show that it has a bivariate Gaussian distribution studied in Exercise 4.2.13. Thus,

$$P(X_1 \geqslant 0,\ X_2 \geqslant 0) = \frac{1}{4} + \frac{1}{2\pi}\text{Arc sin } \rho.$$

Show that, in Exercise 4.2.13 we can find a nonzero linear combination of U_1, U_2, U_{12} independent of (X_1,X_2).

(b) Let (X_1,X_2,X_3) be a centered Gaussian triplet the components of which have variances equal to 1. Calculate $P(X_1 > 0,\ X_2 > 0,\ X_3 > 0)$ as a function of the various correlation coefficients.

E.11. *Perfect gases.* (a) Let v be a vector in $\mathbb{R}^3$ with distribution $N_3(0,\sigma^2 I)$ and u a fixed unit vector. Let θ be the angle between u and v lying between 0 and π. Calculate the distribution of θ and that of $\|v\|$ and verify that these r.v.'s are independent. Calculate the mean of $\|v\|$.

(b) A molecule of mass m arrives at a flat elastic surface with normal vector u, with velocity $v\mathbf{1}_{(0<\theta<\pi/2)}$. It is reflected and its velocity becomes the vector v' symmetrical to $-v$ with respect to u. What is the mean of the change in momentum $m(v - v')$?

(c) n molecules of mass m are contained in a vessel of volume W enclosed by elastic walls. Their positions are independent and uniformly distributed in W. Their velocities are distributed as $N_3(0,\sigma^2 I)$ independently of each other and independent of their positions. Let K be a surface element of area 1 of the wall of the vessel and u a vector normal to K. In a time interval $(t, t + 1)$ (the time unit being small enough...) the velocities are considered to remain constant. A molecule with velocity v will reach K if it lies in the cylinder with base K and axis v. The effects of the various molecules add together and the pressure p on K during this interval is the sum of the expected changes in momentum of the molecules which reach K in this time interval. For Boltzmann's constant k we have: $\sigma^2 = kT/m$, T being the temperature. Prove $pW = nkT$.

5.3. The Central Limit Theorem on $\mathbb{R}^k$

5.3.1. The Limit Distribution for Gaussian Vectors

Gaussian vectors on $\mathbb{R}^k$ have the following important property: every limit in distribution of Gaussian vectors is Gaussian.

Proposition 5.3.10. *In order that X_n, with distribution $N_k(m_n,\Gamma_n)$, converges in distribution, it is necessary and sufficient that there exists an $m \in \mathbb{R}$, and a positive matrix Γ, such that (m_n) tends to m and (Γ_n) tends to Γ. If Γ is zero, the limit distribution is δ_m; in general, it is $N_k(m,\Gamma)$.*

Proof. Let $u \in \mathbb{R}^k$: $\text{Log } E[\exp i \langle u,X_n\rangle] = i\langle u,m_n\rangle - (1/2)^t u\Gamma_n u$. Hence, convergence in distribution of (X_n) is equivalent to convergence, for all u, of $\langle u,m_n\rangle$ and ${}^t u\Gamma_n u$, from Levy's theorem (Section 3.4.2).

5.3.2. The Central Limit Theorem

Theorem 5.3.11. *Let $X_1, ..., X_n$ be an n-sample from a distribution on $\mathbb{R}^k$, with mean m and covariance Γ. Then*

$$\left[\frac{X_1 + ... + X_n - nm}{\sqrt{n}}\right] \xrightarrow{\mathcal{D}} N_k(0,\Gamma)$$

as $n \to \infty$.

Proof. Following theorem 4.4.23, for all $u \in \mathbb{R}^k$ the sequence

$$\left[\frac{<u,X_1> + ... + <u,X_n> - n<u,m>}{\sqrt{n}}\right]$$

converges in distribution to $N(0,\sigma^2(<u,X>))$, with $\sigma^2(<u,X>) = {}^t u\Gamma u$; thus

$$\left[\text{Log } E\left[\exp i \frac{<u,X_1 + ... + X_n> - <u,nm>}{\sqrt{n}}\right]\right]$$

tends to $[-1/2]^t u\Gamma u$. From which the result follows by Levy's theorem.

5.3.3. Convergence of the Multinomial Distributions

Consider an n-sample $X_1, ..., X_n$ from a distribution $p = (p_1, ..., p_k)$ carried by a finite set denoted $(1, ..., k)$. Set $N_i = \Sigma_{j=1}^n 1_{(X_j=i)}$, the number of sample elements taking the value i.

The random vector $N = (N_1, ..., N_k)$ takes values in the set of elements of ^{k} the components of which sum to n:

$$P(N_1 = n_1, ..., N_k = n_k) = \frac{n!}{n_1! ... n_k!} p_1^{n_1} ... p_k^{n_k}.$$

This is the multinomial distribution $M(n,p)$ which we are going to approximate by a Gaussian distribution.

Corollary 5.3.12. *Let $N = (N_1, ..., N_k)$ be a random vector with multinomial distribution $M(n,p)$, $p = (p_1, ..., p_k)$. Let $\sqrt{p}$ be the vector $(\sqrt{p_1}, ..., \sqrt{p_k})$. The sequence of random vectors*

$$\left[\frac{N_1 - np_1}{\sqrt{np_1}}, ..., \frac{N_k - np_k}{\sqrt{np_k}}\right]$$

converges in distribution, as $n \to \infty$, to $N_k(0,\Gamma)$, the distribution of the projection Z of a vector with distribution $N_k(0,I)$ on the orthogonal to $(\sqrt{p})$: $\Gamma = I - \sqrt{p}\ {}^t\sqrt{p}$.

Proof. Let

$$Z_j = \left[\frac{1}{\sqrt{p}_1}(1_{(X_j=1)} - p_1), \ldots, \frac{1}{\sqrt{p}_k}(1_{(X_j=k)} - p_k)\right];$$

$(Z_1, \ldots, Z_k)$ is an n-sample from a centered distribution with covariance Γ; from which the result follows by Theorem 5.3.11. The rank of Γ is at most $k - 1$, the limit distribution being carried by the hyperplane $(\sqrt{p})^{\perp}$.

Exercises 5.3

E.1. A molecule formed by a very long planar chain of n atoms is modelled in the following way: two consecutive atoms are at a distance ε, the vertex angle at each point is assumed random and uniformly distributed on $(0,2\pi)$, independent of the other angles. What is the approximate distribution of the square of the distance between the end points?

E.2. Let $\Lambda_1, \ldots, \Lambda_n$ be independent normal variables. At time t the function $X(t) = \Sigma_{j=1}^n \sin(\Lambda_j t)$ is observed. What is the covariance of $(X(1), \ldots, X(k))$? Give an approximate distribution for this random vector for large n.

5.4. The χ^2 Test

Consider an r.v. (or characteristic) taking values in $\{1, \ldots, k\}$. We want to test "its distribution is $p = (p(1), \ldots, p(k))$," the alternative being "its distribution is different from p." Let $(X_1, \ldots, X_n)$ be an n-sample of the characteristic, and $N_i = \Sigma_{j=1}^n 1_{(X_j=i)}$ the number of elements of the sample taking the value i; $\overline{p}_n(i) = N_i/n$ is the sample estimator of p_i, the sample distribution function being $\overline{p}_n = (\overline{p}_n(1), \ldots, \overline{p}_n(k))$. In Section 1.4 we introduced a "distance" between the probability distributions p and q on $\{1, \ldots, k\}$:

$$\chi^2(p,q) = \sum_{i=1}^{k} \frac{(p(i) - q(i))^2}{p(i)} .$$

For a sample, we will measure the distance of the sample distribution function $\overline{p}_n$ to the distribution p by the χ^2 **goodness of fit**:

$$\chi_n^2(p,\overline{p}_n) = n\chi^2(p,\overline{p}_n) = n \sum_{i=1}^{k} \frac{(p(i) - \overline{p}_n(i))^2}{p(i)}$$

$$= \sum_{i=1}^{k} \frac{(np(i) - N_i)^2}{np(i)} .$$

The distribution of $\chi_n^2(p,\overline{p}_n)$ depends on p. However, if n is large it tends to a limit independent of p (this is the reason for the normalization of $\chi_n^2(p,\overline{p}_n)$ by n). Let us verify this first of all, for $k = 2$:

$$\chi_n^2(p,\overline{p}_n) = n\left[\frac{(p(1) - \overline{p}_n(1))^2}{p(1)} + \frac{(p(1) - \overline{p}_n(1))^2}{1 - p(1)}\right]$$

$$= \frac{n(p(1) - \overline{p}_n(1))^2}{p(1)(1 - p(1))}$$

$$= \frac{(N_1 - np(1))^2}{np(1)(1 - p(1))} ;$$

$(\chi_n^2(p,\overline{p}_n))$, the square of a sequence which converges in distribution to $N(0,1)$, converges to $\chi^2(1)$. For arbitrary k, we have, with the notations of Section 5.3.3,

$$\chi_n^2(p,\overline{p}_n) = \frac{\|Z_1 + \dots + Z_n\|^2}{n} .$$

The norm being a continuous function from $\mathbb{R}^n$ into $\mathbb{R}$, the convergence in distribution of $[(Z_1 + \dots + Z_n/\sqrt{n})]$ to Z implies the convergence of $\chi_n^2(p,\overline{p}_n)$ to $\|Z\|^2$. From Cochran's theorem 5.1.1, $\|Z\|^2$ has a $\chi^2(n - 1)$ distribution.

Theorem 5.4.13. *For each n, let $(X_1, \dots, X_n)$ be an n-sample of a k-valued characteristic with distribution p, and let $\overline{p}_n$ be its empirical distribution. The sequence $(\chi_n^2(p,\overline{p}_n))$ converges in distribution to $\chi^2(k - 1)$.*

Goodness of Fit Test. The sample is used to test the hypothesis H_0 "the distribution of the sample is p." The distribution of $\chi_n^2(p,\overline{p}_n)$ is assimilated to its limit and a test of level close to α is obtained by rejecting H_0 on $\{\chi_n^2(p,\bar{p}_n) > \chi^2_{k-1,\alpha}\}$. The choice of the rejection region is based on the fact that, if $q \neq p$ is the distribution of the sample, the sequence $(\chi_n^2(p,\bar{p}_n))$ tends a.s. to $+\infty$. In fact, from the law of large numbers,

$$\left[\chi^2(p,\bar{p}_n) = \frac{1}{n}\chi_n^2(p,\bar{p}_n)\right]$$

tends a.s. to $\chi^2(p,q)$.

Extensions. We often try to test a hypothesis of the form H_0: "p is in a certain set Θ of probabilities." By assuming H_0 to be true, we have a statistical model and we can use the sample to estimate p: let $\hat{p}_n$ be such an estimator. To test H_0, we can study $\chi_n^2(\hat{p}_n,\bar{p}_n)$. Its distribution depends on the chosen estimator $\hat{p}_n$, and the theory is more complicated. We come back to this in Volume 2. However, we are going to point out two very important practical examples.

Example 1. Test of Independence. We observe an n-sample of a pair of r.v.'s (X,Y), X taking values in $\{1, ..., k\}$ and Y taking values in $\{1, ..., \ell\}$.

Let $p = \{p_{ij} \,;\, 1 \leqslant i \leqslant k,\ 1 \leqslant j \leqslant \ell\}$ be the distribution of (X,Y). If N_{ij} is the number of observations of (i,j) in the n-sample, the sample estimator $\bar{p}_n$ is defined by

$$\bar{p}_n(i,j) = \frac{N_{ij}}{n}.$$

If the characteristics X and Y are independent, p is in the set of product distributions

$$\Theta = \{p;\ p_{ij} = p_{i\cdot}p_{\cdot j} \text{ for } 1 \leqslant i \leqslant k,\ 1 \leqslant j \leqslant \ell\}$$

(the point represents a summation here and not an average as in Section 5.1). $p_{i\cdot}$ is then estimated by the sample estimates $N_{i\cdot}/n$ and $p_{\cdot j}$ by $N_{\cdot j}/n$, thus p_{ij} by $N_{i\cdot}N_{\cdot j}/n^2 = \hat{p}_n(i,j)$. If H_0 is true, $\hat{p}_n$ and $\bar{p}_n$ should be close to one another; the χ^2 **for independence**

$$\chi_n^2(\hat{p}_n,\bar{p}_n) = n \sum_{i,j} \frac{\left[\dfrac{N_{ij}}{n} - \dfrac{N_{i\cdot}}{n}\dfrac{N_{\cdot j}}{n}\right]^2}{\dfrac{N_{i\cdot}}{n}\dfrac{N_{\cdot j}}{n}}$$

should not be very large. We will show (Volume 2) that $\chi^2_n(\hat{p}_n,\bar{p}_n)$ converges in distribution to $\chi^2((k-1)(\ell-1))$. We will give a rule saying that if n parameters are estimated (in a certain way, which will be specified), the number of degrees of freedom of the χ^2 goodness of fit must be decreased. Here we estimate $(k-1)+(\ell-1)$ parameters $p_{i.}$ and $p_{.j}$ (since $\Sigma_i p_{i.} = \Sigma_j p_{.j} = 1$), and the χ^2 to use has $k\ell - 1 - (k-1) - (\ell-1) = (k-1)(\ell-1)$ degrees of freedom. The sequence $((1/n)\chi^2_n(\hat{p}_n,\bar{p}_n))$ converges a.s. to the χ^2 distance from p to the product of the marginal distributions. If H_0 does not hold, $(\chi^2_n(\hat{p}_n,\bar{p}_n))$ tends a.s. to ∞. *A convenient test for the independence of X and Y with a level close to α is that with rejection region*

$$\{\chi^2_n(\hat{p}_n,\bar{p}_n) > \chi^2_{(k-1)(\ell-1),\alpha}\}.$$

Example 2. How Can we Interpret a Problem? Consider the following data. The level of sight of each of the two eyes of 7477 women aged between 30 and 40 years has been classified into 4 groups designated by 1, 2, 3, 4 (in decreasing order from best to worst).

The results are given in the following table. Compare the qualities of the two eyes. Set in this way, the question has several answers. For example; (a) independence of the eyes (obviously false) which leads to the χ^2 test already seen; (b) equality of the distribution for the two eyes; (c) and finally, symmetry of the table, more subtle and convincing than the previous property. To test (c) leads to considering $\Theta = \{p;\ p_{ij} = p_{ji}$ for all $(i,j)\}$ and to test H_0: "$p \in \Theta$." Estimating p knowing that p is in Θ comes down to estimating the 9 p_{ij} with $1 \leqslant i \leqslant j \leqslant 4$ and $(i,j) \neq (4,4)$. We could take the estimator $\hat{p}_n$ defined by

right eye \ left eye	1	2	3	4
1	1520	266	124	66
2	234	1512	432	78
3	117	362	1772	205
4	36	82	179	492

$$\hat{p}_n(i,j) = \hat{p}_n(j,i) = \frac{N_{ij} + N_{ji}}{2n} \quad (i \neq j)$$

and use $\chi^2_n(\hat{p}_n, \overline{p}_n)$, the asymptotic distributions of which is $\chi^2(15 - 9) = \chi^2(6)$.

Exercises 5.4

E.1. *Combination of the values of a characteristic* (see Exercise 1.4.2). If we observe a sample of a qualitative characteristic, combining two possible values i and j into one, denoted $[i,j]$, amounts to putting them together in a single class. To the distribution p then corresponds the distribution which gives the weight $p_i + p_j$ to $[i,j]$, and the same weight as p to the other points. Show that combination decreases the χ^2 goodness of fit to a distribution p. Show that, for the study of a pair of r.v.'s taking a finite number of values, combining two columns (or two rows) in the table of results decreases χ^2 for independence. What can you say about the conclusions of the test for independence based on χ^2?

E.2. Let f be the density of $N_k(0,\Gamma)$, and $f_1, ..., f_k$ its marginal densities. Calculate as a function of Γ

$$\int \frac{[f(x_1, ..., x_k) - f_1(x_1) \dots f_k(x_k)]^2}{f_1(x_1) \dots f_k(x_k)} dx_1 \dots dx_k.$$

E.3. For an n-sample of a pair of characteristics which each take their values in $\{0,1\}$, compare the χ^2 for independence and the sample correlation.

E.4. An n-sample of a pair of characteristics X and Y is studied, both taking values in $\{1, ..., k\}$; $(i,j) \in \{1, ..., k\}^2$ is observed N_{ij} times.

(a) We test hypotheses of symmetry. Show that the statistic

$$\sum_{j=1}^{k} \sum_{i<j} \frac{(N_{ij} - N_{ji})^2}{N_{ij} + N_{ji}}$$

can be used, and give its asymptotic distribution (by assuming the rule used in Example 2) and a suitable test of level close to α of H_0: "the pair (X,Y) is symmetric." Apply this to Example 2 above (on eyesight) for $\alpha = 0.05$. For what values of α is the symmetry acceptable?

(b) Give a test for independence of the two characteristics of level close to α, and apply it to Example 2, for $\alpha = 0.05$.

E.5. In Exercise 4.5.4, construct a χ^2 test of level close to α instead of Kolmogorov's test. Compare the results. Starting from what level would the fit be acceptable?

E.6. *Stuart's statistic: A test for the equality of the marginal distributions.* (a) Let $(X_p,Y_p)_{1 \leq p \leq n}$ be an n-sample from a distribution $q = (q_{ij};\ 1 \leq i \leq k,\ 1 \leq j \leq k)$ on $\{1, \ldots, k\}^2$, of which the marginal distributions are equal. Let $Z_p = \{\mathbf{1}_{(X_p=i)} - \mathbf{1}_{(Y_p=i)}\}_{1 \leq i \leq k-1}$. What is the covariance matrix Γ of Z_p? It is assumed to be invertible and let $\Gamma^{-1} = (\Gamma^{ij})_{1 \leq i,j \leq k-1}$ be its inverse. Let

$$N_{ij} = \sum_{p=1}^{n} \mathbf{1}_{(X_p=i,Y_p=j)}$$

and

$$N_{i\cdot} = \sum_{p=1}^{n} \mathbf{1}_{(X_p=i)}, \quad N_{\cdot j} = \sum_{p=1}^{n} \mathbf{1}_{(Y_p=j)}.$$

What is the asymptotic distribution of

$$\frac{1}{n} \sum_{i=1}^{k-1} \sum_{j=1}^{k-1} (N_{i\cdot} - N_{\cdot i})(N_{j\cdot} - N_{\cdot j})\Gamma^{ij}$$

for large n (see Exercise 5.2.3)?

(b) We want to test the equality of the marginal distribution of q with the help of the preceding n-sample. Let V be the matrix $\{V_{ij}\}_{1 \leq i,j \leq k-1}$ with

$$nV_{ii} = N_{i\cdot} + N_{\cdot i} - 2N_{ii}$$

$$nV_{ij} = -(N_{ij} + N_{ji}), \quad \text{for } i \neq j.$$

Show that, under the hypothesis of equality of the marginal distributions, V converges a.s. to Γ. Let $(V^{ij})_{1 \leq i,j \leq k-1}$ be the inverse of V. Show that "*Stuart's statistic*"

$$\frac{1}{n} \sum_{i=1}^{k-1} \sum_{j=1}^{k-1} (N_{i\cdot} - N_{\cdot i})(N_{j\cdot} - N_{\cdot j})V^{ij}$$

converges to a $\chi^2(k-1)$ distribution as n tends to ∞. Deduce from this a test with level close to α and apply it to Example 2.

5.5. Regression

5.5.1. Least Squares

In order to launch a new product, a marketing and advertising agency wishes to explain the product's monthly consumption Y by a certain number of "explanatory" variables $X^1, \ldots, X^p$. These quantitative variables (taking values in $\mathbb{R}$) are, for example, household income, the price index, and the credit limit. They may possibly have to be transformed: an economic index is often replaced by its logarithm, which is more sensitive to variation and the economic interpretation of which is standard. We try to approximate Y by an affine function $\alpha_1 X^1 + \ldots + \alpha_p X^p + \beta$ in the least squares sense. Assume $Y, X^1, \ldots, X^p$ are in $L^2(\Omega, A, P)$ equipped with scalar product $< , >$ and norm $\| \ \|_2$; $[X] = [1, X^1, \ldots, X^p]$ is the subspace of L^2 generated by $\{1, X^1, \ldots, X^p\}$. We look for $Z = \beta + \Sigma_{i=1}^p \alpha_i X^i$ such that

$$\|Y - Z\|_2 = \inf\{\|Y - A\|_2, \quad A \in [X]\};$$

Z is thus the projection of Y on $[X]$. From which

$$<Y - \beta - \sum_{i=1}^{p} \alpha_i X^i, 1> = E(Y) - \beta - \sum_{i=1}^{p} \alpha_i E(X^i) = 0$$

$$<Y - \beta - \sum_{i=1}^{p} \alpha_i X^i, X^j> = <Y - E(Y) - \sum_{i=1}^{p} \alpha_i (X^i - E(X^i)), X^j>$$

$$= <Y - E(Y) - \sum_{i=1}^{p} \alpha_i (X^i - E(X^i)),\ X^j - E(X^j)> = 0,$$

$$(j = 1, \ldots, p).$$

Let Γ be the covariance matrix of $\{X^1, \ldots, X^p\}$; we thus obtain

$$\begin{cases} \beta = E(Y) - \sum_{i=1}^{p} \alpha_i E(X^i) \\ \begin{bmatrix} \Gamma(Y, X^1) \\ \vdots \\ \Gamma(Y, X^p) \end{bmatrix} = \Gamma_X \begin{bmatrix} \alpha_1 \\ \vdots \\ \alpha_p \end{bmatrix}. \end{cases}$$

The only interesting case is that where Γ_X is invertible, which holds if no affine combination of the X is zero a.s. (From

Theorem 5.2.8, if Γ_X is not invertible, then X, with mean zero, takes its values a.s. in a subspace of dimension strictly less than p).

We have

$$[\alpha_1, ..., \alpha_p] = [\Gamma(Y,X^1), ..., \Gamma(Y,X^p)]\Gamma_X^{-1}.$$

Let then $Z = \beta + \Sigma_{i=1}^p \alpha_i X^i$; $Y - Z$ is uncorrelated with $(X^1, ..., X^p)$ and

$$\begin{aligned} E[(Y - Z)^2] &= E[Y - E(Y)]^2 \\ &\quad - 2E\left[(Y - E(Y))\left[\sum_{i=1}^{p} \alpha_i(X^i - E(X^i))\right]\right] \\ &\quad + E\left[\sum_{i=1}^{p} \alpha_i(X^i - E(X^i))\right] \\ &= \sigma^2(Y) - 2[\alpha_1, ..., \alpha_p]\begin{bmatrix}\Gamma(Y,X^1)\\ \vdots \\ \Gamma(Y,X^p)\end{bmatrix} \\ &\quad + [\alpha_1, ..., \alpha_p]\Gamma_X\begin{bmatrix}\alpha_1\\ \vdots \\ \alpha_p\end{bmatrix} \\ &= \sigma^2(Y) - [\Gamma(X^1,Y), ... \Gamma(X^p,Y)]\Gamma_X^{-1}\begin{bmatrix}\Gamma(X^1,Y)\\ \vdots \\ \Gamma(X^p,Y)\end{bmatrix}. \end{aligned}$$

$E(Y - Z)^2$ represents the *measurement error* or **residual** in the least squares problem. It gives the quality of the approximation.

5.5.2. The Gaussian Case

Assume that $(Y,X^1, ..., X^p)$ is a Gaussian vector; then $Y - Z$ and $(X^1, ..., X^p)$ being uncorrelated is equivalent to independence. Let f be an r.v. on $\mathbb{R}^p$ such that $f(X^1, ..., X^p)$ is square integrable;

$$E([Y - Z - f(X^1, ..., X^p)]^2)$$
$$= E(Y - Z)^2 - 2E[(Y - Z)f(X^1, ..., X^p)]$$
$$+ E[f^2(X^1, ..., X^p)]$$
$$E[(Y - Z)f(X^1, ..., X^p)] = E(Y - Z)E[f(X^1, ..., X^p)] = 0.$$

Thus

$$E([Y - Z - f(X^1, ..., X^p)]^2) \geqslant E(Y - Z)^2.$$

Proposition 5.5.14. *For a Gaussian vector* $(Y,X^1, ..., X^p)$ *the best approximation of* Y *by a function* $f(X^1, ..., X^p)$ *in the least squares sense is the affine function* $E(Y) + \Sigma_{i=1}^p \alpha_i(X^i - E(X^i))$ *calculated above. The difference between* Y *and this approximation is independent of* $(X^1, ..., X^p)$.

5.5.3. Estimation of a Regression

An n-sample is observed from $(Y, X^1, ..., X^p)$, let it be $(Y_i, X_i^1, ..., X_i^p)_{1 \leqslant i \leqslant n}$. If we want to calculate the best approximation, in $\mathbb{R}^n$, of $Y = (Y_1, ..., Y_n)$ by affine combinations of $X^j = (X_1^j, ..., X_n^j)$, $j = 1, ..., p$, we can go back to the preceding case by interpreting the observations as r.v.'s $(Y,X^1, ..., X^p)$ having as distribution their sample distribution. The affine approximation in $\mathbb{R}^n$ consists of looking for $\overline{\alpha} = (\overline{\alpha}_1, ..., \overline{\alpha}_p)$ and $\overline{\beta}$ which minimizes the function

$$(\alpha,\beta) \longmapsto \sum_{i=1}^n \left[Y_i - \beta - \sum_j \alpha_j X_i^j \right]^2.$$

This is obtained by replacing in the preceding formulae the distributions of the r.v.'s by their sample distributions. For $1 \leqslant j \leqslant p$, let $X_.^j$ be the sample mean of X^j and

$$\mathcal{X} = \begin{bmatrix} X_1^1 - X_.^1 & \cdots & X_1^p - X_.^p \\ \cdots & & \cdots \\ X_n^1 - X_.^1 & \cdots & X_n^p - X_.^p \end{bmatrix}.$$

The empirical estimation of the covariance matrix of $(X^1, ..., X^p)$ is then $(1/n)^t \mathcal{X}\mathcal{X}$. The sample covariances $\Gamma_n(X^j,Y)$ of X^j $(1 \leqslant j \leqslant p)$ and Y are calculated by

$$\frac{1}{n}\,{}^{t}X\begin{bmatrix} Y_1 - Y_. \\ \cdots \\ Y_n - Y_. \end{bmatrix} = \begin{bmatrix} \overline{\Gamma}_n(X^1,Y) \\ \cdots \\ \overline{\Gamma}_n(X^p,Y) \end{bmatrix}.$$

If the X^j are affinely independent, the empirical estimators of the affine regression for an n-sample are found:

$$[\overline{\alpha}_1, ..., \overline{\alpha}_p] = [Y_1 - Y_., ..., Y_n - Y_.]X({}^tXX)^{-1}$$

$$\overline{\beta} = Y_. - \sum_{i=1}^{p} \overline{\alpha}_i X^i_. .$$

5.5.4. The Linear Model and Regression on Controlled Factors

Consider the emission of a signal with coordinates x_i^1, x_i^2, x_i^3 transformed into a numerical signal $f(x_i^1,x_i^2,x_i^3)$ by a device. After reception on a noisy line, $Y_i = f(x_i^1,x_i^2,x_i^3) + \xi_i$ is received, where i is the index representing the ith reception and ξ_i the noise, an r.v. with mean zero and distribution independent of i. Likewise consider different doses of fertilizer, in n plots, in order to discover the optimal dose. If we test the dose x_i on the ith plot, we can represent the measured yield in the form $Y_i = \beta + \alpha_1 x_i + \alpha_2(x_i)^2 + \xi_i$, where ξ_i represents the variability (very complex) of the phenomenon assumed to have mean zero. In the two examples, we will have $E(Y_i) = f(x_i^1,x_i^2,x_i^3)$ or $E(Y_i) = \beta + \alpha_1 x_i + \alpha_2(x_i)^2$, $\sigma^2(Y_i) = \sigma^2(\xi_i) = \sigma^2$.

The **Gauss-Markov** or **linear models** are models which are linear with respect to the parameters occuring in the function f. This is the case in the second example above. The parameters x are said to be controlled, they are assumed fixed without error by the experimenter, and thus determine the design of the experiment. The sequence (ξ_i) is called white noise or error, as in Section 5.1.6.

The linear regression model is of the form $Y = \alpha_1 x_i^1 + ... + \alpha_p x_i^p + \beta + \xi_i$, *where, for* $1 \leqslant j \leqslant p$, $x^j = (x_i^j)_{1\leqslant i\leqslant n}$ *is a given point in* $\mathbb{R}^n$ *and where* $(\xi_i)_{1\leqslant i\leqslant n}$ *is a sequence of centered* r.v.'s, *with variance* σ^2 *and pairwise uncorrelated.*

In order to estimate α_1, ..., α_p and β in the least squares sense, it is sufficient to apply the formulae obtained in the empirical case where $(X_i^1, ..., X_i^p)$ was an n-sample of an r.v., since, as we have seen, these empirical formulae correspond to the least squares problem in $\mathbb{R}^n$. They apply by replacing the random X by the controlled x. We are led to use $\overline{m} = \Sigma_{j=1}^{p}\overline{\alpha}_j x^j$ to estimate $m = \Sigma_{j=1}^{p}\alpha_j x^j$: this is an unbiased estimator.

In the case where the r.v. $(\xi_i)_{1\leq i\leq n}$ form an n-sample from $N(0,\sigma^2)$, the problem covered here is an analysis of variance problem (Sections 5.1.6-7). The mean is in the space generated by $(1, x^1, ..., x^p)$ assumed to be of dimension $p + 1$. The projection of Y on this space is $\overline{m} = \Sigma_{j=1}^{p}\overline{\alpha}_j x^j$. This is the unbiased estimator obtained in Section 5.1.6.

Exercises 5.5

E.1. Consider the regression model of Section 5.5.4:

$$Y_i = \alpha + \beta x_i + \xi_i \quad (1 \leq i \leq n),$$

where $(x_i)_{1\leq i\leq n}$ is given in $\mathbb{R}^n$ and where the r.v.'s $(\xi_i)_{1\leq i\leq n}$ are independent with distribution $N(0,\sigma^2)$. Let $\overline{\alpha}$ and $\overline{\beta}$ be the empirical estimators of α and β, and let S^2 be the residual variance. Give the covariance matrix and distribution of $(\overline{\alpha},\overline{\beta},S^2)$. State confidence intervals of a given level for $\overline{\alpha}$ and $\overline{\beta}$. Deduce from this a confidence region of given level for the regression line of Y with respect to X.

E.2. With the data below (p. 238), consider the following regression models:

(1) $$Y_k = b_1 + b_2 \log x_k + \varepsilon_k$$

(2) $$Y_k = b_1 + b_2 x_k + \varepsilon_k$$

where the ε_k are centered, independent, Gaussian variables with variance σ^2.

(a) Discuss the respective value of the two models. Which should we choose?
(b) Consider, for the same data, the model $Y_k = b_1 + b_2 x_k + b_3 x_k^2 + \varepsilon_k$.

Test "$b_3 = 0$" against "$b_3 \neq 0$" at level 0.05. Then compare this model with the first two; the estimation of σ^2 can serve as a criterion.

x	Y
1	0.39
2	1.06
3	0.89
4	1.15
5	1.56
6	1.77
7	0.94
8	0.98
9	1.90
10	1.59
11	1.26
12	1.68
13	1.25
14	1.80
15	1.77
16	1.72

E.3. Interpret the maximum likelihood estimator of θ in the AR1 model (Exercise 5.3.2) as a least squares estimator analogous to that in the linear model.

E.4. *Covariance matrix.* Let $Y_1, \dots, Y_p$ be p square integrable r.v.'s, let Γ be their covariance matrix, and let Γ_1 be that of $(Y_1, \dots, Y_{p-1})$. Denote

$$\Gamma = \begin{bmatrix} \Gamma_1 & \Gamma_{12} \\ {}^t\Gamma_{12} & \sigma_p^2 \end{bmatrix},$$

by isolating the last row and column and let $|\Gamma|$ be the determinant of Γ.

(a) Let Z be the best approximation of Y_p by an affine combination of the other components. Prove, if $|\Gamma_1|$ is nonzero, $|\Gamma| = |\Gamma_1| (\sigma_p^2 - {}^t\Gamma_{12}\Gamma_1^{-1}\Gamma_{12})$ and $E(Z - Y_p)^2 = |\Gamma|/|\Gamma_1|$.

(b) Prove that, if Γ is of rank r, the centered vector $Y - E(Y)$ is a.s. in a subspace of dimension r.

(c) Let $y_i = (y_{i1}, \dots, y_{in})$ for $1 \leqslant i \leqslant k$, be k vectors in $\mathbb{R}^n$;

$$y_{\cdot} = (y_{1\cdot}, \dots, y_{k\cdot}) = \left(\frac{1}{n} \sum_{q=1}^{n} y_{iq}\right)_{1 \leqslant i \leqslant k}$$

and

$$S(y_1, ..., y_k) = \left\{ \frac{1}{n} \sum_{q=1}^{n} (y_{iq} - y_{i.})(y_{jq} - y_{j.}) \right\}_{1 \leqslant i,j \leqslant k}$$

be the *associated sample covariance matrix*. Prove that $S(y_1, ..., y_k)$ is invertible if, and only if, $\{\mathbf{1}_n, y_1, ..., y_k\}$ is a free family of vectors in $\mathbb{R}^n$, thus never for $k \geqslant n$.

Let

$$V(y_1, ..., y_k) = \left\{ \frac{1}{n} \sum_{q=1}^{n} y_{iq} y_{jq} \right\}_{1 \leqslant i,j \leqslant k}.$$

Prove that, V is invertible if, and only if, $\{y_1, ..., y_k\}$ is a free family (never for $k > n$).

(d) Let Y be a random vector of dimension n having a density. Let $y_1, ..., y_k$ be vectors in $\mathbb{R}^n$, which form, with $\mathbf{1}_n$, a free system. For $k < n - 1$, show that $S(y_1, ..., y_k, Y)$ is a.s. invertible. Let $\{y_1, ..., y_k\}$ be a free family in $\mathbb{R}^n$; for $k < n$, show that $V(y_1, ..., y_k, Y)$ is a.s. invertible.

(e) Let $Y_1, ..., Y_k$ be k random vectors of dimension n, $(Y_1, ..., Y_k)$ having a density. Show that $S(Y_1, ..., Y_k)$ is a.s. invertible for $k < n$ and that $V(Y_1, ..., Y_k)$ is a.s. invertible for $k \leqslant n$.

E.5. *Wishart's distribution.* Let M_k^+ be the set of $k \times k$ positive matrices of rank k. Let $m \in \mathbb{R}^k$ and $\Gamma \in M_k^+$. Consider $(X_1, ..., X_n)$ an n-sample from $N_k(m,\Gamma)$, $X_q = (X_{1q}, ..., X_{kq})$ for $1 \leqslant q \leqslant n$ and $\mathcal{X}$ the $n \times k$ matrix the row vectors of which are the vectors $(X_q)_{1 \leqslant q \leqslant n}$.

The sample mean is denoted by $\bar{X} = (X_{1.}, ..., X_{k.})$. Denote by $\bar{\mathcal{X}}$ the $n \times k$ matrix, the column vectors of which are $X_{i.} \mathbf{1}_n$, $1 \leqslant i \leqslant k$. The sample covariance matrix is

$$S = \frac{1}{n}\, {}^t(\mathcal{X} - \bar{\mathcal{X}})(\mathcal{X} - \bar{\mathcal{X}})$$

(verify this). Set also $V = {}^t\mathcal{X}\mathcal{X}$.

(a) Show that $|S|$ is zero for $k \geqslant n$ and that S is a.s. in M_k^+ for $k < n$; that $|V|$ is zero for $k > n$ and that V is a.s. in M_k^+ for $k \geqslant n$ (use Exercise 5.5.4).

(b) Assume $m = 0$; the distribution of V is called the Wishart distribution on n degrees of freedom with parameter Γ. Verify $E(V) = \{E(V_{ij})\}_{1 \leqslant i,j \leqslant k} = n\Gamma$.

(c)_For arbitrary m, what is the distribution of $\overline{X}$? Show that $\overline{X}$ and S are independent and that nS follows a Wishart distribution on n degrees of freedom with parameter Γ.

(d) Let $x \in (\mathbb{R}^k)^n$, $x = (x_1, ..., x_n)$ with, for $1 \leqslant q \leqslant n$, $x_q = (x_{1q}, ..., x_{kq})$. Set

$$\overline{x} = \frac{1}{n} \sum_{q=1}^{n} x_q = (x_{1.}, ..., x_{k.})$$

and

$$s(x) = \frac{1}{n} \left\{ \sum_{q=1}^{n} (x_{iq} - x_{i.})(x_{jq} - x_{j.}) \right\}_{1 \leqslant i,j \leqslant k}.$$

Show that the density of $(X_1, ..., X_n)$ may be written

$$f(x_1, ..., x_n) = \frac{1}{((2\pi)^{k/2}|\Gamma|)^n} \exp\left[-\frac{n}{2} \{ \mathrm{Tr}(\Gamma^{-1} s(x)) + {}^t(\overline{x} - m)\Gamma^{-1}(\overline{x} - m) \} \right].$$

Bibliographic Notes

Gaussian distributions are at the center of probability; see Breiman [1], Renyi, and Ventsez. The linear model is at the center of statistics. Snedecor introduces tests simply; Barra, Lehmann [1], and Rao give mathematical accounts. For various applications see Priece (agronomics); Haberman (medicine); Malinvaud and Theil (econometrics); and Falconer (genetics). To go deeper into the study of tests see Cochran, Coursol, and Scheffe. To study samples of Gaussian vectors, see Anderson and Coursol.

Chapter 6
CONDITIONAL EXPECTATION, MARKOV CHAINS, INFORMATION

Objectives

It is *a priori* natural to approximate an unobservable r.v. by a measurable function of the observations: this is more general than by affine combinations. It is this which leads to conditional expectation. This allows a step by step evolution to be envisaged by a "Markov chain." The distribution of the state at any time, conditional on the past, only depends on the previous state.

If an observation is to be used to provide statistics on an unknown distribution F we are interested in the information which this observation (or its distribution) carries on the distribution F. Various measures of this information are proposed.

6.1. Approximations in the Least Squares Sense by Functions of an Observation

A probability space $(\Omega,\mathcal{A},P)$ is assumed given throughout.

Assume given an observation X, taking values in $(E,\mathcal{E})$, from distribution F_X. For $E = \mathbb{R}^p$, we have studied the problem of multiple regression or approximation of an r.v. by a linear or affine combination of the components of X in the least squares sense. It is natural to extend this and to look for an r.v. f on E, such that $f(X)$ is the closest to Y in the least squares sense. We have seen one case (Proposition 5.5.14) where these two approximations coincide. Assume Y to be square integrable and X to be independent of Y. For all f we have

$$E[Y - E(Y) - f(X)]^2 = \sigma^2(Y) + E[f^2(X)].$$

Thus, if Y is independent of X, no function of X approximates Y better than its expectation.

We have seen (Section 3.3.1) that, for $\mathcal{O} = \sigma(X)$, saying $f \in L^2(E,\mathcal{E},F_X)$ is equivalent to $f(X) \in L^2(\Omega,\mathcal{O},P)$. The problem can thus be reformuled by giving $\mathcal{O}$, a sub σ-algebra of $\mathcal{A}$, and by approximating Y by an r.v. from $L^2(\Omega,\mathcal{O},P)$. $\mathcal{O}$ is to be interpreted as the set of observable events. We will write $L^2(\mathcal{O})$ instead of $L^2(\Omega,\mathcal{O},P)$ and $L^2(\mathcal{A})$ instead of $L^2(\Omega,\mathcal{A},P)$.

Let $\mathcal{N}$ be the set of events of measure zero of $\mathcal{A}$. The observation will be assumed to be described by knowledge of a σ-algebra $\mathcal{O}$ such that $\mathcal{N} \subset \mathcal{O} \subset \mathcal{A}$. Then $L^2(\mathcal{O})$ is a Hilbert subspace of $L^2(\mathcal{A})$. Let $Y \in L^2(\mathcal{A})$. The observable, square integrable random variable Z, i.e. in $L^2(\mathcal{O})$, the closest to Y in the least squares sense, is the projection of Y on $L^2(\mathcal{O})$.

Definition 6.1.1. For a σ-algebra $\mathcal{O}$ satisfying $\mathcal{N} \subset \mathcal{O} \subset \mathcal{A}$, and for $Y \in L^2(\mathcal{A})$, the **expectation of** Y **conditional on** $\mathcal{O}$ is the orthogonal projection of Y on $L^2(\mathcal{O})$ (thus the observable r.v. closest to Y in the least squares sense). This is denoted by $E(Y|\mathcal{O})$ or $E^{\mathcal{O}}(Y)$.

For a given sub σ-algebra $\mathcal{O}$ of $\mathcal{A}$, consider its P-**completion in** $\mathcal{A}$, the σ-algebra generated by $\mathcal{O}$ and $\mathcal{N}$ (as usual, we work up to sets with measure zero), and $E(Y|\mathcal{O})$ is defined as the conditional expectation of Y with respect to its completion in $\mathcal{A}$.

Notes. (a) *The P-completion of* $\mathcal{O}$ *in* $\mathcal{A}$ *is contained in the P-completion of* $\mathcal{O}$ *defined in Exercise* 1.1.19.

(b) *A conditional expectation is an* r.v. *defined* a.s., *and all the relations in which it figures are relations between equivalence classes of* r.v.'s (*we will not state* "a.s.").

Theorem 6.1.2. *Let* $\mathcal{O}$ *be a sub* σ*-algebra of* $\mathcal{A}$ *and let* $Y \in L^2(\Omega,\mathcal{A},P)$.

(a) $\|E(Y|\mathcal{O})\|_2 \leqslant \|Y\|_2$
(b) *$Y \longmapsto E(Y|\mathcal{O})$ is linear from $L^2(\mathcal{A})$ into $L^2(\mathcal{O})$.*
(c) *For $Y \in L^2(\mathcal{O})$ (in particular for Y constant), $E(Y|\mathcal{O})$ coincides with Y.*
(d) *For all $Z \in L^2(\mathcal{O})$, $E[YZ] = E[E(Y|\mathcal{O})Z]$.*
(e) *$E[Y - E(Y|\mathcal{O})]^2 = \inf\{E(Y - Z)^2;\ Z \in L^2(\mathcal{O})\}$.*
(f) *$Y \longmapsto E(Y|\mathcal{O})$ is an increasing function from $L^2(\mathcal{A})$ into $L^2(\mathcal{O})$.*
(g) *For $\mathcal{O}' \subset \mathcal{O}$, (nested σ-algebras), we have*

$$E[Y|\mathcal{O}'] = E[E(Y|\mathcal{O})|\mathcal{O}'].$$

(h) *For Y independent of $\mathcal{O}$ ($\sigma(Y)$ and $\mathcal{O}$ are independent): $E(Y|\mathcal{O}) = E(Y)$.*

Proof. All of the above (except (h)) follow from the standard properties of projections in a Hilbert space. Property (d) signifies that $Y - E(Y|\mathcal{O})$ is orthogonal to $L^2(\mathcal{O})$. It is characteristic of the projection, hence characterizes $E(Y|\mathcal{O})$. Property (e) is the property of best approximation in the least squares sense. Let us prove (f). Assuming Y is positive, for all $Z \in L^2(\mathcal{O})$ we have $(Y - \sup(Z,0))^2 \leqslant (Y - Z)^2$. From (e) this implies that $E(Y|\mathcal{O})$ is positive. By linearity we deduce from this that $Y \leqslant Y'$ implies

$$E(Y' - Y|\mathcal{O}) = E(Y'|\mathcal{O}) - E(Y|\mathcal{O}) \geqslant 0.$$

To show (h) we can write for $Z \in L^2(\mathcal{O})$ and Y independent of $\mathcal{O}$,

$$E[(Y - Z)^2] = E[(Y - E(Y) - Z + E(Y))^2]$$

$$= \sigma^2(Y) + E[(Z - E(Y))^2] \geqslant \sigma^2(Y)$$

and use property (e).

Translation 6.1.3. *Let X be a measurable function from $(\Omega,\mathcal{A},P)$ into $(E,\mathcal{E})$ with distribution F_X. The expectation of $Y \in L^2(\mathcal{A})$ conditional on X is $E(Y|\sigma(X))$, denoted $E(Y|X)$. There then exists a function f of $L^2(E,\mathcal{E},F_X)$ such that $f(X)$ is equal P-a.s. to $E(Y|X)$. It is the function of X closest to Y in the least squares sense.*

The function f is denoted by $x \longmapsto E(Y|X = x)$ and we speak of the **expectation of** Y **when** X **equals** x ..., but N.B.! *We are dealing with a function defined up to* F_X *equivalence.* We are going to see ways of finding a version of it, but for a fixed x the expression $E(Y|X = x)$ only makes sense if F_X charges x.

Example 1. For $Y = f(X)$, we find (fortunately!) for F_X almost all x,

$$E(Y|X = x) = f(x).$$

Example 2. Let A be an atom of (Exercise 3.2.13). $E(Y|\mathcal{O})$ being $\mathcal{O}$ measurable is a.s. constant on A. This constant a is determined by applying property (d) of Theorem 6.1.2 to $\mathbf{1}_A$:

$$E(Y\mathbf{1}_A) = E(Y|A)P(A) = aP(A).$$

On the atom A of $\mathcal{O}$, $E(Y|\mathcal{O}) = E(Y|A)$. This is the idea we met in Section 4.2.1. Hence, if $\mathcal{O}$ is generated by a partition $(A_1, \ldots, A_p, N)$, with $P(A_i) > 0$ for $1 \leqslant i \leqslant p$ and N of measure zero, we have

$$E(Y|\mathcal{O}) = \sum_{i=1}^{p} E(Y|A_i)\mathbf{1}_{A_i}.$$

If X is an observation taking values in $\{1,2, \ldots, p\}$ with distribution F_X, we obtain

$$E(Y|X) = \sum_{i=1}^{p} E(Y|\{X = i\})\mathbf{1}_{(X=i)}$$

and

$$E(Y|X = i) = E(Y|\{X = i\}).$$

Generally, *when* X *is an arbitrary observation the distribution* F_X *of which charges the point* x, *we have*

$$E(Y|X = x) = \frac{E\{Y\mathbf{1}_{(X=x)}\}}{P(X = x)}.$$

Example 3. We are given X and X', two observations taking values in $(E,\mathcal{E})$ and $(E',\mathcal{E}')$. For λ and λ' σ-finite measures on $(E,\mathcal{E})$ and $(E',\mathcal{E}')$ and f an r.v. on $(E \times E', \mathcal{E} \otimes \mathcal{E}')$, assume that (X,X') has distribution $f \cdot \lambda \otimes \lambda' = F$. The marginal distributions are $F_X = f_X \lambda$ and $F_{X'} = f_{X'} \lambda'$ with

$$f_X(\cdot) = \int f(\cdot, y) \lambda'(dy), \quad f_{X'}(\cdot) = \int f(x, \cdot) \lambda(dx).$$

Let

$$\phi \in L^2(E, \mathcal{E}, F_X), \quad \psi \in L^2(E', \mathcal{E}', F_{X'}):$$

$$\begin{aligned} E[\phi(X)\psi(X')] &= \iint \phi(x)\psi(y) f(x,y) \lambda(dx) \lambda'(dy) \\ &= \int \phi(x) f_X(x) d\lambda(x) \left[\int \psi(y) \frac{f(x,y)}{f_X(x)} d\lambda'(y) \right] \\ &= E\left[\phi(X) \int \psi(y) \frac{f(X,y)}{f_X(X)} d\lambda'(y) \right]. \end{aligned}$$

There is no problem in the above for dividing by f_X, which is F_X-a.s. strictly positive. This calculation shows that as soon as a version f of the density of the distribution of (X,X') with respect to $\lambda \otimes \lambda'$ has been chosen, we have, for every $\psi \in L^2(E', \mathcal{E}', F_{X'})$, a version of $E(\psi(X')|X = \cdot)$ by taking the function

$$x \longmapsto \int \psi(y) \frac{f(x,y)}{f_X(x)} d\lambda'(y).$$

Definition 6.1.4. For a pair (X,X') with distribution $f \cdot \lambda \otimes \lambda'$, having chosen a version f of the density, we say that the **density of** X' **conditional on** $(X = x)$ is the function $f_{X'}(\cdot|X = x)$, identical to 0 on the F_X-negligible set $\{x; f_X(x) = 0\}$, and equal, for $f_X(x) \neq 0$, to $f(x, \cdot)/f_X(x)$. The probability defined on $(E', \mathcal{E}')$ which has $f_{X'}(\cdot|X = x)$ for density is the **distribution of** X' **conditional on** $(X = x)$. We have

$$\begin{aligned} P[X' \in \Gamma | X = x] &= \int_\Gamma f_{X'}(y|X = x) d\lambda'(y) \\ &= E[1_\Gamma(X')|X = x]. \end{aligned}$$

Note. *We speak of "the" density or "the" distribution of X' conditional on $(X = x)$, whereas these notions depend on the choice of the version f of the density of (X,X'). In most cases the choice of f is clear and there is no ambiguity.*

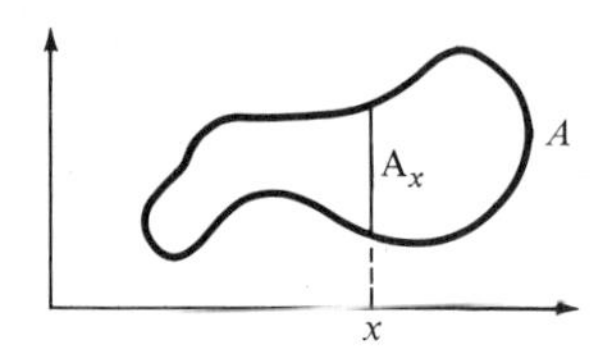

Example 4. Let (X,X') be *a pair of* r.v.'s *having a uniform distribution on a bounded Borel set A.* For $x \in \mathbb{R}$, $A_x = \{y; (x,y) \in A\}$ is the section of A by x. The density of the pair (X,X') is proportional to $\mathbf{1}_A$ and the distribution of X' conditional on $\{X = x\}$ is the uniform distribution on A_x, which corresponds to intuition.

Example 5. Let X and X' be *two independent observations* taking values in $(E,\mathcal{E})$ and $(E',\mathcal{E}')$. Let ϕ be an r.v. on $(E \times E', \mathcal{E} \otimes \mathcal{E}')$ such that $\phi(X,X')$ is square integrable. Then (a.s. for the distribution of X),

$$E[\phi(X,X')|X = x] = E[\phi(x,X')].$$

This is easily shown for ϕ the indicator function of a rectangle, then tricks 3.1.11 and 3.1.13 are used.

Example 6. Let $(Y,X_1, ..., X_p)$ *be a Gaussian vector.* We have seen (Theorem 5.4.14) that $E(Y|X_1, ..., X_p)$ is the affine approximation $E(Y) + \Sigma_{i=1}^p \alpha_i(X_i - E(X_i)) = Z$, and $Y - Z$ is independent of $X_1, ..., X_p$. Let ϕ be a bounded r.v. on $\mathbb{R}$:

$$\begin{aligned} E[\phi(Y)|X_1, ..., X_p) &= (x_1, ..., x_p)] \\ &= E[\phi(Y - Z + Z)|(X_1, ..., X_p) = (x_1, ..., x_p)] \\ &= E\left[\phi(Y - Z + \sum_{i=1}^p \alpha_i(x_i - E(X_i)) + E(Y))\right]. \end{aligned}$$

Now $Y - Z + \Sigma_{i=1}^p \alpha_i(x_i - E(X_i)) + E(Y)$ has a Gaussian distribution with mean $\Sigma_{i=1}^p \alpha_i(x_i - E(X_i)) + E(Y)$, and its variance is

$$\begin{aligned} E(Y - Z)^2 &= \sigma^2(Y) - \sigma^2(Z) \\ &= \sigma^2(Y) - [\Gamma(X_1,Y), ..., \Gamma(X_p,Y)]\Gamma^{-1}\begin{bmatrix} \Gamma(X_1,Y) \\ \vdots \\ \Gamma(X_p,Y) \end{bmatrix} \end{aligned}$$

where Γ is the covariance matrix of $(X_1, \ldots, X_p)$.

Proposition 6.1.5. *For $(Y, X_1, \ldots, X_p)$ a Gaussian vector, the best affine approximation of Y by $(X_1, \ldots, X_p)$ coincides with its conditional expectation*

$$E(Y|X_1, \ldots, X_p) = \sum_{i=1}^{p} \alpha_i[X_i - E(X_i)] + E(Y).$$

The distribution of Y conditional on $(X_1, \ldots, X_p) = (x_1, \ldots, x_p)$ is a Gaussian distribution with mean $\Sigma_{i=1}^{p}\alpha_i(x_i - E(X_i)) + E(Y)$ and with variance independent of $(x_1, \ldots, x_p)$.

A New Interpretation of the Linear Model 5.5.4. In the above framework, an n-sample $(Y, X^1, \ldots, X^p)$ is observed, denoted $(Y_i, X_i^1, \ldots, X_i^p)_{1 \leq i \leq n}$ conditional on $(X_i^1, \ldots, X_i^p) = (x_i^1, \ldots, x_i^p)$, we observe $Y_i = \Sigma_{j=1}^{p}\alpha_j x_i^j + \beta\varepsilon_i$, where ε_i is a centered Gaussian r.v. with variance σ^2. This allows us, in the Gaussian case, to bring together the points of view of Sections 5.5.4 and of 5.5.1 and 5.5.2.

Exercises 6.1

E.1. For all the pairs of r.v.'s (X,Y) of Exercise 3.3.7, determine the distribution and expectation of X conditional on $(Y = y)$ or that of Y conditional on $(X = x)$. What is the regression line of Y on X?

E.2. Let $(X_1, \ldots, X_n)$ be an n-sample from the uniform distribution on $[a,b]$ and $(X_{(1)}, \ldots, X_{(n)})$ its order statistic. Show that the distribution of $(X_{(2)}, \ldots, X_{(n-1)})$ conditional on $(X_{(1)}, X_{(n)}) = (\alpha, \beta)$ is that of an $(n-2)$ sample from a uniform distribution on $[\alpha, \beta]$. Calculate the expectations of X_1 and X_1^2, and the sample mean and variance $\overline{X}$ and S^2 conditional on $(X_{(1)}, X_{(n)})$.

E.3. Let X and Y be two random variables on $(\Omega, \mathcal{A}, P)$; assume X takes values in $\mathbb{N}$, and Y, taking real values, is exponentially distributed with parameter 1. Finally, assume that the distribution of X conditional on $(Y = \lambda)$ is a Poisson distribution with parameter λ. Determine the density of Y, the distribution $F_{X,Y}$ of (X,Y), the marginal distribution F_X of X, and the distribution of Y conditional on $(X = k)$.

E.4. Let (T_1,T_2) be a pair of r.v.'s having the bivariate exponential distribution defined in Exercise 4.2.4. What is the distribution of T_2 conditional on $T_1 = t_1$? Calculate $E(T_2|T_1 = t_1)$.

E.5. Consider a pair of Gaussian r.v.'s, the density f of which is defined (for σ_1, σ_2 constants > 0, and $\rho \in]-1,+1[$) by

$$f(x,y) = \frac{1}{2\pi\sqrt{1-\rho^2}\,\sigma_1\sigma_2} \exp\left[-\frac{1}{2(1-\rho^2)}\left(\frac{x^2}{\sigma_1^2} - 2\rho\frac{xy}{\sigma_1\sigma_2} + \frac{y^2}{\sigma_2^2}\right)\right].$$

Calulate directly the marginal distributions, the means of X and Y, the correlation coefficient of (X,Y), and the distribution of X conditional on $Y = y$. Hence obtain the general result derived for Gaussian vectors.

E.6. *Bayes formula with densities.* In the framework of Example 3 of Section 6.1, prove that for almost all (x,y),

$$f_{X'}(y|X=x) = \frac{f_{X'}(x)f_{X'}(y|X=x)}{\int f_X(u)f_{X'}(y|X=u)\lambda(du)}.$$

Compare with Exercise 1.3.4.

E.7. Let $Y \in L^2(\mathcal{A})$. *The variance of Y conditional on a sub σ-algebra $\mathcal{O}$ of $\mathcal{A}$,* is by definition

$$\sigma^2(Y|\mathcal{O}) = E(Y^2|\mathcal{O}) - E(Y|\mathcal{O})^2.$$

Verify

$$\sigma^2(Y) = E[\sigma^2(Y|\mathcal{O})] + \sigma^2[E(Y|\mathcal{O})].$$

E.8. *The Beta binomial distribution.* (a) Let $a > 0$, $b > 0$, and n be an integer ≥ 1. Consider a pair of r.v.'s (X,Y), Y having a beta distribution with parameters a and b (Exercise 3.3.14), and such that for all $y \in]0,1[$ the distribution of X conditional on $(Y = y)$ is binomial $b(y,n)$. What is the distribution (beta binomial with parameters (a,b,n)) of X? Calculate its mean and variance.

(b) Let $1 \leqslant n \leqslant N$ be two integers and (U,V) a pair of r.v.'s. Assume that U has a beta binomial distribution with parameters (a,b,N) and that, for every integer $\theta \in \{0,1, ..., N\}$, the distribution of V conditional on $(U = \theta)$ is the hypergeometric distribution F_θ, defined on $\{0,1, ..., n\}$ by

$$F_\theta(k) = \frac{\binom{\theta}{k}\binom{N-\theta}{n-k}}{\binom{N}{n}}.$$

What is the distribution of V? What is the distribution of $U - x$ conditional on $V = x$ $(x = 0, ..., n)$? Prove

$$E[U|V = x] = \frac{(N + a + b)x + a(N - n)}{n + a + b}.$$

E.9. *Order statistics.* Let $X = (X_1, ..., X_n)$ be a random vector of dimension n, the distribution of which has density f. What is the distribution of its order statistic $(X_{(1)}, ..., X_{(n)})$? What is the distribution of R, the rank vector of X, conditional on $(X_{(1)}, ..., X_{(n)})$? If X is the n-sample from a distribution on $\mathbb{R}$, rederive the independence of R and its order statistics (Theorem 4.4.29).

E.10. *Association for a Gaussian vector.* Show that a necessary and sufficient condition that a Gaussian vector $(X_1, ..., X_n)$ is associated in the sense of Exercise 4.1.4 is that, for $1 \leqslant i,j \leqslant n$, $\Gamma(X_i,X_j) \geqslant 0$ and $\sigma^2(X_i)\Gamma(X_j,X_k) \geqslant \Gamma(X_i,X_j)\Gamma(X_i,X_k)$.

E.11. *Coefficients of association.* (a) Let (X,Y) be a pair of square integrable r.v.'s the marginal distributions of which are continuous. Let $((X_1,Y_1), (X_2,Y_2))$ be a 2-sample from this distribution. We define the *Kendall coefficient of association* by

$$\tau(X,Y) = 2P[(X_1 - X_2)(Y_1 - Y_2) > 0] - 1.$$

Verify

$$P[(X_1 - X_2)(Y_1 - Y_2) = 0] = 0.$$

Show

(1) $-1 \leqslant \tau(X,Y) \leqslant 1$;
(2) $\tau(X,Y) = \tau(Y,X)$ and $\tau(X,-Y) = -\,\tau(X,Y)$;
(3) if X and Y are independent, $\tau(X,Y)$ is zero;
(4) $\tau(X,Y) = 1$ signifies that there exists an increasing function f, from $\mathbb{R}$ into $\mathbb{R}$, for which $Y = f(X)$; and $\tau(X,Y) = -1$ signifies that there exists a decreasing f, for which $Y = f(X)$. Compare these four properties with those of the correlation coefficient $\rho(X,Y)$. For a Gaussian vector (X,Y), show $\tau = 2/\pi$ Arcsin ρ (see Exercises 4.2.13 and 5.2.11).

(b) Consider $(X_i,Y_i)_{1\leqslant i\leqslant n}$ an n-sample of (X,Y); for $1 \leqslant i < j \leqslant n$ set

$$U_{ij} = \mathbf{1}_{(X_i>X_j)} - \mathbf{1}_{(X_i<X_j)} \quad \text{and} \quad V_{ij} = \mathbf{1}_{(Y_I>Y_j)} - \mathbf{1}_{(Y_i<Y_j)}\ ;$$

$$\overline{\tau} = \frac{\Sigma\ U_{ij}V_{ij}}{(\Sigma U_{ij}^2\ \Sigma V_{ij}^2)^2}$$

(the summations Σ are taken over $1 \leqslant i < j \leqslant n$). Show that $\overline{\tau}$ is an unbiased estimator of τ. Think up reasonable tests of the following hypotheses,

$$H_0\colon \{\tau(X,Y) = 0\} \quad \text{against} \quad H_1\colon \{\tau(X,Y) > 0\},$$

$$H_0'\colon \{\tau(X,Y) = 0\} \quad \text{against} \quad H_1'\colon \{\tau(X,Y) < 0\},$$

$$H_0''\colon \{\rho(X,Y) = 0\} \quad \text{against} \quad H_1''\colon \{\rho(X,Y) > 0\}.$$

Let $(R_1, ..., R_n)$ and $(S_1, ..., S_n)$ be the rank vector of $(X_1, ..., X_n)$ and $(Y_1, ..., Y_n)$, respectively. Associate with them the Spearman correlation (Exercise 1.2.4 and 1.3.11). For this correlation examine the analogue of properties (1)-(4) and use this to define reasonable tests.

E.12. Let $X = (X_1, ..., X_n)$ be an n-sample from $N(0,\sigma^2)$ and $y = (y_1, ..., y_n) \in \mathbb{R}^n$; $\overline{X}$ and $\overline{y}$, $\overline{X}\mathbf{1}_n$ and $\overline{y}\mathbf{1}_n$, respectively.

(a) Give the distributions of $<X,y>$, $\|X\|^2\|y\|^2 - (<X,y>)^2$, $<X - \overline{X}, y - \overline{y}>$, and of $\|X - \overline{X}\|^2\|y - \overline{y}\|^2 - <X - \overline{X}, y - \overline{y}>^2$. Let

$$R(X,y) = \frac{<X-\overline{X},\ y-\overline{y}>}{\|X-\overline{X}\|\ \|y-\overline{y}\|}$$

be the sample correlation. What is the distribution of

$$\sqrt{n-2}\,\frac{R(X,y)}{\sqrt{1-R^2(X,y)}} = T(X,y)?$$

(b) Consider $(X,Y) = (X_i,Y_i)_{1\leq i\leq n}$ an n-sample from a centered Gaussian distribution on $\mathbb{R}^2$. Let $\rho = \Gamma(X_1,Y_1)$. If ρ is zero, what is the distribution of $T(X,Y)$ conditional on $(Y = y)$? What is the distribution of $T(X,Y)$? Show that if ρ is nonzero $T(X,Y)$ tends in probability to $+\infty$ when n tends to ∞. Think up a suitable test of level α of the hypothesis H_0 "X and Y are uncorrelated."

(c) Let $(X_i,Y_i)_{1\leq i\leq n}$ be a sample from a centered distribution on $\mathbb{R}^2$. Discuss the validity for large samples of the test in (b) for no correlation between X and Y.

E.13. Let T be a random vector of dimension k in (Ω,A,P), from the distribution $F = f\cdot\lambda_1 \otimes \dots \otimes \lambda_k$ where $\lambda_1, \dots, \lambda_k$ are σ-finite measures on $\mathbb{R}$. Define

$$\Theta = \{\theta;\ E[\exp <\theta,T>] = \exp \psi(\theta) < \infty\},$$

and the associated exponential model by setting, for $\theta \in \Theta$, $P_\theta = \exp(-\psi(\theta) + <\theta,T>)P$ (Section 3.3.2).

(a) Verify that the distribution of T for P_θ has density $\exp[-\psi(\theta) + <\theta,\cdot>]$ with respect the distribution F.

(b) Let $r < k$ and $U = (T_1, \dots, T_r)$, $V = (T_{r+1}, \dots, T_k)$. Verify that the distributions of U for $\theta \in \Theta$ form an exponential family if $(\theta_{r+1}, \dots, \theta_n)$ is known.

(c) Consider in (Ω,A,P) the density f_v of U conditional on $(V = v)$ associated with the density f. Calculate as a function of f_v the density of U conditional on $(V = v)$ associated with the density $\exp[-\psi(\theta) + <\theta,\cdot>\}f$ of T on (Ω,A,P_θ). Deduce from this that a version of the distribution of U conditional on $(V = v)$, $\theta \in \Theta$, can be chosen which forms an exponential model for every v.

6.2. Conditional Expectation − Extensions

Definition 6.2. For a quasi-integrable r.v. Y on (Ω,A,P) and a sub σ-algebra of A, we call the **expectation of** Y **conditional on** $\mathcal{O}$, the $\mathcal{O}$-measurable and quasi-integrable r.v., unique for a.s. equality **and denoted $E(Y|\mathcal{O})$, such that for all $A \in \mathcal{O}$:**

$$E[Y\mathbf{1}_A] = E[E(Y|\mathcal{O})\mathbf{1}_A].$$

Justification. Denote by $L^0(\mathcal{A})$ or $L^0(\mathcal{O})$ the set of equivalence classes of quasi-integrable r.v.'s $\mathcal{A}$ or $\mathcal{O}$ measurable. The index + designates the set of these r.v.'s which are positive.

(a) Uniqueness is obtained by noticing that if Z and Z' are two elements of $L^0(\mathcal{O})$, the equality $E(Z\mathbf{1}_A) = E(Z'\mathbf{1}_A)$ for all $A \in \mathcal{O}$ implies that $E[(Z - Z')\mathbf{1}_{(N \geqslant Z > Z')}]$ and $E[(Z - Z')\mathbf{1}_{(Z < Z' \leqslant N)}]$ are zero for all $N \in \mathbb{N}$, thus Z and Z' are equal a.s.

(b) For $Y \in L^0_+(\mathcal{A})$, existence is obtained by noticing that $Y \wedge n = \inf(Y,n)$ is in $L^2(\mathcal{A})$. From the Beppo-Levi theorem for $A \in \mathcal{O}$,

$$E(Y\mathbf{1}_A) = \lim_n \uparrow E((Y \wedge n)\mathbf{1}_A) = \lim_n \uparrow E(E(Y \wedge n|\mathcal{O})\mathbf{1}_A).$$

However, $(E(Y \wedge n|\mathcal{O}))$ is an increasing sequence of positive r.v.'s. Its limit, denoted by $E(Y|\mathcal{O})$, is suitable. Since $E(Y)$ and $E(E(Y|\mathcal{O}))$ are equal, $E(Y|\mathcal{O})$ is in $L^1(\mathcal{O})$ if Y is in $L^1(\mathcal{A})$.

(c) For Y quasi-integrable, set $E(Y|\mathcal{O}) = E(Y_+|\mathcal{O}) - E(Y_-|\mathcal{O})$, which makes sense since $E(Y_+|\mathcal{O})$ or $E(Y_-|\mathcal{O})$ is integrable.

(d) For square integrable Y, this definition coincides with that of 6.1.1, due to the characteristic property of projection on $L^2(\mathcal{O})$, stated in (d) of Theorem 6.1.2.

The following properties are easily proved with the usual tricks of integration, by using Theorem 6.1.2.

Theorem 6.2.7. *Let Y be quasi-integrable on $(\Omega,\mathcal{A},P)$ and let $\mathcal{O}$ be a sub σ-algebra of $\mathcal{A}$.*

(a) *$Y \longmapsto E(Y|\mathcal{O})$ is linear and increasing from $L^0(\mathcal{A})$ into $L^0(\mathcal{O})$.*

(b) **Beppo-Levi Theorem.** *Let (Y_n) be an increasing sequence of positive* r.v.'s:

$$\lim \uparrow E(Y_n|\mathcal{O}) = E(\lim \uparrow Y_n|\mathcal{O}).$$

(c) **Fatou's Theorem.** *Let (Y_n) be a sequence of* r.v.'s *in $L^0(\mathcal{A})$. If there exists an* r.v. *Y in $L^1(\mathcal{A})$ which minorizes (Y_n), then*

$$E(\underline{\lim}\, Y_n|\mathcal{O}) \leqslant \underline{\lim}\, E(Y_n|\mathcal{O}).$$

If there exists an r.v. Z *in* $L^1(A)$ *which majorizes* (Y_n), *then*

$$E(\overline{\lim}\, Y_n|\mathcal{O}) \leqslant \overline{\lim}\, E(Y_n|\mathcal{O}).$$

(d) **Lebesgue's Theorem.** *Let* (Y_n) *be a sequence of* r.v.'s *of* $L^0(A)$, *such that* $\sup|Y_n|$ *is integrable and which converges almost surely. Then*

$$E(\lim Y_n|\mathcal{O}) = \lim E(Y_n|\mathcal{O}).$$

(e) *For* $\mathcal{O}'$ *a sub* σ*-algebra of* $\mathcal{O}$, *we have* $E(Y|\mathcal{O}') = E(E(Y|\mathcal{O})|\mathcal{O}')$. *(Thus if* $E(Y|\mathcal{O})$ *is* $\mathcal{O}'$*-measurable, then it is equal to* $E(Y|\mathcal{O}')$*).*

These definitions can be extended to random vectors $X = (X_1, \dots, X_n)$. X is quasi-integrable if its components are, and $E(X|\mathcal{O}) = (E(X_1|\mathcal{O}), \dots, E(X_n|\mathcal{O}))$.

Jensen's Inequality 6.2.8. *Let* C *be a convex set in* $\mathbb{R}^n$ *and* ϕ *a convex function from* C *into* $\mathbb{R}$. *Let* X *be a random vector defined on* (Ω, A), *taking values in* C. *For* X *and* $\phi(X)$ *integrable,*

(1) $E(X|\mathcal{O})$ *takes values in* C;
(2) $E[\phi(X)|\mathcal{O}] \geqslant \phi[E(X|\mathcal{O})]$.

If ϕ *is strictly convex, equality holds on* $\Gamma \in \mathcal{O}$ *if and only if* $1_\Gamma X$ *is* a.s. $\mathcal{O}$*-measurable. The relation* $E[\phi(X)] = E[\phi(E(X|\mathcal{O}))]$ *thus signifies that* X *is* a.s. $\mathcal{O}$*-measurable.*

Sketch of the Proof. We will assume (1), which is clear for $n = 1$ (then it is an interval) and which is usually easy to verify in the examples. We prove (2) by strengthening the hypotheses. Assume C is an open set and ϕ is differentiable with gradient grad ϕ. This gradient, the components of which are the partial derivatives of ϕ, is measurable. We can then write, for all pairs of points (x,y) in C:

$$\phi(x) \geqslant \phi(y) + \langle \text{grad } \phi(y), x - y \rangle$$

and the inequality is strict for ϕ strictly convex and $x \neq y$;

$$\phi(X) \geqslant \phi[E(X|\mathcal{O})] + \langle \text{grad } \phi(E(X|\mathcal{O})), X - E(X|\mathcal{O}) \rangle.$$

For $\Gamma \in \mathcal{O}$, $n \in \mathbb{N}$, and $\Gamma_n = \Gamma \cap \{\|\text{grad } \phi(E(X|\mathcal{O}))\| \leqslant n\}$, we obtain by integrating

$$E[\phi(X)\mathbf{1}_{\Gamma_n}] = E[E(\phi(X)|\mathcal{O})\mathbf{1}_{\Gamma_n}] \geqslant E[\phi(E(X|\mathcal{O}))\mathbf{1}_{\Gamma_n}].$$

Thus $E[\phi(X)|\mathcal{O}] \geqslant \phi(E[X|\mathcal{O}])$, almost surely on $\{\|\text{grad } \phi(E(X|\mathcal{O}))\| \leqslant n\}$ for all n, i.e., a.s. on Ω. The equality $E[\phi(X)|\mathcal{O}] = \phi(E[X|\mathcal{O}])$ on $\Gamma \in \mathcal{O}$ implies that, a.s. on Γ we have

$$\phi(X) = \phi[E(X|\mathcal{O})] + \langle\text{grad } \phi[E(X|\mathcal{O})],\ X - E(X|\mathcal{O})\rangle,$$

i.e., $\mathbf{1}_\Gamma X$, equal to $\mathbf{1}_\Gamma E(X|\mathcal{O})$, is a.s. $\mathcal{O}$-measurable.

Consequence. For $p \in [1,\infty[$ the function $x \longmapsto |x|^p$ is convex, thus for $Y \in L^p(\mathcal{A})$,

$$[E(Y|\mathcal{O})]^p \leqslant E(|Y|^p|\mathcal{O}); \quad \|E(Y|\mathcal{O})\|_p \leqslant \|Y\|_p.$$

It is easy to show that $\|E(Y|\mathcal{O})\|_\infty \leqslant \|Y\|_\infty$ for $Y \in L^\infty(\mathcal{A})$.

Proposition 6.2.9. *$Y \longmapsto E(Y|\mathcal{O})$ is, for $p \in [1,\infty]$, a linear mapping from $L^p(\mathcal{A})$ into $L^p(\mathcal{O})$ such that*

$$\|E(Y|\mathcal{O})\|_p \leqslant \|Y\|_p.$$

Let q be the conjugate of p, such that

$$\frac{1}{p} + \frac{1}{q} = 1.$$

For all $Z \in L^q(\mathcal{O})$ we have

$$E[YZ] = E[E(Y|\mathcal{O})Z].$$

The proof of the last part is standard (Trick 3.2.14).

Exercises 6.2.

How is the expectation of Y conditional on a sub σ-algebra determined? There are rules already used in Section 6.1 based on the use of densities. In general, start by studying what is an $\mathcal{O}$-measurable r.v., then using your imagination "see" which $\mathcal{O}$-measurable r.v. most "resembles" Y (or for $\mathcal{O} = \sigma(X)$, what does Y become when $X = x$ is fixed?) Finally, verify that Definition 6.2.6. holds.

E.1. *Symmetric distributions.* On $(\mathbb{R}^n, \mathcal{B}_{\mathbb{R}^n})$ let P be a 'symmetric' distribution (for $\Gamma \in \mathcal{B}_{\mathbb{R}^n}$, $P(\Gamma) = P(-\Gamma)$).

(a) Let S be the set of all Borel sets Γ symmetric with respect to zero. For quasi-integrable ϕ, determine $E(\phi|S)$.
(b) For $n = 1$, let X be the identity mapping. Determine

$$E(\phi|X^2 = x), \quad E(\phi \mid |X| = x), \text{ and } \quad E(\phi \mid e^{X^2} = x).$$

E.2. *Invariance.* Let $(\Omega, \mathcal{A})$ be a measurable space and G be a *finite* group of bimeasurable bijections on $(\Omega, \mathcal{A})$ (for $g \in G$, g^{-1} is also a measurable bijection). An event $A \in \mathcal{A}$ is said to be G-invariant if, for all $g \in G$, $gA = \{g(\omega); \omega \in A\} = A$.

(a) Show that the set of invariant events $\mathcal{B}_G$ forms a σ-algebra.
(b) A probability P on $(\Omega, \mathcal{A})$ is said to be invariant under G if, for all $g \in G$, $gP = P$. If f is quasi-integrable and P is invariant under G, calculate $E(f|\mathcal{B}_G)$.

E.3. *Exchangeability* (E.2. contd.). (a) On $\mathbb{R}^n$, let τ_n be the group of transformations

$$(x_p)_{1 \leq p \leq n} \overset{\phi_\sigma}{\longmapsto} (X_{\sigma(p)})_{1 \leq p \leq n}$$

for all the permutations σ of $\{1, ..., n\}$. Denote by X the identity function and by $X_1, ..., X_n$ the n coordinate functions. Let P be a probability on $\mathbb{R}^n$ which is τ_n invariant. For quasi-integrable f, calculate $E(f(X_1, ..., X_n)|\mathcal{B}_{\tau^n})$.

(b) A random vector $(X_1, ..., X_n)$ is *exchangeable* (Exercise 4.1.4) when its distribution is invariant under τ_n. Show that an n-sample from a distribution F on $\mathbb{R}$ is exchangeable. For such a sample, let $(X_{(1)}, ..., X_{(n)})$ be the order statistic. Show that if F is the uniform distribution, $(X_{(1)}, X_{(2)} - X_{(1)}, ..., X_{(n)} - X_{(n-1)})$ is exchangeable (Exercise 4.4.10). Let $(R_1, ..., R_n)$ be the associated rank vector assuming F is continuous. Show that it is exchangeable.

(c) Show that if $(X_1, ..., X_n)$ is exchangeable, then for all pairs $1 \leq i,j \leq n$ the correlation coefficient satisfies $\rho(X_i,X_j) \geq -1/(n-1)$.

(d) A random vector taking values in $\mathbb{R}^n$ is said to be "spherically symmetric," if its distribution is invariant under the group G of unitary transformations of $\mathbb{R}^n$. For $k \in \mathbb{N}$, let $a_k \in \mathbb{R}_+\backslash\{0\}$, $p_k \geq 0$ with $\Sigma\, p_k = 1$. Show that a random vector from the distribution $\Sigma\, p_k N_n(0,a_k I_n)$ is spherically symmetric. Show that every spherically symmetric probability is exchangeable. Characterize (by Fourier transforms for example) exchangeable and spherically symmetric vectors X.

(e) The n r.v.'s $X_1, ..., X_n$ are said to be *conditionally independent* with respect to a σ-algebra $\mathcal{C}$ if, for every sequence $f_1, ..., f_n$ of bounded r.v.'s defined on $\mathbb{R}$, we have

$$E[f_1(X_1) \dots f_n(X_n)|\mathcal{C}] = E(f_1(X_1)|\mathcal{C}) \dots E(f_n(X_n)|\mathcal{C}).$$

They have the same distribution conditional on $\mathcal{C}$ if, for every f bounded r.v. on $\mathbb{R}$, $E[f(X_i)|\mathcal{C}]$ does not depend on i. Show that if $X_1, ..., X_n$ are conditionally independent and have the same distribution conditional on the same σ-algebra, then they are exchangeable.

E.4. Let $(X_i)_{1 \leq i \leq n+1}$ be $n + 1$ exponential r.v.'s with parameter $\lambda > 0$. For $1 \leq j \leq n + 1$, set $S_j = \Sigma_{1 \leq i \leq j} X_i$>. Give the distributions of $(X_i)_{1 \leq i \leq n+1}$ and of $(S_j)_{1 \leq j \leq n+1}$. Show that the distribution of $(S_1, ..., S_n)$ conditional on $(S_{n+1} = u)$ is that of the order statistic of n independent r.v.'s uniformly distributed on $[0,u]$ (Exercise 4.4.10); compare with Exercise 4.4.12.

E.5. *Extension of Hölder and Minkowski's inequalities.* Let p and q be conjugate in $[1,\infty]$. Prove that for X and X' in $L^p(\mathcal{A})$ and Y in $L^q(\mathcal{A})$:

$$E(X \cdot Y|\mathcal{O}) \leq [E(X^p|\mathcal{O})]^{1/p}[E(Y^q|\mathcal{O})]^{1/q}$$

$$[E((X + X')^p|\mathcal{O})]^{1/p} \leq [E(X^p|\mathcal{O})]^{1/p} + [E(X'^p|\mathcal{O})]^{1/p}$$

(look at the sketch given in Section 3.2.1 in the unconditional case).

E.6. Assume that $\mathcal{A}$ contains all singleton sets. Let $\mathcal{O}$ be the

σ-algebra generated by these sets and let X be integrable on $(\Omega,\mathcal{A},P)$. Calculate $E(X|\mathcal{O})$.

E.7. Let ϕ be a measurable function from $(\Omega,\mathcal{A},P)$ into $(\Omega',\mathcal{A}')$. Let $\mathcal{B}$ be a sub σ-algebra of $\mathcal{A}'$ and E' the expectation with respect to $\phi(P)$. Prove that if X is a $\phi(P)$ quasi-integrable r.v. on $(\Omega',\mathcal{A}')$, we have for P-almost all ω:

$$E\{X\circ\phi \mid \phi^{-1}(\mathcal{B})\}(\omega) = (E'(X|\mathcal{B}))[\phi(\omega)].$$

E.8. Show that, for a positive r.v. X on $(\Omega,\mathcal{A},P)$ and a sub σ-algebra $\mathcal{O}$ of $\mathcal{A}$, $\{E(X|\mathcal{O}) > 0\}$ is the smallest set of $\mathcal{O}$ containing $\{X > 0\}$ (up to a set of measure zero).

E.9. *Renewal processes.* Let $(\tau_n)_{n\geq 1}$ be a sequence of strictly positive, independent r.v.'s. Define $T_n = \tau_1 + \ldots + \tau_n$ and $\mathcal{G}_n = \sigma(\tau_1, \ldots, \tau_n)$ for $n \geq 1$, $N_t = \Sigma_{n\geq 1}\mathbf{1}_{(T_n \leq t)}$ and $\mathcal{F}_t = \sigma(N_s; s \leq t)$ for $t > 0$; $\mathcal{G}_0$ is trivial and $T_0 = 0$.

(a) Verify the following properties:

(1) An r.v. ϕ is measurable on $(\Omega, \mathcal{F}_t)$ if and only if a sequence $(\phi_n)_{n\geq 0}$ of r.v.'s can be found such that ϕ_n is $\mathcal{G}_n$ measurable for all n and $\phi = \Sigma_{n\geq 0}\phi_n\mathbf{1}_{(T_n \leq t < T_{n+1})}$.

(2) If Y is a quasi-integrable r.v., we have

$$E[Y|\mathcal{F}_t] = E[E(Y|\mathcal{G}_n)|\mathcal{F}_t] \quad \text{on } \{t < T_n\},$$

$$E[Y|\mathcal{F}_t] = \frac{E[Y\mathbf{1}_{(T_n \leq t < T_{n+1})}|\mathcal{G}_n]}{P[T_n \leq t < T_{n+1}|\mathcal{G}_n]} \quad \text{on } \{T_n \leq t < T_{n+1}\}.$$

(b) *The Poisson process* (use Exercise 4.3.6 and compare with the results of 4.4.12 without using these results). Assume that all the r.v.'s (τ_n) are exponentially distributed with parameter λ. Show

$$P[T_n \leq t < T_{n+1}|\mathcal{G}_n] = \mathbf{1}_{(T_n \leq t)}e^{-\lambda(t-T_n)}.$$

Let $T_k^{(t)} + t = T_{N_t+k}$ be the kth instant after t at which an event takes place. Prove that $T_1^{(t)}$ is an r.v. independent of the σ-algebra $\mathcal{F}_t$ and that it has an exponential distribution with parameter λ. Prove that the future σ-algebra generated by $(T_k^{(t)}; k \geq 1)$ coincides with $\sigma(N_{s+t} - N_t; s > 0)$ and is independent of $\mathcal{F}_t$.

6.3. Markov Chains

6.3.1. Transition Probabilities

Let (X,X') be a measurable function taking values in $(E \times E', \mathcal{E} \otimes \mathcal{E}')$, X having distribution F_X. We can define, for all $\Gamma \in \mathcal{E}'$, up to F_X-equivalence, the function $E(1_\Gamma(X')|X = \cdot)$ which is also called the *probability of* $(X' \in \Gamma)$ *conditional on* $(X = \cdot)$ and which is denoted by $P(X' \in \Gamma|X = \cdot)$. For (Γ_n), a sequence of disjoint elements of $\mathcal{E}'$, we have F_X a.s.:

$$\Sigma\, P(X' \in \Gamma_n|X = \cdot) = P(X' \in \cup\, \Gamma_n|X = \cdot).$$

In Examples 3 of Section 6.1, we were able to choose a version of these conditional probabilities, which is, for each x, a probability. This is not always possible since, for a given version, the relation written above is true except for a set of measure zero which depends on (Γ_n). The problem of the existence of such a probability (**lifting problem**) will be considered in Volume 2.

However, cases can be constructed where existence, as above, is evident.

Definition 6.3.10. We call a **transition** from $(E_1,\mathcal{E}_1)$ into $(E_2,\mathcal{E}_2)$ a function from $E_1 \times \mathcal{E}_2$ into $[0,1]$, $(x,\Gamma) \longmapsto \pi(x,\Gamma)$, which satisfies the following properties:

(a) For every $\Gamma \in \mathcal{E}_2$, $\pi(\cdot,\Gamma)$ is an r.v. on $(E_1,\mathcal{E}_1)$.
(b) For every $x \in E_1$, $\pi(x,\cdot)$ is a probability on $(E_2,\mathcal{E}_2)$.

We denote $\pi: (E_1,\mathcal{E}_1) \models\!< (E_2,\mathcal{E}_2)$. We also speak of a σ-finite transition when (b) is replaced by (b'): $\pi(x,\cdot)$ is a σ-finite measure on $(E_2,\mathcal{E}_2)$.

Notations. Let f be a positive (or negative or bounded) r.v. on $(E_2,\mathcal{E}_2)$. Then, for every $x \in E_1$, set

$$\pi f(x) = \int \pi(x,dy)f(y).$$

The function πf thus defined is an r.v. in $(E_1, \mathcal{E}_1)$ (trick 3.2.14).

We can also associate with π the transition $\tilde{\pi}$ from $(E_1,\mathcal{E}_1)$ into $(E_1,\mathcal{E}_1) \times (E_2,\mathcal{E}_2)$ defined by $\tilde{\pi}(x,\cdot) = \partial_x \otimes \pi(x,\cdot)$. For a positive r.v. h on $(E_1,\mathcal{E}_1) \times (E_2,\mathcal{E}_2)$ we have

$$\tilde{\pi}h(x) = \int \pi(x,dy)h(x,y),$$

and, using the same notation for π and $\tilde{\pi}$, we denote this function also by πh.

Finally, composition of transitions is very useful. For π_1: $(E_1,\mathcal{E}_1) \models\!\!< (E_2,\mathcal{E}_2)$ and π_2: $(E_2,\mathcal{E}_2) \models\!\!< (E_3,\mathcal{E}_3)$, we define

$$\pi_1\pi_2: (E_1,\mathcal{E}_1) \models\!\!< (E_3,\mathcal{E}_3)$$

and

$$\pi_1 \otimes \pi_2: (E_1,\mathcal{E}_1) \models\!\!< (E_2,\mathcal{E}_2) \times (E_3,\mathcal{E}_3)$$

by

$$\pi_1\pi_2(x_1,\Gamma) = \int \pi_1(x_1,dx_2)\pi_2(x_2,\Gamma)$$

$$\pi_1 \otimes \pi_2(x_1,\Gamma') = \int \pi_1(x_1,dx_2)\int \pi_2(x_2,dx_3)1_{\Gamma'}(x_2,x_3)$$

$$(\pi_1 \otimes \pi_2 = \pi_1\tilde{\pi}_2).$$

For π: $(E,\mathcal{E}) \models\!\!< (E,\mathcal{E})$, denote by π^n and $\pi^{\otimes n}$ the product of n transitions π. Of course, a probability F on $(E_1,\mathcal{E}_1)$ is a particular transition from $(E_1,\mathcal{E}_1)$ into itself, which allows $F\pi$ and $F \otimes \pi$ to be defined; the operation $\pi_1 \otimes \pi_2$ generalizes the product of probabilities.

Transitions and Conditional Distributions. Let (X_1,X_2) be a pair of observations taking values in $(E_1 \times E_2, \mathcal{E}_1 \otimes \mathcal{E}_2)$. A distribution F_1 of X_1 and a transition π are given such that, for all $\Gamma_2 \in \mathcal{E}_2$, $\pi(X_1,\Gamma_2)$ is a version of $P(X_2 \in \Gamma_2|X_1)$. Standard techniques then give, with the above notation,

$$E[f(X_2)|X_1] = \pi f(X_1); \quad E[h(X_1,X_2)|X_1] = \pi h(X_1);$$

$$E[f(X_2)] = F_1\pi(f); \quad E[h(X_1,X_2)] = F_1 \otimes \pi(h).$$

The distribution of X_2 is $F_1\pi$, that of (X_1,X_2) is $F_1 \otimes \pi$. In the case where X_1 and X_2 are independent, take for π_2 the distribution F_2 of X_2 and the distribution of the pair turns out again to be $F_1 \otimes F_2$.

6.3.2. Markov Chains

Consider the transmission of a message "yes" or "no" in a population. Assume that each individual transmits the received message with probability $1-p$ and the opposite with probability p.

Let X_n be the message transmitted by the nth individual. We have

$$P[X_n = \text{yes}|X_{n-1} = \text{yes}] = P[X_n = \text{no}|X_{n-1} = \text{no}] = 1-p$$

$$P[X_n = \text{yes}|X_{n-1} = \text{no}] = P[X_n = \text{no}|X_{n-1} = \text{yes}] = p.$$

These four numbers determine a transition probability π of {yes, no} into itself, which allows the study of the transmission of the message if the initial message is known. A characteristic is that only the message transmitted by X_{n-1} influences X_n; knowledge of the entire history of the message is not necessary.

Definition 6.3.11. A sequence $(X_n)_{n\geqslant 0}$ of observations taking values in the same state space $(E,\mathcal{E})$, such that, for $\Gamma \in \mathcal{E}$ and $n \geqslant 0$,

$$P[X_{n+1} \in \Gamma|X_0, \ldots, X_n] = P[X_{n+1} \in \Gamma|X_n],$$

is called a **Markov chain** on $(\Omega,\mathcal{A},P)$.

Definition 6.3.12. Let π be a transition from $(E,\mathcal{E})$ into $(E,\mathcal{E})$, and ν a probability on $(E,\mathcal{E})$. The preceding Markov chain is a homogeneous Markov chain with initial distribution ν and transition π if we have

$$P[X_0 \in \Gamma] = \nu(\Gamma)$$

$$P[X_{n+1} \in \Gamma|X_0, \ldots, X_n] = \pi(X_n,\Gamma).$$

Property 6.3.13. *Let f be a positive (or bounded)* r.v. *on* $(E,\mathcal{E})^{p+1}$. *In Definition* 6.3.12,

$$E[f(X_n,X_{n+1}, \ldots, X_{n+p})|X_0, \ldots, X_n]$$
$$= E[f(X_n,X_{n+1}, \ldots, X_{n+p})|X_n]$$
$$= \int \pi(X_n,dx_1)\int \pi(x_1,dx_2) \ldots \int \pi(x_{p-1},dx_p) \, f(X_n,x_1, \ldots, x_p)$$
$$= \int \pi^{\otimes p}(X_n;dx_1, \ldots, dx_p)f(X_n,x_1, \ldots, x_p).$$

This is verified for f, the indicator function of a rectangle, then use Tricks 3.1.13 and 3.2.14. Thus *the distribution of* $(X_{n+1}, \ldots, X_{n+p})$ *conditional on* $(X_0, \ldots, X_n)$ *is* $\pi^{\otimes p}(X_n, \cdot)$. *This is also its distribution conditional on* X_n. *The distribution of* X_{n+p} *conditional on* $(X_0, \ldots, X_n)$ (or on X_n) *is* $\pi^p(X_n, \cdot)$.

Example 1. *A Markov chain on a finite space* $E = \{1,2, \ldots, k\}$. A transition is then defined by a $k \times k$ matrix, $\pi = \{\pi(i,j)\}_{1 \leqslant i,j \leqslant k}$, where $\pi(i,\cdot)$ is a probability

$$\sum_{j=1}^{k} \pi(i,j) = 1 \quad \text{and} \quad 0 \leqslant \pi(i,j).$$

Such a matrix is said to be *stochastic.* It is easily shown that $\{\pi^n(i,j)\}_{1 \leqslant i,j \leqslant k}$ is the matrix π^n, the nth power of π. For

$$\pi = \begin{bmatrix} 1-p & p \\ p & 1-p \end{bmatrix},$$

we obtain by recurrence,

$$2\pi^n = \begin{bmatrix} 1 & 1 \\ 1 & 1 \end{bmatrix} + (1-2p)^n \begin{bmatrix} 1 & -1 \\ 1 & -1 \end{bmatrix}.$$

Then, for $0 < p < 1$, if n tends to ∞, π^n tends to

$$\begin{bmatrix} 1/2 & 1/2 \\ 1/2 & 1/2 \end{bmatrix}.$$

No matter what the initial message, for large n, "yes" and "no" have almost the same probability of being transmitted.

Example 2. *A Random Walk.* For $(Y_n)_{n\geq 0}$, a sequence of independent observations taking values in $(\mathbb{R}^k, \mathcal{B}_{\mathbb{R}^k})$ with the same distribution ν, set $S_n = Y_0 + \dots + Y_n$. Then $(S_n)_{n\geq 0}$ is a Markov chain (random walk) associated with the transition $(x,\Gamma) \longmapsto \nu(\Gamma - x)$. In fact, $Y_{n+1} = S_{n+1} - S_n$ is independent of $(Y_0, \dots, Y_n)$. Thus for $\Gamma \in \mathcal{B}_{\mathbb{R}^k}$ and $\Gamma' \in (\mathcal{B}_{\mathbb{R}^k})^{n+1}$,

$$E[1_\Gamma(S_{n+1})1_{\Gamma'}(S_0, \dots, S_n)|S_0, \dots, S_n]$$

$$= E[1_\Gamma(Y_{n+1} + S_n)1_{\Gamma'}(S_0, \dots, S_n)|S_0, \dots, S_n)]$$

$$= \int d\nu(y)1_\Gamma(y + S_n)1_{\Gamma'}(S_0, \dots, S_n).$$

$$P[S_{n+1} \in \Gamma|S_0, \dots, S_n] = \int d\nu(y)1_\Gamma(y + S_n) = \nu(\Gamma - S_n).$$

Example 3. *The Queue* G/G/1. Customers arrive at the checkout in a shop. The time interval between the arrival of the nth and $(n+1)$th customer is an r.v. A_n. The period during which the nth customer is being served is an r.v. B_n. Assume the r.v.'s (A_n) are finite and independent with distribution α, the r.v.'s (B_n) are finite and independent with distribution β, and these two sequences are independent of one another. Let W_n be the waiting time of the nth customer. For $n \geq 0$, we have

$$W_{n+1} = (W_n + B_n - A_n)_+ \, .$$

If the nth customer arrives at time t, the server is available at time $t + W_n + B_n$ for the $(n+1)$th customer and this customer arrives at time $t + A_n$. Let $\Gamma \in \mathcal{B}_{\mathbb{R}_+}$; (A_n, B_n) is independent of $(W_0, \dots, W_n)$, thus

$$P[W_{n+1} \in \Gamma|W_0, \dots, W_n] = P[(W_n + B_n - A_n)_+$$

$$\in \Gamma|X_0, \dots, X_n)$$

$$= \int d\alpha(a)d\beta(b)P[(W_n + b - a)_+ \in \Gamma].$$

Thus (W_n) is a Markov chain with transition π defined by

$$\pi(x,\Gamma) = \int d\alpha(a)d\beta(b)1_\Gamma((x + b - a)_+).$$

Example 4. *Branching processes.* Consider particles which can produce new particles of the same type before being destroyed. A given particle constitutes generation 0. Each particule produces m particles (m = 0,1,2, ...) with probability p_m = (Σp_m = 1). The direct descendants of the elements of the nth generation constitute the (n+1)th generation, and all particles reproduce independently. This evolution may be described by the probability $p = (p_m)_{m\geq 0}$ on $\mathbb{N}$ and the transition $\pi(k,\cdot) = p^{*k}$. If the number of individuals in generation n is X_n, (X_n) is a Markov chain with transition π and initial state 1.

Exercises 6.3

The asymptotic behavior of Markov chains will be studied in Volume 2. Meanwhile, in this section we shall study important examples of this by means of applications, which will illustrate the theory.

E.1. *A Markov chain with two states.* Consider a transition π from {0,1} into {0,1}. The associated matrix, denoted also by π, is written

$$\pi = \begin{bmatrix} 1-p & q \\ q & 1-q \end{bmatrix}$$

for two numbers p and q from [0,1].

(a) Assume that p and q do not equal 1. Prove

$$\pi^n = \frac{1}{p+q}\left\{\begin{bmatrix} q & p \\ q & p \end{bmatrix} - (1-p-q)^n\begin{bmatrix} -p & p \\ q & -q \end{bmatrix}\right\}.$$

Let μ be the measure defined by $\mu(0) = q/(p+q)$, $\mu(1) = p/(p+q)$. Show

$$\mu\pi = \mu$$

$$\lim_n \pi^n(0,0) = \lim_n \pi^n(1,0) = \mu(0),$$

$$\lim_n \pi^n(0,1) = \lim_n \pi^n(1,1) = \mu(1).$$

For a Markov chain with transition π and initial state 0, calculate the average numbers $N_n(0,0)$ or $N_n(0,1)$ of steps in 0 or in 1 in the first n steps. Study $\lim_{n\to\infty} N_n(0,0)/n$ and $\lim_{n\to\infty} N_n(0,1)/n$.

(b) Describe the Markov chain with initial state 0 when p and q are equal to 1.

E.2. *A Markov chain on a finite space $E = \{1,2, ..., k\}$.* Let π be the associated matrix of transition probabilities on E.

(a) Let $v = (v_1, ..., v_k) \in \mathbb{C}^k$. Using the norm $\|\cdot\|_\infty$ defined by $\|v\|_\infty = \sup_{1\leqslant i\leqslant k}|v_i|$, verify that

$$\|\pi v\|_\infty \leqslant \|v\|_\infty.$$

Prove that there cannot be an eigenvalue with modulus strictly greater than 1 and that the vector **1**, all of the components of which equal 1, is an eigenvector of π.

(b) Let us denote

$$v\pi = \left\{ \sum_{j=1}^{k} v_j \pi(j,i): i = 1, ..., k \right\}$$

the transform of v by the transpose of π. For $v \in \mathbb{R}^k$, prove that $v\pi \geqslant v$ (resp. $v\pi \leqslant v$) implies $v\pi = v$. For $v \in \mathbb{C}^k$, and denoting $|v| = (|v_1|, ..., |v_k|)$, prove that $v\pi = v$ implies $|v|\pi = |v|$.

(c) Let μ be a measure on E, identified by the vector $(\mu(1), ..., \mu(k))$ in $\mathbb{R}^k$. Let S_μ be the set of points charged by μ (its support). The measure μ is said to be invariant under π if $\mu\pi$ is equal to μ. Prove that *there exists a unique probability F invariant under π, if* 1 *is an eigenvalue of order* 1 *of π.*

(d) Let μ and v be two invariant measures with supports S_μ and S_v. Prove that $(\mu - v)_+$ and $\mu - (\mu - v)_+$ are invariant measures and that the support of $\mu - (\mu - v)_+$ is $S_\mu \cap S_v$. Let $i \in E$, charged by at least one invariant measure μ: i is said to be "*recurrent.*" Let $C(i)$ be the intersection of the S_μ, for these invariant measures μ which charge i. Prove that there exists a unique probability G invariant under π, the support of which is $C(i)$; prove that for $j \in C(i)$, $C(i)$ and $C(j)$ coincide. We thus have a partition $E = T \cup C_1 \cup ... \cup C_r$, where T is the set of "*transients*" which are not charged by any invariant measure and where $C_1, ..., C_r$ are r "*recurrent classes.*" For each $\ell = 1, ..., r$ there exists an invariant probability $F^{(\ell)}$, the support of which is C_ℓ. Prove that C_ℓ is a closed set under π, and that for the matrix $(\pi(i,j))_{(i,j)\in C_\ell^2}$, 1 is an eigenvalue of order 1.

(e) Let λ be a root of order k of the minimal polynomial M_π of π. It is known that this is a root of order $k' \geqslant k$ of the characteristic polynomial C_π and that the kernel of $(\pi - \lambda I)^k$ is of dimension k'. Let ν be nonzero in this kernel $\mathrm{Ker}(\pi - \lambda I)^k$. For $|\lambda| < 1$, prove $\pi^n(|\nu|) = O(n^{k-1}|\lambda|^n)$. Study, for $|\lambda| = 1$, the sequence $(\pi^n \nu)$ and prove that it is bounded and that $k = 1$. Prove that there exists p real numbers $\theta_1, \ldots, \theta_p$, $\delta \in]0,1[$ and $p + 1$ matrices $P_0, \ldots, P_p$ such that

$$\pi = P_0 + e^{ni\theta_1}P_1 + \ldots + e^{ni\theta_p}P_p + o(\delta^n).$$

(f) Assume that 1 is the only root of modulus 1 of C_π. If it is a simple root, prove that, F being the invariant probability for all (i,j),

$$\pi^n(i,j) = F(j) + o(\delta^n).$$

If it is a root of order r, prove that there are r recurrent classes $C_1, \ldots, C_r$ and that

$$\pi^n(i,j) = P_0(i,j) + o(\delta^n).$$

Let $F^{(\ell)}$ be the invariant probability concentrated on C_ℓ ($\ell = 1, \ldots, r$). Verify that $P_0(i,\cdot)$ coincides with $F^{(\ell)}$ for all $i \in C_\ell$.

(g) Assume that 1 is a simple root of C_π. Let ν be an eigenvector of π defined up to a factor associated with a complex eigenvalue $e^{i\theta}$ of modulus 1. Let $j \in \{1, \ldots, n\}$ satisfy $|\nu_j| = \sup_{1\leqslant\ell\leqslant n}|\nu_\ell|$. We can take $\nu_j = 1$. Consider

$$\begin{cases} D_0 = \{\ell : \nu_\ell = 1\} \\ D_p = \bigcup_{\ell\in D_0} \{m;\ \pi^p(\ell,m) > 0\} \quad (p \geqslant 1). \end{cases}$$

For $m \in D_p$, prove that $|\nu_m| = 1$. Then (by noticing that the barycenter of complex numbers with modulus 1 only has modulus 1 if they are identical), prove that $D_p = \{m;\ \nu_m = e^{ip\theta}\}$. Deduce from this that these sets (D_p) are disjoint or identical: assume that d amongst them are disjoint. Show that

$e^{i\theta}$ is a dth root of unity. Let F be the unique invariant measure for π. Verify that its support is $D_0 \cup D_1 \cup \ldots \cup D_{d-1}$, that the sets $D_0, \ldots, D_{d-1}$, called *cyclic classes*, are the recurrent classes of π^d, and that there exists a $\delta \in]0,1[$ such that, for $i \in D_p$ and $0 \leqslant r < d - 1$,

$$\pi^{nd+r}(i,j) = \begin{cases} dF(j) + o(\delta^n) & \text{for } j \in D_{p+r} \\ 0 & \text{for } j \notin D_{p+r}. \end{cases}$$

(h) Apply the above to Exercise 6.3.1 and to the following matrices (recurrent classes, invariant measures, cyclic classes in recurrent classes).

$$\begin{bmatrix} 0 & 1 & 0 & 0 \\ 0 & 0 & 1 & 0 \\ 0 & 0 & 0 & 1 \\ 1 & 0 & 0 & 0 \end{bmatrix}; \quad \begin{bmatrix} 0 & 0 & 0 & 1 & 0 & 0 \\ 1/3 & 0 & 0 & 1/3 & 1/3 & 0 \\ 0 & 0 & 0 & 1 & 0 & 0 \\ 1/3 & 0 & 2/3 & 0 & 0 & 0 \\ 0 & 0 & 0 & 0 & 1/2 & 1/2 \\ 0 & 0 & 0 & 0 & 1 & 0 \end{bmatrix}.$$

(i) In the genetic framework set up in Exercise 1.3.6, assume that two alleles A and a of a gene are found in a large population with proportion p and $1 - p = q$. Let $E = \{AA, Aa, aa\}$ be the set of genotypes, and π the transition matrix, which corresponds to the probabilities of the various genotypes of a child conditional on those of its mother. Explain why π is taken to be

$$\pi = \begin{bmatrix} p & q & 0 \\ (1/2)p & 1/2 & (1/2)q \\ 0 & p & q \end{bmatrix}$$

and study the sequence (π^n).

E.3. *Branching processes* (Example 4). Assume in Example 4 that none of the p_k equals 1 and that $p_0 + p_1 < 1$. Let g be the generating function of p, $g(z) = \Sigma_{k=0}^{\infty} z^k p_k$ (for $|z| \leqslant 1$).

(a) Show that, if the sequence (f_n) is defined by the recurrence relations,

$$f_0(z) = z, \quad f_1 = g, \quad f_n = g \circ f_{n-1}$$

the generating function of X_n is f_n.

(b) Assume that the distribution p is square integrable. Let m and σ^2 be its mean and variance. Show

$$E(X_n) = m^n$$

$$\sigma^2(X_n) = \begin{cases} \sigma^2 m^{n-1} \dfrac{m^n - 1}{m - 1} & \text{for } m \neq 1 \\ n\sigma^2 & \text{for } m = 1. \end{cases}$$

(c) Let E be the event "the population becomes extinct": $E = \{\lim_{n\to\infty} X_n = 0\}$. Prove

$$P(E) = \lim_n \uparrow P X_n = 0) = \lim_n \uparrow f_n(0) = \rho.$$

(d) Show that g is strictly convex and that ρ is the smallest root lying between 0 and 1 of the equation $g(x) = x$. Deduce from this that $\rho = 1$ for $m \leqslant 1$, and $0 < \rho < 1$ for $m > 1$.

(e) For n and p integers, prove

$$E[X_{n+p} \mid X_0, \ldots, X_n] = m^p X_n .$$

Setting $Z_n = X_n/m^n$, prove

$$E[Z_{n+p} - Z_n]^2 = \frac{\sigma^2 m^{-n}}{m^2 - m}[1 - m^{-p}] \quad \text{for } m \neq 1.$$

Show that, for $m > 1$, (Z_n) converges in quadratic mean and a.s. to an r.v. Z. Let h, $s \longmapsto E(e^{-sZ})$ be the Laplace transform of the distribution of Z ($s \geqslant 0$). Prove $h(sm) = g[h(s)]$. Show that $\{Z = 0\}$ coincides with E up to a set of measure zero. When it does not become extinct the population thus increases (a.s.) with a speed proportional to (m^n).

(f) For $m \leqslant 1$, study the population total $N = \Sigma_{n\geqslant 0} X_n$. Let G be its generating function. Prove $G(z) = zg[G(z)]$. Show

$$E(N) = \frac{1}{1 - m} \leqslant \infty.$$

E.4. *Recurrent or transient states.* A Markov chain $(\Omega, \mathcal{A}, P, (X_n)_{n\geqslant 0})$ is given. For a state x, assume $\{x\}$ measurable and $P(X_0 = x) = 1$.

(a) Let $S = \inf\{n;\ X_n \neq x\}$. Prove that S has a geometric distribution.

(b) Let

$$\begin{cases} T_1 = \inf\{n;\ n \geq 1,\ X_n = x\} \\ T_k = \inf\{n;\ n > T_{k-1},\ X_n = x\} \text{ on } \{T_{k-1} < \infty\} \end{cases}$$

(the inf of an empty set of $\mathbb{N}$ being $+\infty$); T_k is the time of the kth return to x. Let $n_1, \ldots, n_k$ be k integers; prove

$$P[T_1 = n_1, T_2 = n_1 + n_2, \ldots, T_k = n_1 + \ldots + n_k] = \prod_{i=1}^{k} P(T_1 = n_i)$$

$$P[T_k < \infty] = (P[T_1 < \infty])^k$$

$$E\left[\sum_{n\geq 0} 1_{(X_n = x)}\right] = \sum_{n\geq 0} \pi^n(x,x) = \frac{1}{1 - P[T_1 < \infty]} \leq \infty.$$

Conclusion. The state x can be either "recurrent," it is revisited a.s. an infinite number of times; or "transient," i.e. x is only visited a.s. a finite (integrable) number of times. Verify that the "recurrent" or "transient" states defined in Exercise 6.3.2 are also "recurrent" or "transient" in the sense of this exercise.

E.5. *A random walk on $\mathbb{Z}$.* We are given a Markov chain $(\Omega,\mathcal{A},P,(X_n)_{n\geq 0})$ on $\mathbb{Z}$, with transition π defined for $p \in]0,1[$ by $\pi(n,n+1) = 1 - \pi(n,n-1) = p$. Assume that the initial state is zero.

(a) Check that we are dealing with a random walk.
(b) Calculate $P(X_{2n} = 0)$ for integer n. Show that the state 0 is recurrent if and only if $p = 1/2$ (use the relation

$$\sum_{n=0}^{\infty} \binom{2n}{n} x^n = \frac{1}{\sqrt{1 - 4x}}$$

for $0 \leq x < 1/4$).

E.6. *A random walk on $\mathbb{Z}^2$ and $\mathbb{Z}^3$.* Consider a Markov chain $(\Omega,\mathcal{A},P,(X_n)_{n\geq 0})$ the state space of which is $\mathbb{Z}^2$ or $\mathbb{Z}^3$, the initial state 0 and the transition π, all the terms of which are zero except on $\mathbb{Z}^2$:

$$\pi[(i,j),(i+1,j)] = \pi[(i,j),(i-1,j)] = \pi[(i,j),(i,j+1)]$$
$$= \pi[(i,j),(i,j-1)] = 1/4$$

or on $\mathbb{Z}^3$:

$$\pi[(i,j,k),(i+1,j,k)] = \pi[(i,j,k),(i-1,j,k)] =$$

$$\ldots = \pi[(i,j,k),(i,j,k-1)] = 1/6.$$

(In one step we can move to one of the neighboring points and the possible steps are equiprobable.) Calculate $P(X_n = 0)$ and prove that 0 is recurrent for the random walk on $\mathbb{Z}^2$ and transient on $\mathbb{Z}^3$.

E.7. *Study of the* G/G/1 *queue of Example* 3. Set, for $n \geqslant 1$, $X_n = B_n - A_n$, $S_n = X_1 + \ldots + X_n$, $S_0 = 0$.

(a) Show $W_n = S_n - \inf_{0 \leqslant p \leqslant n} S_p$. Let $I = \inf S_p$. Prove that there are two cases:

Case 1: $P(I = -\infty) = 1$ and there are a.s. an infinite number of customers who do not wait.

Case 2: $P(I = -\infty) = 0$ and the random number of customers who do not wait is integrable.

(b) Show that W_n has the same distribution as $(\sup_{0 \leqslant p \leqslant n} S_p)$.

(c) Show that there are two cases by setting $M = \sup S_n$.

Case I: $P(M = \infty) = 1$, and, for all $a < \infty$, $\lim_{n\to\infty} P(W_n \leqslant a) = 0$. We say "the waiting time tends in distribution towards $+\infty$."

Case II: $P(M < \infty) = 1$, and the sequence (W_n) tends in distribution to the distribution of M.

(d) Assume the r.v. X_n are integrable with mean m. Using the results of Exercise 4.4.13 describe the asymptotic behavior of (W_n) according to whether $m > 0$, $m = 0$, $m < 0$. The results are intuitively obvious.

Prove that (W_n/n) tends a.s. to $\sup(0,m)$. Consider an independent sequence of positive r.v.'s (ε_n), with mean $m' = -m + \eta$ with $\eta > 0$, which is independent of (X_n). By replacing X_n by $X_n^* = X_n + \varepsilon_n$, prove that, for $m > 0$ then for $m \leqslant 0$, the time W_n^* corresponding to (X_p^*) is larger than W_n. Draw conclusions.

E.8. *Autoregressive processes of order* 1 (AR1). Consider a sequence $(\xi_n)_{n\geqslant 1}$ of independent, centered r.v.'s with finite variance σ^2 and X_0 an r.v. independent of this sequence, with distribution ρ. Define for $|\theta| < 1$, a sequence (X_n^θ) by the relations

$$\begin{cases} X_o^\theta = X_0 \\ X_{n+1}^\theta = \theta X_n^\theta + \xi_{n+1} . \end{cases}$$

Show that (X_n^θ) is a Markov chain. Set

$$Y_n^\theta = \theta^n X_0 + \xi_1 + \theta\xi_2 + \ldots + \theta^{n-1}\xi_n .$$

Show that X_n^θ and Y_n^θ have the same distribution. Prove that the sequence (Y_n^θ) converges in L^2 and a.s. to an r.v. Y^θ with distribution μ_θ. Prove that (X_n^θ) converges in distribution to μ_θ. Prove

$$\lim E[X_n^\theta] = 0, \quad \lim E[(X_n^\theta)^2] = \frac{\sigma^2}{1 - \theta^2} .$$

6.4. Information Carried by One Distribution on Another

6.4.1. The Lebesgue and Radon-Nikodym Decomposition

Consider a random phenomenon, the set of realizations of which is $(\Omega,\mathcal{A})$ and the probability of which may be P or Q. If an observation is taken in order to decide which of the probabilities P or Q controls the phenomenon, the decision will be easy when P and Q are 'perpendicular' (denote by $P \perp Q$), i.e., when there exists an $A \in \mathcal{A}$ such that P is concentrated on A and Q on A^c. On the other hand, there would be no point in using an observation from the distribution P to gain information in the distribution Q.

When several probabilities are involved, the question of sets of measure zero, equivalence classes for a.s. equality of r.v.'s, which we have up till now treated lightly, is more fraught with dangers since it depends on the probabilities (it will be necessary to denote Q–a.s., Q–measure zero ....).

Definition 6.4.14. For α and β two σ-finite measures on $(\Omega,\mathcal{A})$, we say that α is **absolutely continuous** with respect to β if the sets of measure zero for β are also sets of measure zero for α. (We also say that β **dominates** α.) They are **perpendicular** if they are concentrated on disjoint sets. These are denoted, respectively, by $\alpha \ll \beta$ and $\alpha \perp \beta$. They are **equivalent** if: $\alpha \ll \beta$ and $\beta \ll \alpha$. This is denoted by $\alpha \sim \beta$. To say that α and β are perpendicular implies that there exists a measurable set N, which is not charged by β and on which α is concentrated. N is defined β a.s. We also say "α is **singular** with respect to β."

Let us assume $\alpha(\Omega)$ finite and $\alpha \leqslant \beta$ ($\alpha(A) \leqslant \beta(A)$ for $A \in \mathcal{A}$). Then almost sure equality for β implies almost sure equality for α and for $f \in L^2(\beta)$, we have

$$\left|\int f\, d\alpha\right| \leqslant \sqrt{\left(\int f^2 d\alpha\right)\left(\int 1\, d\alpha\right)} \leqslant \sqrt{\alpha(\Omega)}\, \sqrt{\int f^2 d\beta}.$$

The mapping $f \longmapsto \int f\, d\alpha$ is a continuous bilinear form on the Hilbert space $L^2(\beta)$. There exists an element $a \in L^2(\beta)$ such that for all $f \in L^2(\beta)$, $\int f\, d\alpha = \int fa\, d\beta$. Taking $f = 1_{[a>1]}$ or $f = 1_{[a<0]}$ in the relations $\int fa\, d\beta \leqslant \int f\, d\beta$ we obtain $0 \leqslant a \leqslant 1$. Taking now α and β to be arbitrary, $\alpha \leqslant \beta + \alpha$ and there exists an r.v. a taking values in $[0,1]$ such that $(1 - a)\alpha = a\beta$. Let

$$N = \{a = 1\}; \quad \alpha = 1_N \alpha + \left(1_{N^c} \frac{1}{1 - a}\right)\beta.$$

For $\alpha \ll \beta$, $\beta(N) = 0$ and $\alpha = a/(1 - a)\beta$.

Theorem 6.4.15. **(Radon-Nikodym and Lebesgue).** *There exists a unique decomposition of α as a sum of a measure singular with respect to β and a measure having a density with respect to β. If α is absolutely continuous with respect to β, it has a density (defined β a.s.), called the Radon-Nikodym derivative of α with respect to β and denoted by $d\alpha/d\beta$.*

This notation is justified from the following property, which follows from the definitions. If γ *dominates* β *which dominates* α, *then*

$$\frac{d\alpha}{d\gamma} = \frac{d\alpha}{d\beta}\frac{d\beta}{d\gamma}.$$

The proof of existence is made when $\alpha(\Omega)$ is finite. This is achieved by writing Ω as an increasing limit of a sequence of events (Ω_n) such that $\alpha(\Omega_n)$ is finite and by summing the traces of α and β on $(\Omega_{n+1}\backslash\Omega_n)$, for $n \geqslant 0$ ($\Omega_0 = \phi$). Uniqueness follows easily from the difference between two candidates since a measure singular with respect to β equal to a measure having a density with respect to β is zero.

6.4.2. The *Dissimilarities* Between Two Probabilities

For quality control, where the manufacturer states that his production follows the distribution P, the consumer's problem is to measure the dissimilarity between the observed distribution Q and P. For estimation problems from a distribution F, we have observed an n-sample and used the sample distribution F_n, which measures the proximity to F. What information does F_n carry on F? A test problem between two hypotheses H_0 and H_1 will be much easier when the dissimilarities of the experimental distributions under H_0 and H_1 are large.

Two probabilities P and Q are always dominated, for example by $P + Q$; let ν dominate P and Q and let $p = dP/d\nu$, $q = dQ/d\nu$. Let ϕ be a function from $\mathbb{R}_+ \times \mathbb{R}_+$ into $\mathbb{R}$. We try to measure the dissimilarity by $\int\phi(p,q)d\nu$. What functions ϕ are adequate?

(a) *The result should not depend on the dominant* ν. It will then be denoted by $\int\phi(P,Q)$. A dominant ν always dominates $P + Q$: $P + Q = f\cdot\nu$ and

$$\int\phi\left[\frac{dP}{d(P+Q)}, \frac{dQ}{d(P+Q)}\right]d(P+Q)$$

$$= \int\phi\left[\frac{dP}{d(P+Q)}, \frac{dQ}{d(P+Q)}\right]f\,d\nu.$$

Now

$$\frac{f\cdot dP}{d(P+Q)} = \frac{dP}{d\nu} \quad \text{and} \quad \frac{f\,dQ}{d(P+Q)} = \frac{dQ}{d\nu}.$$

Thus this expression equals $\int\phi(p,q)d\nu$, *when* ϕ *is homogeneous* ($\phi(tx,ty) = t\phi(x,y)$, for all $t,x,y \geqslant 0$).

(b) $\int\phi(P,P) = 0$ will be assumed by taking ϕ *to be zero on the diagonal* $\{(x,x);\ x \in \mathbb{R}\}$. We would also like $\int\phi(P,Q) \geqslant 0$.

(c) If P and Q are only observed on a sub σ-algebra $\mathcal{B}$ of $\mathcal{A}$, their dissimilarity can only decrease. If ν is a probability, the densities of the restrictions are $E_\nu(p|\mathcal{B})$ (Exercise 6.4.2). Thus we want to ensure

$$E_\nu[\phi(p,q)] \geqslant E_\nu[\phi(E_\nu(p|\mathcal{B}), E_\nu(q|\mathcal{B}))].$$

This will be satisfied *when ϕ is convex* (from Jensen's Inequality 6.2.8).

If X is an observation on $(\Omega,\mathcal{A})$, we can take for densities of $X(P)$ and $X(Q)$ with respect to $X(\nu)$ (Exercise 6.4.3) $E_\nu(p|X = \cdot)$ and $E_\nu(q|X = \cdot)$; hence we again obtain

$$\int\phi(X(P), X(Q)) \leqslant \int\phi(P,Q).$$

Replacing the random phenomenon by an observation X which depends on the phenomenon reduces the dissimilarity.

Taking for ϕ the functions $(x,y) \longmapsto |x - y|$ and $(x,y) \longmapsto (\sqrt{x} - \sqrt{y})^2/2$, we obtain the following two distances between probabilities.

The Distance in Variation. $\|P - Q\| = \int|P - Q| = 2\int(P - Q)_+$. This is most often defined as

$$\|P - Q\| = \sup_{\|f\|\leqslant 1} \left|\int f\, dP - \int f\, dQ\right|.$$

In fact,

$$\begin{aligned}\sup_{\|f\|\leqslant 1} \left|\int f\, dP - \int f\, dQ\right| &= \sup_{\|f\|\leqslant 1} \left|\int f(p - q)d\nu\right| \\ &= \int 1_{(p>q)}(p-q)d\nu + \int 1_{(p<q)}(q-p)d\nu \\ &= \int |p-q|d\nu = 2\int(p-q)_+d\nu,\end{aligned}$$

since

$$\int(p - q)_+d\nu - \int(q - p)_+d\nu = P(\Omega) - Q(\Omega) = 0.$$

The Hellinger Distance. $H^2(P,Q) = (1/2)\int(\sqrt{P} - \sqrt{Q})^2 = 1 - \int\sqrt{PQ}$. This lends itself well to the passage to product measures (hence to the study of samples). For $P_1 = p_1\nu_1$, $Q_1 = q_1\nu_1$, $P_2 = p_2\nu_2$, $Q_2 = q_2\nu_2$, we have

$$P_1 \otimes P_2 = p_1 \otimes p_2 \cdot \nu_1 \otimes \nu_2;$$

$$Q_1 \otimes Q_2 = q_1 \otimes q_2 \cdot \nu_1 \otimes \nu_2.$$

We often set $\rho(P,Q) = \int\sqrt{PQ}$;

$$\rho(P_1 \otimes P_2, Q_1 \otimes Q_2) = \rho(P_1,Q_1)\rho(P_2,Q_2).$$

However, we have seen that many statistical problems are asymmetric. It is necessary to measure the dissimilarity between one measure and a reference measure. This is what led us to use in Sections 1.4 and 5.4 the χ^2 **distance** from P to Q,

$$\chi^2(P,Q) = \int \frac{(q-p)^2}{q} d\nu,$$

assuming q is ν a.s. strictly positive (Q equivalent to ν). The **Kullback information of** P **on** Q is often used, defined by

$$K(P,Q) = \int p \operatorname{Log}\frac{p}{q}\, d\nu \quad \text{for } P << Q$$

$$= \infty \quad \text{otherwise.}$$

It is related to the function $(x,y) \longmapsto -x \operatorname{Log}(y/x)$, which is convex and 1-homogeneous on $\{(x,y);\ x > 0\}$. This function is extended by setting $0 \operatorname{Log} 0 = 0 = 0 \operatorname{Log}(0/0)$. The hypothesis $P << Q$ implies that q is nonzero, P-a.s.

In this case we can write $p/q = dP/dQ$ (Q-a.s. hence P-a.s.) and

$$K(P,Q) = E_P\left[\operatorname{Log}\frac{dP}{dQ}\right].$$

This expression may be infinite.

Kullback information is positive; it is zero if and only if P *and* Q *are equal.* Assuming $P << Q$, we use the fact that the function $x \longmapsto x \operatorname{Log} x + 1 - x = \phi(x)$ is positive and equals zero only for $x = 1$:

$$\int \left(\frac{p}{q} \operatorname{Log}\frac{p}{q} + 1 - \frac{p}{q}\right) q\, d\nu = K(P,Q) \geq 0.$$

And $K(P,Q) = 0$ implies $p/q = 1$, Q-a.s., thus $P = Q$. Let $\mathcal{O}$ be a sub σ-algebra of $\mathcal{A}$ and $P|\mathcal{O}$, $Q|\mathcal{O}$ the traces of P and Q on $\mathcal{O}$:

$$K(P|\mathcal{O}, Q|\mathcal{O}) \leq K(P,Q).$$

This is clear for $K(P,Q)$ infinite. Assume, therefore, in order to check this, $P << Q$. Then from Exercise 6.4.2, $P|\mathcal{O}$ is absolutely continuous with respect to $Q|\mathcal{O}$, and its density is

$$E_Q\left[\frac{dP}{dQ} \mid \mathcal{O}\right],$$

an $\mathcal{O}$ measurable r.v. defined Q-a.s. Recall the expression

$$K(P,Q) = \int\left[\frac{dP}{dQ} \operatorname{Log} \frac{dP}{dQ} + 1 - \frac{dP}{dQ}\right] dQ = E_Q\left[\phi\left(\frac{dP}{dQ}\right)\right].$$

The function ϕ is strictly convex, the conditional Jensen's inequality may be applied to it:

$$E_Q\left[\phi\left(\frac{dP}{dQ}\right) \mid \mathcal{O}\right] \geqslant \phi\left[E_Q\left(\frac{dP}{dQ} \mid \mathcal{O}\right)\right], \quad Q\text{-a.s.};$$

and there is equality Q-a.s. if and only if dP/dQ is $\mathcal{O}$ measurable Q-a.s. On integrating we obtain the stated inequality.

On the other hand, *for $P \ll Q$, $K(P|\mathcal{O}, Q|\mathcal{O})$ coincides with $K(P,Q)$ if and only if dP/dQ is $\mathcal{O}$-measurable.* If T is a measurable observation on $(\Omega,\mathcal{A})$, $d(T(P))/d(T(Q))$ is $E_Q[dP/dQ|T = \cdot]$, $T(Q)$ a.s. By taking $\mathcal{O} = \sigma(T)$ above we obtain: $K(T(P),T(Q)) \leqslant K(P,Q)$ with equality in the case $P \ll Q$ if and only if dP/dQ is Q-a.s. $\sigma(T)$ measurable.

Kullback information has the specific property of being additive for independent observations:

$$K(P_1 \otimes P_2, Q_1 \otimes Q_2) = K(P_1,Q_1) + K(P_2,Q_2).$$

Exercises 6.4

E.1. Let f and g be two positive r.v.'s and let μ be a σ-finite measure on $(\Omega,\mathcal{A})$. What is the Lebesgue decomposiion of $f \cdot \mu$ with respect to $g \cdot \mu$?

E.2. Let $\mathcal{B}$ be a sub σ-algebra of $\mathcal{A}$. Let P and Q be two probabilities on $(\Omega,\mathcal{A})$ such that $P = p \cdot Q$. Prove that, on $(\Omega,\mathcal{B})$, we have $P = E_Q[p|\mathcal{B}] \cdot Q$. Let T be a measurable function from $(\Omega,\mathcal{A})$ into $(E,\mathcal{E})$; determine dP/dQ on $(\Omega,\sigma(T))$ and, for the distribution $T(P)$ and $T(Q)$ of T, $dT(P)/dT(Q)$.

E.3. Prove that two σ-finite measures α and β on $(\Omega,\mathcal{A})$ are equivalent if and only if α is absolutely continuous with respect to β and $d\alpha/d\beta$ is nonzero β-a.s. Prove

$$\frac{d\beta}{d\alpha} = \frac{1}{d\alpha/d\beta}.$$

E.4. With the notations of Section 6.4.2,

(a) Set $P\wedge Q = \int p\wedge q \, d\nu$ with $p\wedge q = \inf(p,q)$. Prove the relations

$$\|P - Q\| = 2 - 2P\wedge Q$$

$$P\wedge Q = \inf_{0\leqslant f\leqslant 1} \left| \int f \, dP + \int (1-f) dQ \right|$$

$$\|P - Q\| = 2 \sup_{A\in \mathcal{A}} |P(A) - Q(A)|$$

(b) Prove

$$H^2(P,Q) \leqslant \frac{1}{2}\|P - Q\| \leqslant H(P,Q)[2 - H^2(P,Q)]^{1/2}$$

$$K(P,Q) \geqslant -2 \operatorname{Log} (1 - H^2(P,Q)).$$

E.5. *Kullback information for bounded measures.* Let $\mathcal{U}$ be the set of positive random variables u in $(E,\mathcal{E})$, such that u and $1/u$ are bounded. Let α and β be two bounded measures on $(E,\mathcal{E})$. Set

$$J = \sup_{u\in\mathcal{U}} \left[\int \operatorname{Log} u \, d\alpha + \int (1-u) d\beta \right].$$

(a) Assume $\alpha << \beta$ and $\operatorname{Log}(d\alpha/d\beta)$ is in $L^1(\alpha)$. Let $a = d\alpha/d\beta$. Verify $J \leqslant \int (\operatorname{Log} a) d\alpha + \beta(E) - \alpha(E) = I$.

(b) For $A \in \mathcal{E}$, set $u = e^k 1_A + 1_{A^c}$ to conclude

$$k\alpha(A) + (1 - e^k)\beta(A) \leqslant J.$$

For $J < +\infty$, deduce from this $\alpha << \beta$.

(c) Assume $\alpha = a\cdot\beta$ and $a \in \mathcal{U}$. Verify $J = I$.

(d) Assume $\alpha = a\cdot\beta$. Let $a_n = \sup(a, 1/n)\wedge 1/n$. Prove that

$$\lim_{n\to\infty} \left[\int \operatorname{Log} a_n d\alpha + \int (1 - a_n) d\beta \right]$$

equals $\int \operatorname{Log} a \, d\alpha$ if $\operatorname{Log} a$ is in $L^1(\alpha)$ and $+\infty$ otherwise.

Consequence:

$$J = \begin{cases} \int \text{Log } a \, d\alpha + \beta(E) - \alpha(E) & \text{for } \alpha = a \cdot \beta \\ & \text{and Log } a \in L^1(\alpha) \\ +\infty & \text{otherwise.} \end{cases}$$

E.6. *Kullback information.* (a) Take $E = \{1,2, ..., n\}$, $P = (p_1, ..., p_n)$ a probability on E, and Q the uniform probability on E. Calculate $K(P,Q)$ as a function of the entropy of P (Exercise 1.3.13).

(b) Calculate $K(F_1,F_2)$ for $(F_1,F_2) = (F_i)_{i=1,2}$ with $F_i = N(m_i,\sigma^2)$, $\sigma > 0$ known;

$$F_i = N(m,\sigma_i^2), \quad m \in \mathbb{R} \text{ known};$$

$$F_i = N(m_i,\sigma_i^2), \quad m_i \in \mathbb{R}, \ \sigma_i^2 > 0;$$

$$F_i = b(p_i), \quad p_i \in]0,1[;$$

$$F_i = p(\lambda_i), \quad \lambda_i > 0.$$

(c) Let (X,Y) be a centered, Gaussian random vector of dimension 2, with correlation coefficient ρ. Denote its distribution by $F_{X,Y}$ and its marginal distributions by F_X and F_Y. Prove

$$K[F_{X,Y}, F_X \otimes F_Y] = -\frac{1}{2}\text{Log}(1 - \rho^2);$$

$$K(F_{X,Y}, F_X \otimes F_Y) + K(F_X \otimes F_Y, F_{X,Y}) = \frac{\rho^2}{1 - \rho^2}.$$

Bibliographic Notes

Conditional expectation is covered in Metivier, Breiman [3], and in Neveu [1] in a very complete manner.

Amongst works on Markov chains and various processes, let us cite Blanc, LaPierre and Fortet, Rényi, Karlin and Taylor, Ross, Hoel, Port and Stone (Vol. III), Borovkov (the theory of queues), Feller (Vols. I and II), Kemeny and Snell. Dissimilarities are developed in Kullback.

Chapter 7
DOMINATED STATISTICAL MODELS AND ESTIMATION

Objectives

In this chapter we begin a more systematic account of the statistical ideas which we have handled for the Bernoulli and Gaussian distributions and for samples of independent observations. We bring out the concepts which have, in the preceding examples, led to some results.

Next we show how to measure the information carried by a statistical experiment on the parameter on which it depends. Finally we tackle the idea of maximum likelihood and show how it allows some estimators and very useful tests to be constructed simply.

7.1. Dominated Statistical Models

7.1.1. Examples of Experiments Dependent on a Parameter

In the preceding chapters (2, 5, 6) we have studied the following experiments.

(a) *An n-sample from a distribution F on* $(E,\mathcal{E})$. The observations are, for example, defined on $(\Omega,\mathcal{A}) = (E,\mathcal{E})^n$, given the probability $P_F = F^{\otimes n}$. If nothing further is specified, F plays the role of an unknown parameter varying in the set of probabilities on $(E,\mathcal{E})$. Some very weak restrictions can be imposed. For example, F has a finite mean, $(E,\mathcal{E})$ being Euclidean (for the case of empirical methods of estimation and nonparametric tests (Section 4.4.5)). However, F is often constrained to vary in a family of probabilities, denoted $(F_\theta)_{\theta \in \Theta}$, where Θ is a very much 'smaller' set. This is the case for Bernoulli trails with parameter $\theta \in [0,1]$, or the distributions $\mathcal{N}(m,\sigma^2)$ with $\theta = (m,\sigma^2)$, $\theta \in \mathbb{R} \times]0,\infty[$. In writing this, we have made the convention of denoting P_θ for P_{F_θ} and have replaced a subset of the set of probabilities on $(\Omega,\mathcal{A})$ by a subset of $\mathbb{R}^k$. This is possible if θ is *identifiable*, i.e. if the mapping $\theta \longmapsto F_\theta$ is bijective.

(b) *The first* $(n + 1)$ *observations of a Markov chain.* We observe $(X_0, ..., X_n)$ the first steps of a Markov chain with initial distribution ρ and transition π on $(E,\mathcal{E})$. We can take for the space of realizations $(\Omega,\mathcal{A}) = (E,\mathcal{E})^{n+1}$, and for probability $P_{\rho,\pi} = \rho \otimes \pi^{\otimes n}$. If no other restrictions are made, the parameter here is (ρ,π).

7.1.2. Dominated Statistical Models

Definition 7.1.1. A measurable space $(\Omega,\mathcal{A})$ together with a family $(P_\theta)_{\theta \in \Theta}$ of probabilities is called a **statistical (or experimental) model**; Θ is the set of values of the parameter θ.

The set Θ being arbitrary, there are difficult problems related to all the questions of sets of measure zero for the various P_θ, except when they are "dominated."

Definition 7.1.2. The above model is **dominated** if there exists a σ-finite measure μ (said to be **dominant**) such that for all $\theta \in \Theta$, P_θ is dominated by μ (P_θ is absolutely continuous with respect to μ). We will denote $P_\theta = p_\theta \cdot \mu$.

Examples. (a) The model $(E,\mathcal{E},\delta_\theta)_{\theta \in E}$ is only dominated for countable E.

(b) Let μ be a σ-finite measure on $(E,\mathcal{E})$ and, for all $\theta \in \Theta$, $F_\theta = \ell(\theta,\cdot) \cdot \mu$, where $\ell(\theta,\cdot)$ is a positive r.v. on $(E,\mathcal{E})$, having integral with respect to μ equal to 1. An n-sample from F_θ has distribution P_θ, the density of which with respect to $\mu^{\otimes n}$ is $L(\theta) = \ell(\theta,X_1) ... \ell(\theta,X_n)$, still denoting the coordinates by $X_1, ..., X_n$; $\mu^{\otimes n}$ is then a dominant measure.

(c) Let there be a Markov chain on $(E,\mathcal{E})$ the initial distribution ρ of which is given, and the transition $\pi_\theta = \pi(\theta,\cdot,\cdot)$ of which satisfies for arbitrary (θ,x):

$$\pi(\theta,x;\cdot) = \ell(\theta,x,\cdot)\pi(x;\cdot),$$

for a given σ-finite transition π from E into E. If the first $n + 1$ states are observed, we can take $(\Omega,\mathcal{A}) = (E,\mathcal{E})^{n+1}$ and $P_{\theta,\rho} = \rho \otimes \pi_\theta^{\otimes n}$. Then $\mu = \rho \otimes \pi^{\otimes n}$ is a dominant measure and a density of $P_{\theta,\rho}$ with respect to μ is

$$L(\theta) = \ell(\theta,X_0,X_1)\ell(\theta,X_1,X_2) \dots \ell(\theta,X_{n-1},X_n).$$

7.1.3. Privileged Dominants

Let $(P_\theta)_{\theta\in\Theta}$ be a family of dominated probabilities. The set $D(\Theta)$ of dominant measures is preordered by the order of absolute continuity. We shall study the minimal elements of $D(\Theta)$.

Proposition 7.1.3. *Consider the "convex hull" of* Θ, *denoted by* conv Θ, *the set of all convex combinations* $\Sigma a_n P_{\theta_n}$, (P_{θ_n}) *a sequence of probabilities and* (θ_n) *being a sequence of elements of* Θ. *Then there exists a probability* $P \in$ conv Θ *which dominates* $(P_\theta)_{\theta\in\Theta}$.

Given this proposition, if μ dominates $(P_\theta)_{\theta\in\Theta}$, μ dominates P and every measure which dominates $(P_\theta)_{\theta\in\Theta}$ and is minimal is equivalent to P. A P-negligible set is negligible for every P_θ. Conversely, since P is in conv Θ, every set having measure zero for each P_θ has P-measure zero.

Definition 7.1.4. For a dominated family $(P_\theta)_{\theta\in\Theta}$ a **privileged dominant** is a probability P which is a combination of a countable subfamily "equivalent" to $(P_\theta)_{\theta\in\Theta}$. A set is P-negligible if and only if it has measure zero for all P_θ. A set is said to be **negligible**, and measurable functions are said to be **equivalent**, if that is the case for P.

In the model $(\Omega,\mathcal{A},(P_\theta)_{\theta\in\Theta})$ the term a.s. will from now on be relative to a privileged dominant P; an r.v. and its equivalence class for P-a.s. equality are then often identified. An r.v. (defined P-a.s.) will be said to be integrable (or square integrable) if that is the case for each of the probabilities P_θ, $\theta \in \Theta$.

Proof of the Proposition. Let μ be a σ-finite measure which dominates $(P_\theta)_{\theta\in\Theta}$; μ can be replaced by $Z\cdot\mu$ for Z a strictly positive and integrable r.v. We can thus assume that μ is a probability. Let $\mathcal{C}\subset\mathcal{A}$ be the set of events C, charged by μ and such that there exists a probability $p_C \in \operatorname{conv}\Theta$ with density $p_C\mu$-a.s. strictly positive on C. Let (C_n) be a sequence of $\mathcal{C}$ such that $(\mu(C_n))$ tends to $\sup_{C\in\mathcal{C}}\mu(C)$. Set

$$C_0 = \bigcup_{n=1}^{\infty} C_n, \quad p_0 = \sum_{n\geq 1} 2^{-n} p_{C_n}, \quad P_0 = p_0\mu.$$

On C_0, p_0 is μ-a.s. strictly positive and C_0 is charged by μ; P_0 is in $\mathcal{C}$. Let $P_\theta = p_\theta\cdot\mu$, $\theta\in\Theta$; $(P_\theta + P_0)/2 \in \operatorname{conv}\Theta$, thus $\{p_\theta + p_0 > 0\}$ and C_0 coincide μ-a.s. because of the maximality of $\mu(C_0)$. Let A be an event of P_0-measure zero; $A\cap C_0$ has μ-measure zero, thus has P_θ-measure zero, and $A\subset(A\cap C_0)\cup C_0^c$ has P_θ-measure zero. The probability P_0 dominates P_θ for every $\theta\in\Theta$.

7.1.4. Likelihood

We are going to choose in a dominated model a likelihood, a particular version of the density; the most regular possible. We have already made such a choice in the preceding chapters.

Definition 7.1.5. A **likelihood** of the dominated model of Definition 7.1.1 is a function $(\theta,\omega)\longmapsto L(\theta,\omega)$ such that, for all θ, $L(\theta,\cdot)$ is a version of the density of P_θ with respect to μ. Denote $L(\theta)$ for $L(\theta,\cdot)$.

When the choice of a regular likelihood (continuous, infinitely differentiable, etc.) is clear, it will be made without further discussion, and by abuse of language we speak of "the" likelihood as if it was unique. This is the way in which we chose the densities in Section 3.3.3. If $\omega\in\Omega$ is charged by μ, every likelihood equals $P_\theta(\omega)/\mu(\omega)$ at the point ω. In particular for Ω countable and μ the measure which gives to each point the weight 1, $L(\theta,\omega)$ equals $P_\theta(\omega)$ at each point ω.

7.1.5. Conditional Expectation in a Dominated and Sufficient Model

Assume known an observation T on $(\Omega,\mathcal{A})$, thus the σ-algebra $\mathcal{B} = \sigma(T)$; or assume observed a sub σ-algebra $\mathcal{B}$ of $\mathcal{A}$. We know that, in the model $(\Omega,\mathcal{A},P_\theta)$, this allows a positive r.v. Y to be approximated by its conditional expectation, defined P_θ-a.s. If the model is dominated by a probability P (we will denote by E_θ and E the expectations for P_θ and P), *we associate to each choice of likelihood a version of the conditional expectations defined* P_θ-a.s.; this results from:

Proposition 7.1.6. *For* $\mathcal{B}$ *a sub* σ*-algebra of* $\mathcal{A}$, *we have, with the preceding notations,* P-a.s.:

$$E_\theta[Y|\mathcal{B}] = \frac{E[L(\theta)Y|\mathcal{B}]}{E[L(\theta)|\mathcal{B}]} \quad \text{on} \quad E[L(\theta)|\mathcal{B}] > 0$$

$$= 0 \quad \text{on} \quad E[L(\theta)|\mathcal{B}] = 0.$$

Proof. The event $\Gamma = \{E(L(\theta)|\mathcal{B}) = 0\}$ is not charged by P_θ. In fact, $P_\theta[\Gamma] = E[L(\theta)1_\Gamma] = 0$. We can thus take $E_\theta(Y|\mathcal{B})$ to be arbitrary on this set. For $A \subset \Omega\backslash\Gamma$, and $\mathcal{B}$ measurable, we have

$$E_\theta\left[1_A \frac{E(L(\theta)Y|\mathcal{B})}{E(L(\theta)|\mathcal{B})}\right] = E\left[1_A L(\theta) \frac{E(L(\theta)Y|\mathcal{B})}{E(L(\theta)|\mathcal{B})}\right]$$

$$= E[1_A E(L(\theta)Y|\mathcal{B})] = E[L(\theta)Y\, 1_A]$$

$$= E_\theta[1_A Y].$$

Sufficiency. A probability P_θ is characterized by its trace $P_{\theta|\mathcal{B}}$ on $\mathcal{B}$ and by the expectation of bounded r.v.'s conditional on $\mathcal{B}$. In fact, for $A \in \mathcal{A}$,

$$P_\theta(A) = E_\theta[E_\theta(1_A|\mathcal{B})] = \int E_\theta(1_A|\mathcal{B})dP_{\theta|\mathcal{B}}.$$

When a version of $E_\theta(1_A|\mathcal{B})$ independent of θ can be found, P_θ is characterized by its trace on $\mathcal{B}$. We have seen, for example, that the distribution of an n-sample from a Bernoulli distribution $(X_1, ..., X_n)$ conditional on $(X_1 + ... + X_n)$

does not depend on its parameter. Thus the distribution of $(X_1 + \dots + X_n)$ allows that of $(X_1, \dots, X_n)$ to be calculated (Section 2.3.3).

Definition 7.1.7. For a model $(\Omega, A, P_\theta)_{\theta \in \Theta}$ dominated by a σ-finite measure μ, a σ-algebra B (resp. an observation T) is **sufficient** if, for all $A \in A$, we can find a version of $E_\theta(1_A|B)$ independent of θ and defined μ-a.s., (resp. of $E_\theta(1_A|T)$). This version is μ-a.s. denoted by $E[1_A|B]$. (We can then also talk of $E(Y|B)$ for an r.v. $Y \geqslant 0$ or bounded.)

Theorem 7.1.8. *The following propositions are equivalent:*

(1) *(resp. T) is sufficient;*
(2) *For every privileged dominant P, dP_θ/dP is* a.s. B *(resp. $\sigma(T)$) measurable;*
(3) *For every dominant measure μ, we can find an* r.v. h, *which does not depend on Θ, and G_θ a B measurable* r.v. *(resp. an* r.v. g_θ *on the set of values of T) such that*

$$\frac{dP_\theta}{d\mu} = h \cdot G_\theta \quad \left[resp.\ \frac{dP_\theta}{d\mu} = h \cdot g_\theta(T)\right].$$

Proof. (a) Let $P = \sum_n a_n P_\theta$ be a privileged dominant and let B be sufficient:

$$E_P[1_A|B] = \sum_{n=1}^{\infty} a_n E_{\theta_n}[1_A|B] = E[1_A|B]$$

$$P_\theta(A) = E_P\left[\frac{dP_\theta}{dP} 1_A\right] = E_\theta[E[1_A|B]]$$

$$= E_P\left[\frac{dP_\theta}{dP} E(1_A|B)\right]$$

$$= E_P\left[E(1_A|B) E\left(\frac{dP_\theta}{dP} \mid B\right)\right]$$

$$= E_P\left[1_A\, E\left(\frac{dP_\theta}{dP} \mid B\right)\right].$$

Thus $E(dP_\theta/dP \mid B)$ is a B measurable version of dP_θ/dP.

(b) As soon as a probability P satisfies (3) (or (2)), (1) is obtained from Proposition 7.1.6. Thus, if a measure satisfies (3) a P-privileged dominant which it dominates satisfies (2).

Examples. (a) For an n-sample from an exponential distribution with parameter $\theta > 0$, or a Bernoulli distribution with parameter $\theta \in [0,1]$, $X_1 + \dots + X_n$ is sufficient.

(b) For an n-sample from an $N(m,\sigma^2)$ distribution ($m \in \mathbb{R}$, $\sigma > 0$), the likelihood is

$$L(m,\sigma^2) = \frac{1}{(2\pi\sigma^2)^{n/2}} \exp\left\{\frac{-1}{2\sigma^2}\left[\sum_{i=1}^{n} X_i^2 - 2m \sum_{i=1}^{n} X_i + nm^2\right]\right\}.$$

The statistic $(\Sigma_{i=1}^n X_i^2, \Sigma_{i=1}^n X_i)$ (or $\bar{X}$, S^2) is sufficient.

(c) For an n sample from a uniform distribution on the interval $[a,b]$ of $\mathbb{R}$ $(-\infty < a < b < +\infty)$, $(\inf_{1 \leqslant i \leqslant n} X_i, \sup_{1 \leqslant i \leqslant n} X_i)$ is sufficient. The likelihood may be written as

$$L(a,b) = \frac{1}{(b-a)^n} \prod_{i=1}^{n} \mathbf{1}_{[a,b]}(X_i)$$

$$= \frac{1}{(b-a)^n} \mathbf{1}_{[a \leqslant \inf_{1 \leqslant i \leqslant n} X_i \leqslant \sup_{1 \leqslant i \leqslant n} X_i \leqslant b]} .$$

(d) The first $n + 1$ states $(X_0, \dots, X_n)$ of a Markov chain are observed, with arbitrary initial distribution ρ and transition π; $\theta = (\rho,\pi)$. Let $N_n^{ij} = \Sigma_{p=0}^{n-1} \mathbf{1}_{(X_p=i, X_{p+1}=j)}$ be the number of jumps from state i to state j. The density of $\rho \otimes \pi^{\otimes n}$ with respect to the measure which gives mass 1 to each point of E^{n+1} is, for countable E,

$$\rho(X_0)\pi(X_0,X_1) \dots \pi(X_{n-1},X_n) = \rho(X_0) \prod_{i,j} [\pi(i,j)]^{N_n^{ij}} .$$

Hence, $(X_0,(N_n^{ij})_{1 \leqslant i,j \leqslant k})$ is sufficient.

For an n-sample from an arbitrary distribution F on E, we would obtain likewise that $(N_n^i)_{1 \leqslant i \leqslant k}$ is sufficient, where N_n^i is the number of components of the n-sample equal to i; in other words, the sample distribution is sufficient.

Minimal Sufficient σ-Algebra. For a choice of likelihood, the σ-algebra $\sigma(L(\theta); \theta \in \Theta)$ is sufficient, but it depends on the choice which has been made of the version $L(\theta)$ of dP_θ/dP. On the other hand, its P-completion in $\mathcal{A}$ for a privileged dominant P (Definition 6.1.1) does not depend on the likelihood chosen. This σ-algebra $\mathcal{M}$ is also sufficient. Every sufficient σ-algebra which contains all the events of P-measure zero of $\mathcal{A}$ contain $\mathcal{M}$. We say that $\mathcal{M}$ is "*minimal sufficient.*"

When an observation T, measurable on $(\Omega,\mathcal{A})$ is such that its P-completion in $\mathcal{A}$ of $\sigma(T)$ contains $\mathcal{M}$, T is a sufficient statistic. T is minimal sufficient if the P-completion of $\sigma(T)$ in $\mathcal{A}$ is $\mathcal{M}$. The distribution of T on $(\Omega,\mathcal{A},P_\theta)$ depends on the parameter θ. On the other hand, (Exercise 7.1.4), a statistic independent of T has a distribution which does not depend on θ, and its information does not tell us anything about θ.

7.1.6. The Exponential Model

In Section 3.3.2 we introduced the exponential models. Let us recall the definition here.

Definition 7.1.9. The exponential model. Let $(\Omega,\mathcal{A},\mu)$ be a measure space, μ being σ-finite and T a random vector taking values in $\mathbb{R}^k$. Assume that Θ is a subset of $\mathbb{R}^k$ defined by $\Theta = \{\theta;\ e^{\psi(\theta)} = \int e^{<\theta,T>}d\mu < \infty\}$. We can then define an **exponential model** by its likelihood $\exp(-\psi(\theta) + <\theta,T>)$ for $\theta \in \Theta$. This is the model $(\Omega,\mathcal{A},P_\theta)_{\theta\in\Theta}$, for $dP_\theta/d\mu = \exp(-\psi(\theta) + <\theta,T>)$.

In the rest of this chapter, we will say "*the exponential model*" in speaking of this model and assuming $\mathring{\Theta}$ nonempty.

Note. *We have seen (Section 3.3.2) that Θ is convex and we have proved that, in the interior $\mathring{\Theta}$ of Θ, the function ϕ is infinitely differentiable. We assume $\mathring{\Theta}$ is nonempty and often study an exponential model by replacing Θ by its interior.*

In the exponential model the probabilities P_θ are all equivalent, any P_{θ_0} is privileged dominant, and

$$\frac{dP_\theta}{dP_{\theta_0}} = \exp<\theta - \theta_0,T>.$$

Let us show that T is minimal sufficient: $\sigma(\exp\langle\theta,T\rangle;\ \theta \in \Theta)$ *is contained in* $\sigma(T)$ *and, for* $u \in \mathbb{R}^k$,

$$\langle u,T\rangle = \lim_{h\to 0} \frac{\exp\langle\theta + uh,T\rangle - \exp\langle\theta,T\rangle}{h \exp\langle\theta,T\rangle}$$

is $\sigma(\exp\langle\theta,T\rangle;\ \theta \in \mathring{\Theta})$ *measurable. The* σ*-algebras* $\sigma(T)$ *and* $\sigma(\exp\langle\theta,T\rangle;\ \theta \in \Theta)$ *coincide.*

Proposition 7.1.10. *For the exponential model of Definition* 7.1.9, *T is a minimal sufficient statistic.*

Examples. (a) Look back at Examples (a) and (b) of the preceding section and the case of the n-sample from a distribution on a finite space E seen in (d).

(b) *An n-sample from an exponential model* $(\Omega,\mathcal{A},P_\theta) = (E,\mathcal{E},F_\theta)^n$, with $F_\theta = \exp[-\phi(\theta) + \langle\theta,T\rangle]$. μ *is again exponential* (Θ unchanged). In fact, if $X_1, \ldots, X_n$ are the n coordinate functions of Ω, we have

$$P_\theta = F_\theta^{\otimes n} = \exp[-n\phi(\theta) + \langle\theta,T(X_1) + \ldots + T(X_n)\rangle]\mu^{\otimes n},$$

and $T(X_1) + \ldots + T(X_n)$ *is minimal sufficient.*

7.1.7. How Can we Improve an Estimator?

Definition 7.1.11. Estimator. Let α be a function from Θ into $\mathbb{R}$. We try to estimate $\alpha(\theta)$; an **estimator** U is an r.v. on $(\Omega,\mathcal{A},(P_\theta)_{\theta\in\Theta})$ defined possibly a.s. Its **quadratic risk**, which we try to minimize, is, if the parameter equals θ, $E_\theta[U - \alpha(\theta)]^2$. If U is integrable its **bias** b is the function $\theta \longmapsto b(\theta) = E_\theta(U)$. If the bias is zero, U is said to be **unbiased** and its quadratic risk coincides with the variance.

Assume given, a square integrable estimator U, and a sufficient statistic T; $E[U|T]$ is also an estimator and the conditional expectation is a projection. From Pythagoras' theorem, for all θ,

$$E_\theta[U - \alpha(\theta)]^2 = E_\theta[E(U|T) - \alpha(\theta)]^2 + E_\theta[U - E(U|T)]^2.$$

Theorem 7.1.12. *If U is a square integrable estimator of* $\alpha(\theta)$, *and if T is a sufficient statistic,* $E(U|T)$ *is an estimator with* *lower risk than that of U (strictly lower if U is not a function of T).*

Consider an unbiased estimator U of $\alpha(\theta)$; this is improved by replacing it by $\Psi(T) = E(U|T)$, which is also unbiased. In the class of square integrable unbiased estimators, it is thus sufficient to consider those which are functions of T. Let $\Psi_1(T)$ and $\Psi_2(T)$ be two estimators of this form; $(\Psi_1 - \Psi_2)(T)$ has zero mean for P_θ, θ *arbitrary*. In numerous cases, this is sufficient to assure the equality of $\Psi_1(T)$ and $\Psi_2(T)$.

Definition 7.1.13. A statistic T, which is a measurable function from $(\Omega,\mathcal{A})$ into $(E,\mathcal{E})$, is **complete** (resp. b-**complete**) for the model if every r.v. ϕ on $(E,\mathcal{E})$ such that $\phi(T)$ is integrable (resp. bounded) and has integral zero for P_θ, θ arbitrary, is zero (P-a.s.).

Note. *In spite of the unfortunate clash between the vocabularies, this idea has nothing to do with the completion of the σ-algebra $\sigma(T)$ with respect to a given measure (Exercises* 3.1.19 *or* 6.1).

Theorem 7.1.14. *Let U be an unbiased, square integrable estimator of $\alpha(\theta)$ and T a complete, sufficient statistic. The unbiased estimator $E(U|T)$ is then the unique unbiased estimator which is a function of T; its variance is, for all θ, less than the variance of U. The variance of this estimator is, for all θ, lower than that of every unbiased estimator. It is* **uniformly of minimum variance among unbiased estimators** (*this will be abbreviated by* MVUE – *minimum variance unbiased estimator*).

Some Criteria. (a) *In the exponential model, T is sufficient and complete.* Let f be an r.v. on $\mathbb{R}^k$ such that $f(T)$ is integrable and centered for P_θ, for arbitrary θ. Let F be the image measure of μ by T:

$$0 = E_\theta[f(T)] = \int f(T)\exp(-\psi(\theta) + \langle\theta,T\rangle)d\mu$$

$$= \exp(-\psi(\theta))\int f(t)(\exp\langle\theta,t\rangle)dF(t).$$

The Laplace transforms of the measures f_+F and f_-F are equal on Θ : $f_+F = f_-F$; f is zero F_- a.s. and $f(T)$ is zero P-a.s.

(b) *If T is a sufficient and b-complete statistic, then T is minimal sufficient.* The σ-algebra $\sigma(T)$, complete for P, contains the minimal sufficient σ-algebra $\mathcal{M}$. For $C \in \sigma(T)$ and all θ, $E_\theta[1_C - E(1_C|\mathcal{M})]$ is zero, and $1_C - E[1_C|\mathcal{M}]$ is zero. Thus 1_C is a.s. $\mathcal{M}$-measurable and T is measurable for the P-completed, σ-algebra $\mathcal{M}$: T is minimal. The converse is false (Exercise 7.1.3).

Exercises 7.1

E.1. Let $(X_1, ..., X_n)$ be an n-sample from $\mathcal{N}(m,\sigma^2)$. Give a minimum variance unbiased estimator of m, when $\sigma = \sigma_0$ is known, then when σ is unknown. Give a minimum variance unbiased estimator for σ^2 when $m = m_0$ is known, then when m is unknown.

E.2. *A Gaussian vector.* Consider the observation of a Gaussian vector $(X_1, ..., X_n)$ from an $\mathcal{N}_n(m,\Gamma)$ distribution for $m \in \mathbb{R}^n$, and Γ a positive definite symmetric matrix. Give a minimal sufficient statistic

(a) when Γ is known and m is the parameter;
(b) when m is known and Γ is the parameter;
(c) when m and Γ are unknown;
(d) for $\Gamma = \sigma^2 I_n$, and m in a given vector subspace E of $\mathbb{R}^n$, σ^2 and m unknown.

In Proposition 5.1.5 show that, for all $v \in \mathbb{R}^n$, $<v,X_E>$ is an MVUE of $<v,m>$ and that the residual variance is an MVUE of σ^2. Show that for every unbiased estimator Z of m, we have $E(\|X_E - m\|^2) \leq E(\|Z - m\|^2)$.

E.3. *Uniform distributions.* (a) Let $(X_1, ..., X_n)$ be an n-sample from $\mathcal{U}_{[0,\theta]}$, $\theta \in]0,\infty[$; give a minimal sufficient complete statistic. Improve the unbiased estimator $\overline{X}$ of the mean $\theta/2$ and give an MVUE of $\theta/2$.

(b) Let $(X_1, ..., X_n)$ be an n-sample from $\mathcal{U}_{[\theta-1/2,\theta+1/2]}$, $\theta \in \mathbb{R}$. Show that $X_{(n)} - X_{(1)}$ is a distribution-free statistic (its distribution does not depend on θ). Show that $(X_{(1)},X_{(n)})$ is minimal sufficient, but not complete.

E.4. *Distribution-free statistic.* Let $(\Omega,\mathcal{A},P_\theta)_{\theta\in\Theta}$ be a dominated model and P a privileged dominant. Let T be a sufficient statistic and Z a measurable function from $(\Omega,\mathcal{A})$ into $(E',\mathcal{E}')$.

(a) Show that, if Z is independent of T under P, it is a distribution-free statistic (its distribution $Z(P_\theta)$ does not depend on θ).
(b) Show that, if Z is distribution-free and if T is b-complete, Z is independent of T under P and P_θ.
(c) Give an example to show that (b) is false if T is not assumed b-complete (see, for example, Exercise 7.1.3).

E.5. *The Hypergeometric distribution.* For N and n fixed, consider $\Theta = \{0,1, ..., N\}$ and the family of hypergeometric distributions on $\Omega = \{0,1, ..., n\}$, $(P_\theta)_{\theta \in \Theta}$,

$$P_\theta(k) = \frac{\binom{\theta}{k}\binom{N-\theta}{n-k}}{\binom{N}{n}}, \quad k \leqslant \inf(n,\theta).$$

We observe X with distribution P_θ. Show that X is sufficient. Let ϕ be an r.v. on Ω for which $E_\theta[\phi(X)]$ is zero for all θ. Show that ϕ is zero (we can show by recurrence that $\phi(k)$ is zero for every integer k). Give a minimum variance unbiased estimator of θ; calculate its variance (see Exercise 1.3.9).

E.6. *An n-sample from a distribution on* $\{1,2, ..., k\}$. Consider the set Θ of distributions on $\{1,2, ..., k\}$. We observe an n-sample from $\theta \in \Theta$, $(X_1, ..., X_n)$. Construct an exponential submodel and give a minimal sufficient statistic. Give a minimum variance unbiased estimator of $\theta(1)$.

E.7. *A Markov chain on* $\{1,2, ..., k\}$. Consider the set Θ of pairs (ρ,π) where ρ is a given distribution on $E = \{1, ..., k\}$ and π is an unknown transition from E into E. We observe $(X_0, X_1, ..., X_n)$, the first $n + 1$ states of a Markov chain with initial distribution ρ and transition π. Construct an exponential submodel; give a minimal sufficient, complete statistic.

E.8. *Order statistics.* Let $(X_1, ..., X_n)$ be the n coordinate functions of $\mathbb{R}^n$.

(a) To $\theta = (\theta_1, ..., \theta_n) \in \mathbb{R}^n$, we associate the distribution F_θ with density proportional to $x \longmapsto \exp(\theta_1 x + \theta_2 x^2 + ... + \theta_n x^n - x^{2n}]$. Show that the statistic

$$\left[\sum_{i=1}^{n} X_i, \sum_{i=1}^{n} X_i^2, \ldots, \sum_{i=1}^{n} X_i^n\right]$$

is complete for the family of probabilities $(F_\theta^{\otimes n})_{\theta \in \mathbb{R}^n}$ on $\mathbb{R}^n$.

(b) Prove that the order statistic $(X_{(1)}, \ldots, X_{(n)})$ is a sufficient and complete statistic for the family of distributions of n-samples from a distribution on $\mathbb{R}$ having a density (see Exercise 3.1.15 and 4.4.3). Deduce from this that $\bar{X}$ is the MVUE of the mean for the family of distributions having a density and a mean.

E.9. Let $(X_1, \ldots, X_n)$ be an n-sample from an exponential distribution with parameter $\theta > 0$. We want to estimate $P(X_1 \leqslant t)$ for a given $t > 0$. Show that the sample estimator $(1/n)\Sigma_{i=1}^n 1_{(X_i \leqslant t)}$ is unbiased. What is its variance? Determine an estimator with minimum variance amongst the unbiased estimators.

7.2. Dissimilarity in a Dominated Model

7.2.1. Distances in a Statistical Model

The statistician has available observations that he wishes to use to take a decision on the value of the parameter θ which regulates the phenomenon: a test problem, an estimation problem, ... This will be very much easier if the two probabilities to be distinguished are very dissimilar.

The space Θ is sometimes given with *a metric space structure.* This structure is implicit in the case where Θ is finite or countable (it is the discrete metric: $d(\theta,\theta')$ is zero for $\theta = \theta'$ and equals 1 otherwise). It is explicit if Θ is, for example, an open set of $\mathbb{R}^k$ (the case of samples from Gaussian distributions

$$\Theta = \{(m,\sigma^2) \in \mathbb{R} \times]0,\infty[\}).$$

Θ can also be equipped with distances between probabilities defined in Section 6.4 (variation distance, Hellinger distance).

The tradition is to distinguish between *parametric* and *nonparametric model*, but this distinction is not always very clear. We can consider as "parametric" a problem where Θ is (with a judicious choice of parameter) an open set in $\mathbb{R}^k$ together with Euclidean distance.

7.2.2. Dissimilarities in a Dominated Model

Let $(\Omega,\mathcal{A},P_\theta)_{\theta\in\Theta}$ be a dominated model, P being a privileged dominant. The Kullback information $K(P_\theta,P)$ of P_θ on P is $E_\theta[\mathrm{Log}(dP_\theta/dP)]$. From Section 6.4.2 and Theorem 7.1.8 a sub σ-algebra $\mathcal{B}$ of $\mathcal{A}$ (or a statistic T) is minimal if, and only if, $K(P_\theta|\mathcal{B},\ P|\mathcal{B})$ (or $K(T(P_\theta),T(P))$ coincides with $K(P_\theta,P)$: *a sufficient statistic is a statistic which does not diminish the Kullback information of* P_θ *on* P. In the case where the measures $(P_\theta)_{\theta\in\Theta}$ are all equivalent, each one of them is a privileged dominant; to every pair $(\theta,\theta')\in\Theta^2$ associate $K(P_\theta,P_{\theta'})$, denoted by $K(\theta,\theta')$.

Example. *For the exponential model,*

$$K(\theta,\theta') = \psi(\theta') - \psi(\theta) + \langle\theta - \theta',E_\theta(T)\rangle.$$

We have seen (Section 3.3.2) that, if θ is in the interior of Θ, $E_\theta(T)$ equals grad $\psi(\theta)$. Thus in this case,

$$K(\theta,\theta') = \psi(\theta') - \psi(\theta) - \langle\theta' - \theta,\ \mathrm{grad}\ \psi(\theta)\rangle.$$

For θ in the interior of Θ, the function ψ has derivatives of all orders. When θ' is close to θ, a second-order approximation of $K(\theta,\theta')$ is obtained with the help of the second derivative of ψ:

$$K(\theta,\theta') = \frac{1}{2}\, d^2\psi(\theta)[\theta' - \theta] + o(\|\theta' - \theta\|^2),$$

by setting

$$d^2\psi(\theta)[h_1, \ldots, h_k] = \sum_{i=1}^{k}\sum_{j=1}^{k} h_i h_j \frac{\partial^2\psi}{\partial\theta_i\partial\theta_j}(\theta) = {}^t h I(\theta) h.$$

The second derivative matrix $I(\theta) = \{\partial^2\psi/\partial\theta_i\partial\theta_j\}$ appears as a local element of information. In what follows we try to generalize this to other dominated models.

7.2.3. Local Dissimilarity: Fisher's Information

For a problem of estimation of the parameter $\theta \in \Theta$, Θ always assumed to be an open set in $\mathbb{R}^k$, the important thing is the variability of the distributions in the neighborhood of θ. From this comes the idea of local information.

Consider a dominated model, with Θ open in $\mathbb{R}^k$. Assume given a dominant probability P and a likelihood L; denote $L(\theta) = L(\theta; \cdot)$. We formulate hypotheses on the function $\theta \longmapsto L(\theta,\omega)$, it being always understood that they are true for each ω. Assume first of all that this function is differentiable.

If the order of differentiation and integration can be changed, by differentiating the function $\theta \longmapsto E[L(\theta)]$ identically equal to 1, we obtain

$$E[\text{grad } L(\theta)] = 0;$$

$$E_\theta\left[\frac{\text{grad } L(\theta)}{L(\theta)}\right] = E_\theta[\text{grad Log } L(\theta)] = 0.$$

Assume $L(\theta)$ is in $L^2(P)$; let $Y \in L^2(P)$. If the order of differentiation and integration can again be changed (as before where Y was constant), we have

$$\begin{aligned}\text{grad } E_\theta[Y] &= \text{grad } E[L(\theta)\cdot Y] = E[\text{grad } L(\theta)\cdot Y] \\ &= E_\theta[\text{grad}(\text{Log } L(\theta))\cdot Y] \\ &= E_\theta[\text{grad}(\text{Log } L(\theta))\cdot(Y - E_\theta(Y))].\end{aligned}$$

Let $u \in \mathbb{R}^k$, we thus have

$$<u,\text{grad } E_\theta(Y)> = E_\theta[<u,\text{grad Log } L(\theta)>(Y-E_\theta(Y))].$$

From which, by Schwarz's inequality,

$$E_\theta[Y - E_\theta(Y)]^2 \geqslant \frac{<u,\text{grad } E_\theta(Y)>^2}{E_\theta[<u,\text{grad Log } L(\theta)>^2]}\,.$$

Definition 7.2.15. We call the **Fisher information matrix**, the matrix

$$\begin{aligned}I(\theta) &= \left\{E_\theta\left[\frac{\partial}{\partial\theta_i} \text{Log } L(\theta) \cdot \frac{\partial}{\partial\theta_j}\text{Log } L(\theta)\right]\right\}_{1\leqslant i,j\leqslant k} \\ &= \{I_{ij}(\theta)\}_{1\leqslant i,j\leqslant k}\,.\end{aligned}$$

For $k = 1$, $I(\theta) = E_\theta[L'(\theta)/L(\theta)]^2$. The matrix $I(\theta)$ is positive; it is the covariance matrix for P_θ of grad Log $L(\theta)$. Let us look then, for the inequality above, for the vector u maximizing the right-hand term. In view of the form of the inequality, we can normalize u by $\langle u, \text{grad } E_\theta(Y)\rangle = 1$ and apply the method of Lagrange multipliers. Setting $v = \text{grad } E_\theta(Y)$, we are led, for a multiplier $\lambda \in \mathbb{R}$, to take u satisfying $I(\theta)u = \lambda v$ and $\langle u,v\rangle = 1$.

Assume $I(\theta)$ is invertible: $u = \lambda(I(\theta))^{-1}v$:

$$1 = \langle u,v\rangle = \lambda\, {}^t v(I(\theta))^{-1}v$$

$$\begin{aligned} {}^t u I(\theta) u &= \lambda^2\, {}^t v(I(\theta))^{-1} I(\theta)(I(\theta))^{-1} v \\ &= \lambda^2\, {}^t v(I(\theta))^{-1} v = \frac{1}{{}^t v(I(\theta))^{-1} v}. \end{aligned}$$

Hence for the above choice of u, we obtain

$$E_\theta(Y - E_\theta(Y))^2 \geqslant {}^t(\text{grad } E_\theta(Y))[I(\theta)]^{-1}(\text{grad } E_\theta(Y)).$$

This results in the following minimization:

Theorem 7.2.16. Cramer-Rao Inequality. *Consider the hypotheses:*

(a) $\theta \longmapsto L(\theta)$ *is differentiable on* Θ *and* grad Log $L(\theta)$ *is centered and square integrable for* P_θ;

(b) Y *is in* $L^2(\Omega, \mathcal{A}, P_\theta)$ *and*

$$\text{grad } E(L(\theta)\cdot Y) = E(\text{grad } L(\theta)\cdot Y);$$

(c) $I(\theta)$ *is invertible. Then:*

$$E_\theta[Y - E_\theta(Y)]^2 \geqslant {}^t(\text{grad } E_\theta(Y))[I(\theta)]^{-1}(\text{grad } E_\theta(Y)).$$

Consequence. If Y is an unbiased estimator of a real differentiable function g of the parameter θ, $E_\theta(Y)$ equals $g(\theta)$ for arbitrary θ. The quadratic error is bounded below:

$$E_\theta(Y - g(\theta))^2 \geqslant {}^t(\text{grad } g(\theta))[I(\theta)]^{-1}(\text{grad } g(\theta)).$$

For $k = 1$,

$$E_\theta(Y - g(\theta))^2 \geqslant \frac{[g'(\theta)]^2}{I(\theta)}.$$

If we have a basis $b(\theta) = E_\theta[Y - g(\theta)]$, we obtain for $k = 1$:

$$E_\theta(Y - g(\theta))^2 \geqslant (b(\theta))^2 + \frac{(b'(\theta) + g'(\theta))^2}{I(\theta)}.$$

The quadratic error of an unbiased estimator is thus minorized by a positive number, which is smaller when Fisher's information is large and when the function varies only a little in the neighborhood of θ.

An unbiased estimator is said to be **efficient** *if the lower bound of the Cramer–Rao inequality is attained.* There is not always an efficient unbiased estimator (Exercise 7.2.3).

Particular Case. For the **exponential model**, we have for $\theta \in \mathring{\Theta}$, the interior of Θ, grad Log $L(\theta)$ = −grad $\psi(\theta) + T$. This vector is centered and its covariance matrix, the Fisher information matrix, is

$$I(\theta) = \left\{\frac{\partial^2\psi}{\partial\theta_i\partial\theta_j}(\theta)\right\}_{1\leqslant i,j\leqslant k} \qquad \text{(Section 3.3.2).}$$

For $k = 1$, this means that T *is an unbiased efficient estimator of* $\psi'(\theta)$: indeed, its variance is $\psi''(\theta)$ equal to $(\psi''(\theta))^2/I(\theta)$. Thus the sample mean is an efficient estimator of the mean for an n-sample from $\{b(p)\}_{p\in]0,1[}$, or from $\{N(m,1)\}_{m\in\mathbb{R}}$, or from $\{p(\lambda)\}_{\lambda>0}$, or from $\{E(\lambda)\}_{\lambda>0}$. This is not the case for an n-sample from an arbitrary distribution (Exercise 7.1.3); in return for weak hypotheses of regularity, the existence of efficient estimators is a characteristic of exponential families.

In this case the matrix $I(\theta)$ is the matrix related to the local dissimilarity, which we had obtained in Section 7.2.2. By noticing that $\partial^2\psi/\partial\theta_i\partial\theta_j$ coincides here with $-\partial^2 L(\theta)/\partial\theta_i\partial\theta_j$, we have a particular case of the following proposition.

Proposition 7.2.17. *Add to the hypotheses of* 7.2.16 *the following hypothesis* (d):

$$\begin{cases} \theta \longmapsto L(\theta) \text{ is twice differentiable} \\ E\left[\dfrac{\partial^2}{\partial\theta_i\partial\theta_j} L(\theta)\right] = \dfrac{\partial^2}{\partial\theta_i\partial\theta_j} E(L(\theta)) = 0 \quad (1 \leqslant i,j \leqslant k). \end{cases}$$

Then

$$\frac{\partial^2}{\partial\theta_i\partial\theta_j}(\text{Log } L(\theta)) = \left[\frac{\partial^2}{\partial\theta_i\,\partial\theta_j}L(\theta)\right]\frac{1}{L(\theta)}$$

$$-\left[\frac{\partial}{\partial\theta_i}L(\theta)\right]\left[\frac{\partial}{\partial\theta_j}L(\theta)\right]\frac{1}{(L(\theta))^2}\ ;$$

$$E_\theta\left[\frac{1}{L(\theta)}\ \frac{\partial^2}{\partial\theta_i\partial\theta_j}(L(\theta))\right] = E\left[\frac{\partial^2}{\partial\theta_i\partial\theta_j}(L(\theta))\right] = 0.$$

Hence,

$$I_{ij}(\theta) = -E\left[\frac{\partial^2}{\partial\theta_i\partial\theta_j}\text{Log } L(\theta)\right].$$

7.2.4. Estimation of a Function of the Parameter Taking Values in $\mathbb{R}^q$

In the case where the function $g(\theta)$ to be estimated takes values in $\mathbb{R}^q$, $q > 1$, we have a natural extension of the Cramer–Rao inequality. The choice of a quadratic loss function on $\mathbb{R}^q$ is somewhat arbitrary. Let $C(\theta)$ be a family of positive, invertible matrices; take as a loss ${}^t(Y - g(\theta))C(\theta)(Y - g(\theta))$, if Y estimates $g(\theta)$. Let us denote by Γ_θ the covariance matrix for P_θ. If Y is unbiased, the risk is

$$E_\theta[{}^t(Y - g(\theta))C(\theta)(Y - g(\theta))] = \text{Tr}[C(\theta)\Gamma_\theta(Y)].$$

It may therefore be seen that, *for arbitrary* $C(\theta)$, *an unbiased estimator* Y *is better (has a smaller risk) than* Z, *also unbiased, when* $\Gamma_\theta(Z) - \Gamma_\theta(Y)$ *is a positive matrix.* In this sense a bound for $\Gamma_\theta(Y)$ is obtained by:

Proposition 7.2.18. *Under the conditions of Theorem* 7.2.16, *let* g *be differentiable from* Θ *into* $\mathbb{R}^q$ *with Jacobian matrix* $\nabla_\theta g$. *At the point* θ *if* Y *is an unbiased estimator of* $g(\theta)$, *the matrix* $\Gamma_\theta(Y) - (\nabla_\theta g)[I(\theta)]^{-1}\,{}^t(\nabla_\theta g)$ *is positive.*

Proof. It is sufficient to apply Theorem 7.2.16 to the variable $\langle u,Y\rangle$ with expectation $\langle u,g(\theta)\rangle$ for all $u \in \mathbb{R}^q$.

For the exponential model and $\theta \in \mathring{\Theta}$, the estimator T of grad $\psi(\theta)$ has covariance matrix $I(\theta)$, equal to the lower bound of Proposition 7.2.18.

7.2.5. How Does the Information Increase When the Number of Experiments Increases?

(a) **Independent Experiments.** Consider an open set Θ of $\mathbb{R}^k$, $(E,\mathcal{E})$ a measurable space and $(F_\theta)_{\theta\in\Theta}$ a family of probabilities on $(E,\mathcal{E})$ dominated by a measure μ. Let L be a likelihood satisfying the hypotheses of Theorem 7.2.16 and $I(\theta)$ its Fisher information matrix.

We observe an n-sample $(X_1, \ldots, X_n)$ from F_θ: $(\Omega,\mathcal{A},P_\theta) = (E,\mathcal{E},F_\theta)^n$ is a model dominated by $\mu^{\otimes n}$ and we can associate with it the likelihood $(\theta,x_1, \ldots, x_n) \longmapsto \prod_{i=1}^n L(\theta,x_i)$. Then the hypotheses of Theorem 7.2.16 remain true for an information matrix $I_n(\theta) = nI(\theta)$. *The Cramer–Rao bound decreases as $1/n$ when the size of the sample increases.*

(b) **Markov Chains.** Recall example (c) of Section 7.1.2. Assume that Θ is an open interval of $\mathbb{R}$ and that $\theta \longmapsto \ell(\theta,x,y)$ is differentiable for all θ. Let $\dot{\ell}(\theta,x,y) = \partial/\partial\theta\, \ell(\theta,x,y)$. Assume

(a) $$\int \dot{\ell}(\theta,x,y)\Pi(x;dy) = 0 = \frac{\partial}{\partial\theta}\int \ell(\theta,x,y)\Pi(x;dy).$$

(b) $$\int \left[\frac{\dot{\ell}(\theta,x,y)}{\ell(\theta,x,y)}\right]^2 \rho(dx)\pi^n(\theta,x;dy) < \infty \quad \text{for all } n \text{ and all } \theta.$$

Then

$$E_{\theta,\rho}\left[\frac{\dot{\ell}(\theta,X_{n-1},X_n)}{\ell(\theta,X_{n-1},X_n)}\,\Big|\, X_0, \ldots, X_{n-1}\right] = 0.$$

Fisher's information may be written

$$I_n(\theta,\rho) = E_{\theta,\rho}\left[\sum_{i=1}^n \frac{\dot{\ell}(\theta,X_{i-1},X_i)}{\ell(\theta,X_{i-1},X_i)}\right]^2$$

$$= \sum_{i=1}^n \int \left[\frac{\dot{\ell}(\theta,x,y)}{\ell(\theta,x,y)}\right]^2 \rho(dx)\pi^n(\theta,x;dy).$$

This increases with n. If ρ is a probability μ_θ, invariant under $\Pi_\theta(\mu_\theta\pi_\theta = \mu_\theta)$, then $I_n(\theta,\mu_\theta) = nI_1(\theta,\mu_\theta)$. This result is similar to that found for independent observations.

Exercises 7.2.

E.1. Let f be the density of a distribution F on $\mathbb{R}$; assume that there exists an r.v., denoted by f', integrable with respect to Lebesgue measure such that, for all x, $f(x) = \int_{-\infty}^{t} f'(t)dt$. *Then the "Fisher's information on the function f"* is (setting $0/0 = 0$):

$$I(f) = \int \left(\frac{f'}{f}\right)^2 (x)f(x)dx.$$

(a) Calculate as a function of $I(f)$, Fisher's information for the family of distributions which are *translates* of F, with densities $x \mapsto f(x - \theta)$ $(\theta \in \mathbb{R})$? Then calculate Fisher's information for the family of distributions derived from F by "*change of scale*", with densities $x \mapsto e^{\theta} f(xe^{\theta})$, $(\theta \in \mathbb{R})$?

(b) Let f_1 and f_2 be two densities of the preceding type. Show that the convolution $f_1 * f_2$ has smaller information than $I(f_1)$.

(c) Calculate $I(f)$, when f is the $N(0,1)$ density; the $E(\lambda)$ density; the $U_{[0,1]}$ density; the triangular distribution $x \mapsto (1 - |x|)1_{\{|x| \leq 1\}}$. In each of these cases describe the family of distributions associated by (a) and (b).

E.2. *Properties of Fisher's information.* (a) Let X be a measurable function taking values in $(E,\mathcal{E})$ on the model $(\Omega,\mathcal{A},P_\theta)_{\theta \in \Theta}$ and P a privileged dominant of the model. Show that $X(P)$ is a privileged dominant of the image model $(E,\mathcal{E},X(P_\theta))_{\theta \in \Theta}$.

(b) Take Θ to be an open set in $\mathbb{R}$ and assume given likelihoods L and L_X of the model and its image under X, satisfying the hypotheses of Theorem 7.2.16. Let $I(\theta)$ and $I_X(\theta)$ be their Fisher information. Show: $0 \leq I_X(\theta) \leq I(\theta)$. Show that "$I_X(\theta)$ is zero" is equivalent to "X is a distribution free statistic"; and that "$I_X(\theta)$ is equal to $I(\theta)$" is equivalent to "X is sufficient."

E.3. Let S^2 be the sample variance of an n-sample from $N(m,\sigma^2)$ for $m \in \mathbb{R}$ and $\sigma^2 > 0$. What is its quadratic risk for estimating σ^2? and that of cS^2 for $c > 0$?

Show that S^2 is not an admissible estimator of σ^2, that it is an unbiased estimator with minimum variance, but that nevertheless it is not efficient.

E.4. *Superefficiency.* Let $\bar{X}$ be the sample mean of an n-sample from $N(0,1)$. Set $\hat{X} = \bar{X}1_{\{|\bar{X}|>n^{-1/4}\}}$. Calculate the quadratic error of $\hat{X}$ and the limit, as $n \to \infty$, of this error multiplied by n. Deduce from this that, for $\theta = 0$ and n large, $\hat{X}$ is better than $\bar{X}$, although $\bar{X}$ is efficient. Does this contradict the Cramer-Rao inequality?

E.5. *Fisher's information for a recurrent Markov chain in a finite space.* The framework is that of example (b) of Section 7.2.5, but here E is a finite space. Assume that, for all θ, E is a recurrent class of the Markov chain (Exercise 6.3.2). Show that for every initial distribution ρ, $(I_n(\theta,\rho)/n)$ tends to a number $I(\theta)$ independent of ρ.

E.6. *The Hellinger Arc.* Let P_0 and P_1 be two inequalities,

$$P = \frac{P_0 + P_1}{2}, \quad f_0 = \frac{dP_0}{dP}, \quad f_1 = \frac{dP_1}{dP}.$$

Assume that P_0 and P_1 are not perpendicular. Show that for $t \in]0,1[$, $\phi(t) = \int f_1^t f_0^{1-t} dP$ is a.s. > 0 and finite. Let P_t be the probability having density $(\phi(t))^{-1} f_1^t f_0^{1-t}$ with respect to P. We call the "Hellinger arc constructed on P_0 and P_1" the family $\{P_t: t \in [0,1]\}$. Show that it is an exponential family; prove

$$\lim_{t\uparrow 1} \phi'(t) = K(P_1,P_0), \quad \lim_{t\downarrow 0} \phi'(t) = -K(P_0,P_1)$$

(the Kullback informations may be infinite). Let P_{θ_0} and P_{θ_1} be two probabilities from an exponential family $(P_\theta)_{\theta\in\Theta}$ for $\Theta \subset \mathbb{R}$. Show tht $\{P_\theta;\ \theta \in [\theta_0,\theta_1]\}$ can be identified with the Hellinger arc constructed on P_{θ_0} and P_{θ_1}.

E.7. *Autoregressive processes of order* 1. Consider on $(\mathbb{R},\mathcal{B}_{\mathbb{R}})^{n+1}$ the measure P_ρ which is the product of an arbitrary probability ρ and of n identical distributions F having a density f. Let $\theta \in \mathbb{R}$ and let ϕ_θ be the function from $\mathbb{R}^{N+1}$ into $\mathbb{R}^{n+1}$ defined by

$$\begin{cases} \phi_\theta(x_0,x_1, \ldots, x_n) = (y_0,y_1, \ldots, y_n) \\ y_0 = x_0;\ y_p = \theta y_{p-1} + x_p, \quad 1 \leqslant p \leqslant n. \end{cases}$$

Let $P_{\theta,\rho} = \phi_\theta(P_\rho)$ and let $(X_0, \ldots, X_n)$ be the $n + 1$ coordinates of $\mathbb{R}^{n+1}$.

(a) Prove that $(\mathbb{R}^{n+1}, \mathcal{B}_{\mathbb{R}^{n+1}}, (X_p)_{0 \leq p \leq n}, P_{\theta,\rho})$ is a Markov chain. Show that, if λ is Lebesgue measure on $\mathbb{R}^n$, $\rho \otimes \lambda$ dominates the family $P_{\theta,\rho}$. Calculate the density $L_n(\theta)$ of $P_{\theta,\rho}$ with respect to $\rho \otimes \lambda$.

(b) Assume f is strictly positive, differentiable with derivative f', and

$$\int f'(y)dy = 0, \qquad \int \frac{f'^2(y)}{f(y)} dy = I < \infty.$$

Assume that the distributions ρ and F have finite second moments. Let $U_n(\theta) = \text{grad } L_n(\theta)$; prove that $U_n(\theta)$ is a centered r.v. and is square integable with respect to $P_{\theta,\rho}$. Let $I_n(\theta,\rho)$ be the Fisher information of this observation. Prove

$$I_n(\theta,\rho) = I E_{\theta,\rho}(X_0^2 + \dots + X_{n-1}^2).$$

(c) Show that the chain $(X_0, \dots, X_n)$ studied here has for $P_{\theta,\rho}$ the same distribution as $(X_0^\theta, \dots, X_n^\theta)$ in Exercise 6.3.8, by taking the ξ_n with density f. Show that for $|\theta| < 1$ and every initial distribution, $(I_n(\theta,\rho)/n)$ tends to $\sigma^2 I/(1 - \theta^2)$ if n tends to ∞.

7.3. Likelihood

The use of likelihood is without doubt the central idea of statistics. This use often leads to simple estimators of empirical type and for certain test problems, to optimal solutions. We shall see in Volume 2 its importance for other types of problems. However it is not a panacea; there are numerous pitfalls.

7.3.1. Estimation by Maximum Likelihood

Consider an n-sample $(X_1, \dots, X_n)$ from a discrete distribution F, concentrated on a countable set D. Its sample distribution

$$\overline{F}_n = \frac{1}{n} \sum_{i=1}^{n} \delta_{X_i}$$

is a measure which is also concentrated on D. The Kullback information carried by $\overline{F}_n$ on F is

$$K(\overline{F}_n,F) = \sum_{x\in D} \overline{F}_n(x)\text{Log } \overline{F}_n(x) - \sum_{x\in D} \overline{F}_n(x)\text{Log } F(x)$$
$$= -H(\overline{F}_n) - \frac{1}{n}\text{Log } F(X_1) \dots F(X_n).$$

The function $H(\overline{F}_n)$ (the entropy of $\overline{F}_n$, Exercise 1.3.13) depends only on the result of the experiment.

If, starting from the observations, we want to estimate which distribution F is, from a family of distributions $\mathcal{F}$ on D, we can look for the one which, in a given sense, is the closest to $\overline{F}_n$: if possible, $\hat{F} \in \mathcal{F}$ is chosen such that

$$K(\overline{F}_n,\hat{F}) = \inf\{K(\overline{F}_n,F);\ F \in \mathcal{F}\},$$

where $\hat{F}(X_1) \dots \hat{F}(X_n) = \sup\{F(X_1) \dots F(X_n);\ F \in \mathcal{F}\}$. Hence it is necessary to maximize the likelihood of this model, dominated by the measure $\Sigma_{x\in D^n} \delta_x$.

Example. If $D = \{1, \dots, k\}$ is a set of k elements, the distribution F is a vector $p = (p_1, \dots, p_k)$ from the subset of $\mathbb{R}^k$, $\Theta = \{p;\ p_j \geqslant 0$ for $1 \leqslant j \leqslant k$ and $\Sigma p_j = 1\}$. Let: $N_j = \Sigma_{i=1}^n 1_{\{j\}}(X_i)$, $j = 1, \dots, k$. The logarithm of the likelihood is $N_1\text{Log } p_1 + \dots + N_k\text{Log } p_k = L(p)$. The method of Lagrange multipliers can be applied to find the maximum of $L(p)$ by varying p in the open set $\{0 < p_j < 1,\ j = 1, \dots, k\}$ under the constraint $\Sigma_{j=1}^k p_j = 1$. We find $\hat{p}_j = N_j/n$ $(j = 1, \dots, k)$. Since every physical measurement only ever consists of a finite number of significant figures, the importance of finite models and the foundation they give to maximum likelihood are clear. We are now going to generalize the definition, but without using $K(\overline{F}_n,F)$, which is very often infinite: we will use likelihood.

Definition 7.3.19. Let $(\Omega,\mathcal{A},P_\theta)_{\theta\in\Theta}$ be a dominated statistical model, and T a sufficient statistic taking values in $(E,\mathcal{E})$. Assume given a measurable function $(\theta,t) \longmapsto L(\theta,t)$ of $\Theta \times E$, such that $L(\theta,T)$ is a likelihood.

A **maximum likelihood estimator** of θ is an estimator $\hat{\theta}(T)$, where $\hat{\theta}$ is a function from E into Θ such that, for all $t \in E$,

$$L(\hat{\theta}(t),t) = \sup_{\theta\in\Theta} L(\theta,t).$$

In such general terms, these definitions do not make much sense... . L is not defined uniquely; for $\hat{\theta}$, we have in general neither existence nor uniqueness. Things are clear in the case of discrete distributions. Before making detailed criticism, let us look at *an interesting case: that of the* **exponential model.** Let us set, for $t \in \mathbb{R}^k$ and $\theta \in \Theta$,

$$L(\theta,t) = \exp(-\psi(\theta) + <\theta,t>).$$

The function $\theta \longmapsto \text{Log } L(\theta,t)$ is convex on Θ and its Young transform h (Section 3.3.5) is defined by $L(\theta,t) = \sup_{\theta\in\Theta}(-\psi(\theta) + <\theta,t>)$. If an r.v. $\hat{\theta}$ on $\mathbb{R}^k$ can be found such that, for all t, $h(t) = -\psi[\hat{\theta}(t)] + <\hat{\theta}(t),t>$, then $\hat{\theta}(T)$ is a maximum likelihood estimator of θ.

We know (Section 3.3.2) that ψ is infinitely differentiable on the interior $\overset{\circ}{\Theta}$ of Θ, that grad ψ is bijective from $\overset{\circ}{\Theta}$ into $C =$ grad $\psi(\overset{\circ}{\Theta})$, and that if $\hat{\theta}$ is its inverse, $L(t) = \text{Log } L(\hat{\theta}(t),t)$. Hence, if the value $t \in C$ is observed of the sufficient statistic T, θ is estimated by $\hat{\theta}(t)$, which maximizes $\theta \longmapsto L(\theta,t)$. *Thus if T takes its values in C*, $\hat{\theta}(T)$ is a maximum likelihood estimator of θ. For $t \notin C$, the function $\theta \longmapsto L(\theta,t)$ can have no maximum on Θ, and there is then no maximum likelihood estimator. Let us still assume that T takes values in C. As we have seen, $T = \text{grad } \psi(\hat{\theta}(T))$ is an efficient unbiased estimator of grad $\psi(\theta)$; we can estimate grad $\psi(\theta)$ by substituting for θ its maximum likelihood estimator.

This is a very natural idea. If we are trying to estimate a function $g(\theta)$ with parameter θ and we have a good estimator $\hat{\theta}$ of θ, we can try to estimate $g(\theta)$ by $g(\hat{\theta})$. This idea has been used in the case of empirical estimators. If $\hat{\theta}$ is a maximum likelihood estimator, the estimator $g(\hat{\theta})$ thus obtained corresponds, if g is bijective, to the maximum likelihood estimator for the reparametrization $\theta \longmapsto g(\theta)$. Take in the above $\Omega = U$, an open set in $\mathbb{R}^k$, T the identity r.v., and a dominated model of distributions on U with densities $L(\theta,\cdot)$, $\theta \in \Theta$. Let ϕ be a diffeomorphism from U into another open set V of $\mathbb{R}^k$. From Section 3.3.3, the likelihood $L_{\phi(T)}$ of $\phi(T)$ satisfies, for $T \in U$,

$$L(\theta,t) = |J_\phi|(t)L_{\phi(T)}(\theta,\phi(t)),$$

$$\sup_{\theta\in\Theta} L(\theta,t) = \sup_{\theta\in\Theta} L_{\phi(T)}(\theta,\phi(t))|J_\phi|(t).$$

If it exists, the maximum likelihood estimator $\hat{\theta}$ is also the maximum likelihood estimator of the observation $\phi(T)$.

Example. Let $(X_1, ..., X_n)$ be an n-sample from the Log-normal distribution (Section 3.3.4) with parameter (m,σ^2): Log X is distributed as $N(m,\sigma^2)$. The logarithm of the likelihood of the n-sample is

$$(x_1, ..., x_n) \longmapsto \sum_{i=1}^{n} \text{Log} \frac{1}{|x_i|} - \frac{1}{2\sigma^2} \sum_{i=1}^{n} [\text{Log } x_i - m]^2 - \frac{n}{2} \text{Log} \frac{\sigma^2}{2\pi} .$$

The maximum likelihood estimator of (m,σ^2) is thus

$$\left[\hat{m} = \frac{1}{n} \sum_{i=1}^{n} \text{Log } X_i, \quad \sigma^2 = \frac{1}{n} \sum_{i=1}^{n} (\text{Log } X_i - \hat{m})^2\right],$$

the estimator corresponding to the Gaussian n-sample (Log X_1, ..., Log X_n).

The mean and variance of X are

$$\exp\left[m + \frac{\sigma^2}{2}\right] = \mu \quad \text{and} \quad \mu^2(\exp(\sigma^2) - 1) = \tau^2.$$

By substituting the maximum likelihood estimators, we obtain the estimators:

$$\exp\left[\hat{m} + \frac{\hat{\sigma}^2}{2}\right] = \hat{\mu} \quad \text{and} \quad \hat{\mu}^2(\exp(\hat{\sigma}^2) - 1) = \hat{\tau}^2.$$

Criticism of Estimation by the Method of Maximum Likelihood. This method has the advantage of allowing, in many cases, easy calculation of an estimator. It has many shortcomings. A minor shortcoming is that it gives biased estimators. A more serious shortcoming is that there is, in general, neither existence nor uniqueness. Also, we may obtain a nonadmissible estimator.

Examples. (a) The likelihood for an n-sample from a uniform distribution on $[\theta,\theta + 1]$ is

$$(x_i)_{1 \leq i \leq n} \longmapsto 1_{[\theta \leq \inf_i x_i < \sup_i x_i \leq \theta+1]} .$$

For $x_{(1)} = \inf_i x_i$ and $x_{(n)} = \sup_i x_i$, we observe (a.s.) $x_{(n)} - x_{(1)}$ less than 1. The likelihood equals 1 for $\theta \in [x_{(n)} - 1, x_{(1)}]$, and $\hat{\theta}(x_1, ..., x_n)$ can be given an arbitrary value in this interval.

(b) Let X be an r.v., the distribution of which is a mixture of normal distributions, with density

$$x \longmapsto \frac{1}{2\sqrt{2\pi}} \exp\left[-\frac{1}{2}(x-m)^2\right] - \frac{1}{2\sigma\sqrt{2\pi}} \exp\left[-\frac{1}{2}\frac{(x-m)^2}{\sigma^2}\right].$$

The parameter is $(m,\sigma^2) \in \mathbb{R} \times]0,\infty[$. Taking $m = x$, the likelihood equals

$$\left(1 + \frac{1}{\sigma}\right)\frac{1}{2\sqrt{2\pi}},$$

which tends to $+\infty$ if $\sigma \to 0$: the maximum likelihood, infinite, is not attained.

(c) For an n-sample from the uniform distribution on $[0,\theta]$, the likelihood is

$$(x_i)_{1 \leqslant i \leqslant n} \longmapsto \frac{1}{\theta^n} 1_{[\sup x_i \leqslant \theta]} .$$

The maximum likelihood estimator is $X_{(n)}$. It is easy to calculate that

$$E_\theta[X_{(n)} - \theta]^2 = \frac{2\theta^2}{(n+1)(n+2)},$$

$$E_\theta\left[\frac{n+2}{n+1}X_{(n)} - \theta\right]^2 = \frac{\theta^2}{(n+1)^2}.$$

The estimator $X_{(n)}$ is therefore inadmissible.

7.3.2. The Likelihood Ratio Test

Let $\Theta = \Theta_0 \cup \Theta_1$ be a partition of Θ. Within the framework of Definition 7.3.19, we can define a test, based on the following intuitive idea: an hypothesis is accepted if it is quite likely. The rejection region of Θ_0 is taken to be of the form

$$\left\{\sup_{\theta\in\Theta_0} L(\theta,T) < C \sup_{\theta\in\Theta_1} L(\theta,T)\right\}$$

for a given constant $C \geqslant 0$.

Such a test is easy to construct. It will be seen in Volume 2 that it has good asymptotic properties. For samples of fixed size, interesting properties of this test are obtained only in certain cases which will be tackled in Sections 8.3 and 8.4.

Exercises 7.3

E.1. *Statistics on a Poisson process.* Successive breakdown times $(T_n)_{n \geqslant 0}$ are observed assuming $T_0 = 0$. Setting $\tau_n = T_n - T_{n-1}$ $(n \geqslant 1)$, τ_n is assumed to be exponentially distributed with parameter $\theta > 0$ and the sequence (τ_n) is assumed independent. A model of this situation may be constructed by taking $\Omega = \mathbb{R}_+^{\mathbb{N}}$ and τ_n the nth coordinate function. Set $G_n = \sigma(\tau_1, ..., \tau_n)$, $A = \vee G_n$. (Ω, A) is given the appropriate unique probability P_θ (cf. Section 6.3.1).

(a) Prove that, for $\theta \neq \theta'$, P_θ and $P_{\theta'}$ are perpendicular.

(b) In this part, the observation $(T_1, ..., T_n)$ is the first n breakdown times. Show that on (Ω, G_n),

$$P_\theta = \theta^n e^{(1-\theta)T_n} P_1.$$

Give a sufficient statistic having observed $(T_1, ..., T_n)$. What is the Fisher information of $(T_1, ..., T_n)$? What is the maximum likelihood estimator of $1/\theta$ for this observation? Is it unbiased? efficient? What is the maximum likelihood estimator of θ for this observation? Show that it is biased. Give an unbiased estimator proportional to this and calculate its variance: it is not efficient. Show that an unbiased efficient estimator of θ cannot be found.

(c) For $t \geqslant 0$, set $N_t = \Sigma_{n \geqslant 1} 1_{(T_n \leqslant t)}$. The process $(\Omega, A, P_\theta, (N_t)_{t \geqslant 0})$ is a Poisson process with parameter θ [Exercise 4.4.12]. Let $F_t = \sigma(N_s;\ s \leqslant t)$ be the σ-algebra of observations prior to t. Prove that on (Ω, F_t):

$$P_\theta = \theta^{N_t} e^{(1-\theta)t} P_1.$$

Give a sufficient statistic, having observed the breakdown times prior to t. What then is the maximum likelihood estimator of θ? Is it efficient?

E.2. *Breakdown times.* n machines are started *at the same time*, the breakdown times of which occur at independent exponentially distributed times with mean θ.

(a) In order to estimate θ, the first r breakdown times are observed (i.e., the order statistic $(T_{(1)}, \ldots, T_{(r)})$. What is the density of $(T_{(1)}, \ldots, T_{(r)})$? What is the maximum likelihood estimator $\hat{\theta}$? Prove that $2r\hat{\theta}/\theta$ has a $\chi^2(2r)$ distribution, that $\hat{\theta}$ is an unbiased, efficient estimator, and that it is a complete sufficient statistic (see Exercise 4.4.10(d)).

(b) The observations are now stopped at a fixed time t; N_t equals r on $\{T_{(r)} \leqslant t < T_{(r+1)}\}$. $T_{(1)}, \ldots, T_{(N_t)}$ is observed. Show that the model is still dominated. Calculate the maximum likelihood estimator.

E.3. *The Pareto Distribution.* An n-sample $(X_1, \ldots, X_n)$ is observed from a one-sided Pareto distribution with density

$$x \longmapsto \frac{\alpha r^{\alpha}}{x^{\alpha+1}} 1_{[x>r]}, \quad (\alpha > 0,\ r > 0).$$

(a) Give a sufficient statistic for the study of (α, r).
(b) Assume α is known. What is the maximum likelihood estimator of r? What is its distribution? Is it unbiased?

E.4. Let $(X_1, \ldots, X_n)$ be an n-sample from a Poisson distribution with parameter $\sqrt{\theta}$. We want to estimate θ. Let $T = \Sigma_{i=1}^n X_i$. Prove that $T(T-1)/n^2$ is an unbiased estimator of θ. Can an unbiased estimator with smaller variance be found?

Give the maximum likelihood estimator of θ and its variance. Calculate the Cramer-Rao bound for the variance of an unbiased estimator of θ. Comment on this.

E.5. *Estimation on an n-sample from $N_k(m,\Gamma)$.* The notations are those of Exercise 5.5.4 and 5.5.5. Take $n > k$ for questions (a) – (c).

(a) Let us take $\Theta = \mathbb{R}^k \times M_k^+$, and study an n-sample from $N_k(m,\Gamma)$, with $(m,\Gamma) \in \Theta$. Show that we have a dominated model and that we can take a likelihood, the logarithm of which is

$$L(X,m,\Gamma) = -\frac{nk}{2}\text{ Log } 2\pi - \frac{n}{2}\text{ Log}|\Gamma| - \frac{n}{2}\text{Tr}(\Gamma^{-1}S)$$
$$-\frac{n}{2}\,{}^t[\bar{X} - m]\Gamma^{-1}[\bar{X} - m].$$

Let S be the sample covariance and $\lambda_1, \ldots, \lambda_k$ the eigenvalues of the symmetric matrix $\Gamma^{-1}S$. Prove, for m and $\hat{m}$ in $\mathbb{R}^k$, $\Gamma \in M_k^+$:

$$L(X,\hat{m},S) - L(X,m,\Gamma)$$
$$= \frac{n}{2}\left\{\left[\sum_{i=1}^{k} \lambda_i - k - \sum_{i=1}^{k} \text{Log } \lambda_i\right] + {}^t[\bar{X} - m]\Gamma^{-1}[\bar{X} - m]\right.$$
$$\left. - {}^t[\bar{X} - \hat{m}]S^{-1}[\bar{X} - \hat{m}]\right\}.$$

Verify that the quantity in the first square brackets is positive, and is zero if and only if S and Γ coincide. Determine the maximum likelihood estimator of m when Γ is known; of Γ when m is known; then of (m,Γ) when both are unknown.

(b) Now take $\Theta = \mathbb{R} \times M_k^+$; for $(\mu,\Gamma) \in \Theta$, X is an n-sample from $N_k(\mu 1_k,\Gamma)$. What is the maximum likelihood estimator of μ when Γ is known?

(c) Take $\Theta = \mathbb{R}^k \times]0,\infty[$ and for $(m,\sigma^2) \in \Theta$, X is an n-sample from $N_k(m,\sigma^2 I_k)$. Give a maximum likelihood estimator of (m,σ^2).

(d) Assume m is zero and $n \leq k$. Show that there then does not exist a maximum likelihood estimator of Γ (taking values in M_k^+).

E.6. *Likelihood ratio tests for samples of Gaussian vectors.* The notations are those of Exercises 5.5.4, 5.5.5, and 7.3.5. An n-sample from $N_k(m,\Gamma)$ is observed for $(m,\Gamma) \in \Theta$, $\Theta = \mathbf{R}^k \times M_k^+$, $n > k$. Let $\Theta_0 \subset \Theta_1 \subset \Theta$. To test H_0 "$\theta \in \Theta_0$" against H_1 "$\theta \in \Theta_1 \backslash \Theta_0$," the likelihood ratio Δ is used, with

$$\text{Log } \Lambda = \sup\{L(X,m,\Gamma);\ (m,\Gamma) \in \Theta_0\}$$
$$- \sup\{L(X,m,\Gamma);\ (m,\Gamma) \in \Theta_1\}.$$

In what follows, tests with rejection regions $\{\Lambda \leq C\}$ are taken, for a given constant $C < 1$.

(a) Let $m_0 \in \mathbb{R}^k$. Assume that the covariance Γ is known. Give a test of level α of "$m = m_0$" against "$m \neq m_0$." This is

Hotelling's test.

(b) For a *test of sphericity*, "$\Gamma \in \Delta = \{\sigma^2 I_k;\ \sigma > 0\}$" against "$\Gamma \in M_k^+ \backslash \Delta$," show that we obtain

$$\Lambda = \frac{|S|^{n/2}}{(\mathrm{Tr}\ S/k)^{nk/2}} .$$

Under what condition does Λ equal 1?

E.7. *The maximum likelihood estimator in an exponential model.* It is necessary to see by examples the difficulties not dealt with in Section 7.3.1 (we shall use the same notations). Also refer to Section 3.3.5.

(a) Consider a Pareto distribution with density proportional to $x \longmapsto e^{\theta x}(a/x^{a+1})1_{[x>1]}$ on $\mathbb{R}$; $a > 1$ is a constant. We have, on $\Theta =]-\infty,0]$,

$$e^{\psi(\theta)} = \int_1^\infty e^{\theta x} \frac{a}{x^{a+1}}\, dx < \infty.$$

Study on $]-\infty,0[$, ψ, its derivative ψ', its Young transform h; draw their graphs. Show that $\hat{\theta}$ is a maximum likelihood estimator where $\hat{\theta}$ equals 0 on $[(a-1)/a, \infty[$ and equals ψ'^{-1} on $[1, a/a-1]$; it is not a sufficient statistic.

(b) Assume $\Theta = \mathbb{R}^k$. Prove that the maximum likelihood exists and is unique. Show that it is a sufficient statistic.

(c) Consider the family $N_k(m,\Gamma)$, $\theta = (m,\Gamma)$, and $\Theta = \mathbb{R} \times M_k^+$, M_k^+ being the set of positive definite $k \times k$ matrices. Show that it is an exponential family, but that an n-sample does not give a maximum likelihood estimator of Γ except for $n > k$ (Exercise 7.3.5).

Bibliographic Notes

The foundations of statistics are covered in Hoel, Port and Stone [2], Breiman [2], Bickel–Doksum and Raoult. More complex, and containing numerous examples, let us cite Cox and Hinkley [1], [2], and for the exponential model Barndorff–Neilsen; Lehmann [1] has in its first chapters a very complete account. Some mathematical developments in a very general framework are covered in LeCam.

Chapter 8
STATISTICAL DECISIONS

Objectives

In this section we follow up the research, started in Chapter 7, on the general principles which have been used in the previous chapters for statistical problems. Various criteria are given for guiding a statistical decision (estimation, or more often testing). It will be worthwhile finally to go back to Chapters 2, 4, and above all, 5.

8.1. Decisions

Even if it does not represent all aspects of statistics, the decision formalism is fairly intuitive and has allowed clear mathematical developments.

8.1.1. A Two-Person Game

A statistician S plays an opponent J. The various possible games of J are elements of a set Θ ; those of S are elements of a set of strategies (or actions) A. The payoff matrix gives a loss $C(\theta,a)$ to the statistician, if he plays a while J plays θ (J then wins $C(\theta,a)$): $C(\theta,a)$ is positive or negative. If S does not know what J is going to do and if he is careful, he thinks only of the maximum loss $\sup_{\theta} C(\theta,a)$: in order to minimize this maximum loss he will try to determine $\inf_a \sup_\theta C(\theta,a)$ and if the inf over a is attained for a_m, he will play a_m, the minimax strategy (which minimizes the maximum risk).

In any case, he will only play a if it is an *admissible strategy*, i.e., there is no a' such that the function $C(\cdot,a')$ is greater than or equal to $C(\cdot,a)$ everywhere, the inequality being strict at one point at least (see Exercise 2.3.5).

8.1.2. Statistical Decisions

The statistician S's opponent is Nature. He observes a phenomenon, the distribution of which depends on a parameter $\theta \in \Theta$, in order to undertake an **action** $a \in A$, the **cost** of which depends on the state of nature. Let $C(\theta,a)$ be this cost. The function C, assumed here to be positive, is the statistician's loss. The set A of actions is imposed on the statistician. The set Θ is often given a measurable structure, which will be taken as understood.

Before deciding, the statistician carries out a statistical experiment $(\Omega, \mathcal{A}, P_\theta)_{\theta \in \Theta}$ (which he is often free to organize). His decision, a function of the result of the experiment, will be a measurable function δ from $(\Omega,\mathcal{A})$ into A assumed given a σ-algebra $\mathcal{U}$. The cost is then an r.v. $C(\theta,\delta)$. If the statistician chooses the **decision rule** (or **strategy**) δ, his **risk** or average cost will be, when the parameter is θ:

$$R(\theta,\delta) = \int dP_\theta(\omega)C(\theta,\delta(\omega)) = E_\theta(C(\theta,\delta)).$$

We often limit ourselves to a class $\mathcal{D}$ of decisions; for example to those decisions which have finite risk for all θ. The class $\mathcal{D}$ could be a class of decision functions of an observation X, which amounts to replacing the model by its image under X.

As for the game, the prudent statistician will try to minimize $\sup_\theta R(\theta,\delta)$ following a strategy called **minimax**. The natural partial order associated with risk leads to the following definitions:

Definition 8.1.1. In the set $\mathcal{D}$ of proposed decisions, a decision δ is **as good** as δ' if for all θ, $R(\theta,\delta)$ is less than $R(\theta,\delta')$; it is **better** than δ' if the inequality is strict for at least one θ. A decision δ is **optimal** if it is as good as all the others; it is **admissible** if there does not exist a better decision.

In general, we look, if possible, for an optimal strategy in a class $\mathcal{D}$ in use. In any case we should only accept admissible strategies. Nevertheless, there exist nonadmissible "natural" strategies which are used, in spite of this (Exercise 8.1.2). As in Section 2.3.8, note

Proposition 8.1.2. *An admissible decision with constant risk is minimax.*

Note. *Admissibility and minimax depend on the loss function chosen.*

8.1.3. The Case of Estimation

To estimate θ, the set of actions is the set Θ. The properties of the probability P_θ are used; if two values of θ are associated with the same probability, it is useless to try to distinguish them by statistical means. Hence the model is assumed to be **identifiable** (Section 7.1.1).

More generally, if g is a function from Θ into U, an open set in $\mathbb{R}^k$, to estimate $g(\theta)$ is to take $A = U$ and to attribute a value a to $g(\theta)$. If g is not bijective, we have the same restrictions as above.

A decision here is called an **estimator**; its cost will most often be taken equal to $\|g(\theta) - a\|^2$. We have seen in Section 2.3 numerous admissible estimators based on the observation of a Bernoulli n-sample.

In this case we will restrict ourselves most often to the class $\mathcal{D}_2$ of estimators T which are square integrable for all P_θ, or to the class $\mathcal{D}_2^b$ of **unbiased** estimators T of $\mathcal{D}_2$, i.e., those such that $E_\theta(T)$ is equal to $g(\theta)$ for all θ. *In the class* $\mathcal{D}_2^b$ *an efficient estimator* (Section 7.2.3) *is optimal; but it is not necessarily optimal in the class* $\mathcal{D}_2$ (Exercise 7.7.4).

8.1.4. Tests

Let $\Theta_0 \cup \Theta_1 \cup \ldots \cup \Theta_k$ be a partition of Θ. We want to choose the set of the partition in which θ lies. This is a **problem of choice amongst several hypotheses.** Take $A = \{1, \ldots, k\}$, the set of subsets of A and, for $1 \leqslant i \leqslant k$, the error cost $C(\theta,i)$ to be zero for $\theta \in \Theta_i$ and equal to a positive number C_{ji} for $\theta \in \Theta_j$ and $i \neq j$. Denote hypothesis H_i "$\theta \in \Theta_i$."

The essential use of choice between two hypotheses is a **test** problem. Instead of taking $A = \{0,1\}$, take more generally $A = [0,1]$ in order to allow random decisions (analogous to those used for the games in Exercise 2.3.5). An r.v. ϕ defined on $(\Omega, \mathcal{A})$, taking values in $[0,1]$, is called a **test function** (or simply a **test**). It is used as follows: for $\phi = 0$, choose H_0; for $\phi = 1$, choose H_1; if ϕ equals p, $0 < p < 1$, a random draw is made from a Bernoulli distribution with parameter p, *independent of the observation*, which leads to deciding H_1 (with probability p) if 1 is drawn, and H_0 (with probability $(1 - p)$) if 0 is drawn. When ϕ takes values in $\{0,1\}$, the test is called **deterministic** (or **nonrandomized**): ϕ is an indicator function 1_D for $D \in \mathcal{A}$. The region D is the **rejection region** of the test. It often occurs that for a given r.v. T, D is of the form $(T \geqslant s)$ and s is then the test **threshold.**

To the decision $p \in [0,1]$ corresponds the average cost in the Bernoulli random draw

$$C(0,p) = pC_{01} + (1 - p)C_{00} = pC_{01}$$

$$C(1,p) = pC_{11} + (1 - p)C_{10} = (1 - p)C_{10} .$$

The risk of test ϕ is then

$$E_\theta[C(\theta,\phi)] = \begin{cases} C_{01}E_\theta(\phi) & \text{for } \theta \in \Theta_0 \\ C_{10}E_\theta(1 - \phi) & \text{for } \theta \in \Theta_1 . \end{cases}$$

We often take $C_{01} = C_{10} = 1$. As we have seen (Sections 2.4.1, 4.5, 5.1, 5.4, and 5.5), the hypotheses H_0 and H_1 often do not play a symmetric role for the statistician. In signal theory H_0 is the hypothesis of alarm, H_1 that of non-alarm; in the field of scientific experimentation, H_0 is the generally accepted hypothesis and H_1 an hypothesis leading to a new theory. The quantity $\sup_{\theta \in \Theta_0} E_\theta(\phi)$ is called the **size** of the test ϕ. A test is said to be of **level** α if its size is less than or equal to α ; α thus controls the probability of wrongly rejecting H_0. $E_\theta(\phi)$ for $\theta \in \Theta_0$ is also called the **type I error**. The function $\theta \longmapsto E_\theta(\phi)$, defined on Θ_1, is called the **power** of the test; $1 - E_\theta(\phi)$ is the probability of a false alarm or

type II error. "Level" is often used in place of "size" (Section 5.1).

Decisions on tests are chosen according to the following rule: α is fixed, the class $\mathcal{D}_\alpha$ of tests of level α is considered, and we try to maximize the power in this class. To say that a test ϕ is **uniformly more powerful** than ϕ' signifies $E_\theta(\phi) \geq E_\theta(\phi')$ for all $\theta \in \Theta_1$. Notice that the relations "better" and "more powerful" are not equivalent, since they are associated with two different preorderings. It is mainly to construct tests of a given size that we are led to make auxiliary random draws (Sections 4.5.3 and 8.3.2).

8.1.5. Confidence Regions

A variation on the ideas of testing and estimation is that of a confidence region. A decision S takes values in $A = \mathcal{D}(\Theta)$, the set of subsets of Θ; the cost of decision S, if the parameter equals θ, is taken equal to $1_S(\theta)$. A **confidence region** is a decision rule, i.e., a function $\omega \longmapsto S(\omega)$ from Ω into $\mathcal{D}(\Theta)$; if the parameter equals θ, the risk $P_\theta[\omega; \theta \notin S(\omega)] = P_\theta[\theta \notin S]$ is associated with it, by assuming that $\{\omega; \theta \notin S(\omega)\}$ is in A. If, for all θ, $P_\theta(\theta \notin S)$ is majorized by α, S is said to be a **confidence region of level** $1 - \alpha$: we have seen examples of them for the Bernoulli (Section 2.4.5) and Gaussian (Sections 5.1.5, 5.1.6) distributions.

Let S be such a confidence region of level $1 - \alpha$. For all $\theta_0 \in \Theta$, it allows a test of "$\theta = \theta_0$" against "$\theta \neq \theta_0$" to be defined, of size α. It is sufficient to take the rejection region $D(\theta_0) = \{\omega; \theta_0 \notin S(\omega)\}$.

This relation also allows us to move from a deterministic test of "$\theta = \theta_0$" with rejection region $D(\theta_0)$ to a confidence region, by setting $S(\omega) = \{\theta; 1_{D(\theta)}(\omega) = 0\}$. We can also associate with S an estimator chosen from S (for example, if S is an interval, we could choose its midpoint as an estimator).

Example. Two independent samples are compared, $(X_1, \ldots, X_n)$ a sample from the distribution F on $\mathbb{R}$ and $(Y_1, \ldots, Y_n)$ a sample from the distribution F translated by θ. Y has the distribution of $X + \theta$, for θ a real parameter. To carry out a two-sided test of "$\theta = \theta_0$" against "$\theta \neq \theta_0$" we test if $(X_1 + \theta_0, \ldots, X_n + \theta_0)$ and $(Y_1, \ldots, Y_n)$ have the same distribution; we can thus (as in Section 4.4.2) introduce W_{θ_0}, the sum of the ranks of the Y's in $(X_1 + \theta_0, ..., X_n + \theta_0, Y_1, ..., Y_n)$. If F is continuous and $\theta = \theta_0$, W_{θ_0} is a Wilcoxon statistic distributed independently of F and θ_0. Let c and d be constants such that

$$P_{\theta_0}(c \leqslant W_{\theta_0} \leqslant d) = 1 - \alpha.$$

For each θ_0, $D(\theta_0) = \{W_{\theta_0} \notin [c,d]\}$ is the rejection region of a test of size α of "$\theta = \theta_0$" against "$\theta \neq \theta_0$." From this we obtain a confidence region S of level $1 - \alpha$, a set of θ_0's where this test would accept "$\theta = \theta_0$": $S = \{\theta;\ c \leqslant W_\theta \leqslant d\}$. Here $\theta \longmapsto W_\theta(\omega)$ is an increasing function and $S(\omega)$ is an interval $(\underline{\theta}(\omega),\overline{\theta}(\omega))$. θ could then be estimated by $(\underline{\theta} + \overline{\theta})/2$.

Exercises 8.1

E.1. *The Bluff.* Consider the following game:

(a) A and B each put a in the pot.
(b) A draws a card which wins with probability P, and loses otherwise; the two players know the value of P.
(c) If A's card is a winner, he increases the stakes by b (without saying anything). If A's card is a loser, A chooses between two strategies: he lays down his card and loses a or he bluffs and stakes b.
(d) If A stakes b, B can either lay down his card and lose a, or can follow by staking b. In the latter case, A's card is looked at: if he has bluffed, B wins $(a + b)$, otherwise B loses $(a + b)$.

(1) Following the players' strategies, there are four possible cases without the extra draw; construct a table of A's winnings.
(2) We are also interested in strategies in which the players make an extra draw. Show that for $P \geqslant (2a+b)/(2a+2b)$, A should always bluff and B should always lay down his cards. On the other hand, for $P < (2a+b)/(2a+2b)$, this is not the case. Determine

the minimax strategy for A and the minimax strategy for B. Has this game a value (in the sense of Exercise 2.4.5)?

E.2. *Inadmissibility of the sample mean for* $N(\theta, I_k)$, $k \geq 3$. Let X be a random vector of dimension $k \geq 3$, with unknown mean θ and covariance matrix I, the $k \times k$ identity matrix. $X = \theta + Y$, where Y is a vector with k independent, $N(0,1)$ components. In what follows, denote by θX the scalar product $\langle\theta, X\rangle$ and $X^2 = \langle X, X\rangle$. The question is to compare the usual estimator of θ, $\bar{\theta}(X) = X$, with estimators of the form $\hat{\theta}(X) = (1 - b/(a + X^2))X$ with $a > 0$, $b > 0$, and to prove that if b is sufficiently small and a sufficiently large that $\hat{\theta}$ is strictly better than $\bar{\theta}$.

(a) Demonstrate

$$E_\theta[\hat{\theta}(x) - \theta]^2 < k - 2bE_\theta\left[\frac{Y(Y + \theta) - b/2}{a + (Y + \theta)^2}\right].$$

(b) By using the identity $1/(1 + x) = 1 - x + x^2/(1 + x)$, applied to $x = 2\theta Y/(a + Y^2 + \theta^2)$, show, when a tends to infinity:

$$E_\theta\left[\frac{Y(Y + \theta) - b/2}{a + (Y + \theta)^2}\right] > \frac{k - 2 - b/2}{a + \theta^2} + \frac{1}{a + \theta^2}\, O\left(\frac{1}{\sqrt{a}}\right)$$

uniformly in θ. Deduce from this that $\bar{\theta}$ is not admissible.

(c) Deduce the same result for an n-sample and $\bar{\theta}$ the sample mean.

8.2. Bayesian Statistics

8.2.1. Bayesian Games

Statistician S plays against an opponent J who chooses the θ he is going to play with the help of a roulette; θ is thus random with distribution ν. If S knows J's method, he knows that to play a will lead to an average (or Bayesian) loss $\int d\nu(\theta)C(\theta,a)$. He will thus choose a to minimize this average cost.

8.2.2. The Posterior Distribution for a Gaussian Vector

The output from a machine being set to produce rods of length θ is, due to the imperfections present, a sequence of rods of length $X_1, \ldots, X_n$ independently distributed as $N(\theta,\sigma^2)$; σ^2, a characteristic of the machine, is known and we want information on θ. The statistician knows how the machine has been set; it was required to be set at value m, but this was not able to be done precisely. This leads to assuming that θ is an r.v. with distribution $N(m,\tau^2)$: thus the **prior distribution** of θ is known.

In other words, $(\theta,X_1, \ldots, X_n)$ is a random vector, the distribution of θ is $N(m,\tau^2)$, and the distribution of $X = (X_1, \ldots, X_n)$ conditional on θ is $N^{\otimes n}(\theta,\sigma^2)$. The observation of X gives information on θ contained in its distribution conditional on X, or **posterior distribution** of θ. The density of (θ,X) is, for a suitable constant K,

$$(t,x_1, \ldots, x_n) \longmapsto K \exp - \left[\frac{(t-m)^2}{2\tau^2} + \sum_{i=1}^{n} \frac{(x_i - t)^2}{2\sigma^2}\right]$$

$$= K \exp\left[\frac{-1}{2\sigma^2\tau^2}\left[t^2(\sigma^2 + n\tau^2) - 2t\left(\tau^2 \sum_{i=1}^{n} x_i + \sigma^2 m\right) + \sigma^2 m^2 + \tau^2 \sum_{i=1}^{n} x_i^2\right]\right]$$

The distribution of θ conditional on $X = (x_1, \ldots, x_n)$ has a density which, up to a factor dependent on $x_1, \ldots, x_n$, equals

$$t \longmapsto \exp\left[-\frac{1}{2\sigma^2\tau^2}\left[t^2(\sigma^2 + n\tau^2) - 2t\left(\tau^2 \sum_{i=1}^{n} x_i + \sigma^2 m\right)\right]\right]$$

Hence it is

$$N\left(\frac{\tau^2 \sum_{i=1}^{n} x_i + \sigma^2 m}{\sigma^2 + n\tau^2}, \frac{\sigma^2\tau^2}{\sigma^2 + n\tau^2}\right).$$

Proposition 8.2.3. *Let $(\theta,X_1, \ldots, X_n)$ be a random vector, for which θ is distributed as $N(m,\tau^2)$ and $(X_1, \ldots, X_n)$ has a distribution conditional on θ equal to $N^{\otimes n}(\theta,\sigma^2)$. The distribution of θ conditional on $(X_1, \ldots, X_n) = (x_1, \ldots, x_n)$ is then*

$$N\left[\frac{\tau^2 \sum_{i=1}^{n} x_i + \sigma^2 m}{n\tau^2 + \sigma^2}, \frac{\sigma^2\tau^2}{n\tau^2 + \sigma^2}\right].$$

How is θ to be estimated after this experiment? The estimator which will naturally be chosen is

$$\hat{\theta} = \frac{\tau^2 \sum_{i=1}^{n} X_i + \sigma^2 m}{n\tau^2 + \sigma^2}$$

the mean of the posterior distribution.

8.2.3. Which Group Does an Observed Individual Come From?

Two populations of cells of type A and type B are available. In a large sample the proportion of A is π_A and that of B is π_B. The cell diameters are measured, the distribution of which is $N(m,\sigma^2)$ for A and $N(m',\sigma^2)$ for B, with $m > m'$; π_A, π_B, m, m' and σ^2 are known. A cell is chosen at random. Let us denote by θ the population to which it belongs and by X its diameter. Starting from the observation X, we try to test "$\theta \in A$" against "$\theta \in B$", the costs of misclassification being a, for $\theta \in A$, and b, for $\theta \in B$. If it is decided to make a random draw and to choose B with probability p, the cost will be ap, for $\theta \in A$, and $b(1 - p)$, for $\theta \in B$. Hence if decision $\phi(X)$ is chosen, where ϕ is an r.v. from $\mathbb{R}$ into [0,1] and $\phi(X)$ is the probability of deciding B after having observed X, it is necessary to minimize:

$$aP(\theta \in A)E[\phi(X)|\theta \in A] + bP(\theta \in B)E[(1 - \phi(X))|\theta \in B]$$

$$= a\pi_A \frac{1}{\sigma\sqrt{2\pi}} \int \phi(x)\exp\left[-\frac{(x - m)^2}{2\sigma^2}\right] dx$$

$$+ b\pi_B \frac{1}{\sigma\sqrt{2\pi}} \int (1 - \phi(x))\exp\left[-\frac{(x - m')^2}{2\sigma^2}\right] dx.$$

The solution is the indicator function ϕ of

$$\left\{x;\ a\pi_A \exp\left[-\frac{(x-m)^2}{2\sigma^2}\right] < b\pi_B \exp\left[-\frac{(x-m')^2}{2\sigma^2}\right]\right\}.$$

Hence the chosen test is nonrandomized, which consists of deciding $\theta \in B$ for

$$X < \frac{m+m'}{2} - \frac{\sigma^2}{m-m'} \operatorname{Log} \frac{a\pi_A}{b\pi_B}.$$

8.2.4. Bayesian Statistical Models

Definition 8.2.4. A Bayesian statistical model is the following structure:

(a) A measure set $(\Theta, \mathcal{C}, \nu)$ of **parameters**; ν is the **prior distribution** of the parameter;
(b) a measurable set $(\Omega, \mathcal{A})$ of observations;
(c) a transition probability $(\theta, A) \overset{P}{\longmapsto} P_\theta(A)$ from $(\Theta, \mathcal{C})$ into $(\Omega, \mathcal{A})$, P_θ being the distribution of the observations conditional on the fact that the parameter equals θ.

In other words, $(\Theta, \mathcal{C}) \times (\Omega, \mathcal{A})$ is given the probability $P_\nu = \nu \otimes P$; the distribution of an observation conditional on θ is $P_\theta(\cdot)$, its marginal distribution is νP.

A **never ending controversy** livens up the statistician's world on the foundation of this framework. A **Bayesian** statistician considers it impossible to view nature without reasoning starting from the intuitive idea (or based on his past experience) which he has: thus he decides to start from this *a priori* idea (the prior distribution of the parameter) in order to organize his experiment.

A **non-Bayesian** statistician refuses this *a priori*; he will sometimes tolerate the Bayesian model in examples such as Sections 8.2.2 and 8.2.3, or for the theory of games (Section 8.2.1), where nature is the opposing player. Above all, he will use the Bayesian model as a mathematical tool, which will give results applicable to the non-Bayesian framework. Such a philosophical quarrel is not the aim of this book: we shall not take sides!

8.2.5. Bayesian Decisions Before Any Experiments

The statistical problem being that posed in Section 8.1.2, what decision is to be taken without experimenting, with the help of the prior distribution? If action a is chosen, the loss is $C(\theta,a)$: the average loss is $\int d\nu(\theta)C(\theta,a)$ (assuming that this integral exists). With these hypotheses we have:

Definition 8.2.5. The **Bayesian risk** for the prior distribution ν is the number $\inf_{a\in A}\int d\nu(\theta)C(\theta,a) = \rho(\nu)$. If this lower bound is attained by an action $d(\nu)$ of A, $d(\nu)$ is said to be the **Bayesian decision.**

The Case of Estimation. $\Theta = A$ is an open set in $\mathbb{R}^k$, and $C(\theta,a)$ equals $\|\theta - a\|^2$. Assume that $\int\|\theta\|^2 d\nu(\theta)$ is finite. It is necessary to minimize ϕ, with

$$\phi(a_1, ..., a_k) = \int \sum_{i=1}^{n} (\theta_i - a_i)^2 d\nu(\theta)$$

$$\frac{\partial\phi}{\partial a_i}(a_1, ..., a_k) = 2\left[a_i - \int\theta_i\, d\nu(\theta)\right].$$

The gradient of ϕ vanishes for $a = \int\theta\, d\nu(\theta)$; the *Bayesian estimator based on the prior distribution* ν *is* $\int\theta\, d\nu(\theta)$ and the Bayesian risk is $\int\|\theta - \int\theta\, d\nu(\theta)\|^2 d\nu(\theta)$. For arbitrary Θ and g a measurable function from Θ into $\mathbb{R}$ such that $\int\|g(\theta)\|^2 d\nu(\theta)$ is finite, we obtain similarly the Bayesian estimator $\int g(\theta)d\nu(\theta)$.

The Case of Tests Between Two Hypotheses. The set Θ being the union of two disjoint sets Θ_0 and Θ_1, we test "$\theta \in \Theta_0$" or "$\theta \in \Theta_1$." Here $A = \{0,1\}$, the cost is defined by two strictly positive numbers C_{01} and C_{10}, and

$$C(\theta,0) = \begin{cases} 0 & \text{for } \theta \in \Theta_0 \\ C_{10} & \text{for } \theta \in \Theta_1 \end{cases}$$

$$C(\theta,1) = \begin{cases} C_{01} & \text{for } \theta \in \Theta_0 \\ 0 & \text{for } \theta \in \Theta_1 , \end{cases}$$

Set $\nu = \nu(\Theta_1) = 1 - \nu(\Theta_0)$. Then the average cost of deciding 1 is $(1 - \nu)C_{01}$; the average cost of deciding 0 is νC_{10}. From which

$$\rho(\nu) = \inf\{(1 - \nu)C_{01}, \nu C_{10}\};$$

$$d(\nu) = \begin{cases} 0 & \text{for } \nu < C_{01}/(C_{01} + C_{10}) \\ 1 & \text{for } \nu > C_{01}/(C_{01} + C_{10}) \\ \text{either} & \text{for } \nu = C_{01}/(C_{01} + C_{10}). \end{cases}$$

8.2.6. Bayesian Decisions After the Experiment, For a Dominated Model

Consider a dominated statistical experiment $(\Omega, \mathcal{A}, P_\theta)_{\theta \in \Theta}$, in the Bayesian framework. Assume that there exists a σ-finite μ on $(\Omega, \mathcal{A})$ and a likelihood L, a positive r.v. on $(\Theta, \mathcal{C}) \times (\Omega, \mathcal{A})$, such that, for all θ, $L(\theta, \cdot)$ is a version of $dP_\theta/d\mu$. In other words, with the notations of Section 6.3.1, $(\Theta, \mathcal{C}) \times (\Omega, \mathcal{A})$ is given the probability $P_\nu = \nu \otimes P_\theta$, the density of which is L with respect to $\nu \otimes \mu$. The distribution of θ conditional on a result of the experiment, $\omega = x$, or the **posterior distribution** ν_x of θ, thus has, with respect to ν, the density

$$\theta \longmapsto \frac{L(\theta,x)}{\int L(\theta,x)d\nu(\theta)}$$

defined for all x not in the set

$$\{x; \int L(\theta,x)d\nu(\theta) = 0\}$$

of μ-measure 0 (Section 6.1, Example 3).

Proposition 8.2.3 gives the posterior distribution of θ after having observed $(x_1, \dots, x_n)$ in an n-sample from $N(\theta,\sigma^2)$, when the prior distribution of θ is $N(m,\tau^2)$. For $\Theta = [0,1]$ given a prior distribution $\beta(a,b)$ $(a > 0,\ b > 0)$, the posterior distribution after having observed $(x_1, \dots, x_n)$ in an n-sample from $b(\theta)$ is $\beta(a + x,\ b + n - x)$ (see Section 2.3.4).

The average risk of a decision δ is

$$\begin{aligned} R(\nu,\delta) &= \int dP_\nu(\theta,x)C(\theta,\delta(x)) \\ &= \int d\nu(\theta) \int L(\theta,x)d\mu(x)C(\theta,\delta(x)) \end{aligned}$$

$$= \int d\mu(x) \int d\nu(\theta) L(\theta,x) C(\theta,\delta(x))$$

$$= \int d\mu(x) (\int L(\theta,x) d\nu(\theta)) [\int C(\theta,\delta(x)) d\nu_x(\theta)].$$

Hence if for μ-almost all x, a Bayesian decision $d(\nu_x)$ can be found and if $x \longmapsto d(\nu_x)$ is measurable from $(\Omega,\mathcal{A})$ into $(A,\mathcal{U})$, the average risk is minimized by taking $\delta(x) = d(\nu_x)$.

Proposition 8.2.6. *Let there be given a dominated Bayesian model, with a measurable likelihood L on $(\Theta,\mathcal{C}) \times (\Omega,\mathcal{A})$, and a prior distribution ν. The posterior distribution after having observed $\omega = x$ is the distribution ν_x with density*

$$\theta \longmapsto \frac{L(\theta,x)}{\int L(\theta,x) d\nu(\theta)}$$

with respect to ν. If, for μ-almost all x, there is a Bayesian decision $d(\nu_x)$ and if $x \overset{\delta}{\longmapsto} d(\nu_x)$ is measurable from $(\Omega,\mathcal{A})$ into $(A,\mathcal{U})$. This decision δ minimizes the average risk. This is the Bayesian decision after the experiment.

The Case of Estimation. For Θ an open set in $\mathbb{R}^k$, the **Bayesian estimator** of θ will therefore be

$$x \longmapsto \int \theta \, d\nu_x(\theta) = \frac{\int \theta L(\theta,x) d\nu(\theta)}{\int L(\theta,x) d\nu(\theta)} .$$

The stated estimator has been dealt with in Section 8.2.2. Let g be an r.v. on Θ ; we obtain similarly the Bayesian estimator of $g(\theta)$,

$$\frac{\int g(\theta) L(\theta,x) d\nu(\theta)}{\int L(\theta,x) d\nu(\theta)}.$$

The Case of Tests of "$\theta \in \Theta_0$" Against "$\theta \in \Theta_1$." We then calculate

$$\nu_x(\Theta_1) = \frac{\int_{\Theta_1} L(\theta,x) d\nu(\theta)}{\int_{\Theta} L(\theta,x) d\nu(\theta)} .$$

We decide "$\theta \in \Theta_1$" for $\nu_x(\Theta_1) > C_{01}/(C_{01} + C_{10})$. In other words, *the Bayesian test of "$\theta \in \Theta_0$" against "$\theta \in \Theta_1$"* is (having observed $\omega = x$): *decide "$\theta \in \Theta_1$" if the* **likelihood ratio**

$$\frac{\int_{\Theta_1} L(\theta,x)d\nu(\theta)}{\int_{\Theta_1} L(\theta,x)d\nu(\theta)}$$

exceeds C_{01}/C_{10} *and* "$\theta \in \Theta_0$" *if this ratio is less than* C_{01}/C_{10} (if there is equality there is no preference for either decision).

Particular Case. For a test between two simple hypotheses $\Theta = \{\theta_0,\theta_1\}$, we test "$\theta = \theta_0$" against "$\theta = \theta_1$" and a likelihood ratio test is obtained, where the rejection region of "$\theta = \theta_0$" is of the form

$$\left\{x;\ \frac{L(\theta_1,x)}{L(\theta_0,x)} \geqslant k\right\}$$

for a particular value of k. Such a test, where ν does not appear, is acceptable to a non-Bayesian statistician; we will come back to the study of this in a non-Bayesian framework.

Note. *By using a privileged dominant, we see that if* T *is a sufficient statistic, the posterior distribution of* ν *has a* T*-measurable density. A Bayesian decision is thus* T*-measurable. The observation can be replaced by that of a sufficient statistic.*

8.2.7. Bayesian Decisions for Non-Bayesians

In Sections 2.3.4 and 2.3.5, we have constructed a whole host of admissible estimators of the parameter of a Bernoulli distribution, starting from various prior distributions. Prior distributions appear here as tools for constructing estimators or other decision rules. It is no longer a question here of intuition about the state of nature, but of tricks of calculation. Prior distributions are taken, for which the calculation of the posterior distributions is easy; for example *a family of distributions said to be* **conjugate** *distributions for the experiment*, closed under replacement of a distribution by its posterior distribution (Exercises 8.2.6, 8.2.14, and 8.2.15).

Is the Bayesian decision $x \longmapsto d(\nu_x)$ admissible? We know that, for every other decision δ, we have

$$\int d\nu(\theta)\int dP_\theta(x)C(\theta,\delta(x)) \geqslant \int d\nu(\theta)\int dP_\theta(x)C(\theta,d(\nu_x)).$$

It is then impossible that δ is better than the Bayesian decision, which is therefore admissible under various, easy to verify hypotheses, for example:

– *if the Bayesian decision is unique up to equivalence in the sense of Definition* 7.1.4, *the model being dominated;*

– *if* Θ *is countable and if* ν *charges all the points of* Θ;

– *if* Θ *is an open set in* $\mathbb{R}^k$, *if* ν *charges every open set contained in* Θ *and if* $\theta \longmapsto \int dP_\theta(x)C(\theta,\delta(x))$ *is, for every decision, a continuous function.*

The Case of Tests. For a test between two simple hypotheses, *a likelihood ratio test is always admissible.*

The Case of Estimation. Let Θ be an open set in $\mathbb{R}^k$, and ν a measure, the support of which is $\overline{\Theta}$. If the Bayesian estimator of θ associated with ν has a continuous risk, it is admissible in the class of estimators, the risk of which, $\theta \longmapsto \int dP_\theta(x)\|\theta - \delta(x)\|^2$, is continuous.

Exercises 8.2

E.1. In the framework of Section 8.2.2 show that the (marginal) distribution of X_1 is $N(m,\sigma^2 + \tau^2)$. Are the r.v.'s $(X_1, \ldots, X_n)$ independent for the probability P?

E.2. *Bayesian estimation and the median* (Exercise 3.2.11). Let Θ be an interval of $\mathbb{R}$. Here $\theta \in \Theta$ is estimated by using the cost function $C(\theta,a) = |\theta - a|$.

(a) Give a Bayesian estimator for a prior distribution in the absence of experiment. Determine the Bayesian estimators of θ associated with this loss and compare them to those associated with quadratic loss in the following cases:

– $\theta > 0$ is estimated by observing an r.v. X with distribution $U_{[0,1]}$ with an exponential prior distribution with parameter 1.

– $\theta \in \mathbb{R}$ is estimated by observing an n-sample from $N(\theta,1)$ with the prior distribution $N(m,\tau^2)$; or with the uniform prior distribution on $\{1, 2, 4\}$.

E.3. What is the Bayes estimator for an n-sample from $b(\theta)$ if the prior distribution of the parameter θ is $U_{[0,1]}$ and if the loss function is $(\theta,a) \longmapsto (\theta - a)^2/\theta(1 - \theta)$. What is its risk? Show that it is minimax.

E.4. We want to estimate $\theta \in [0,1]$ with the loss $(\theta,a) \longmapsto \theta(1 - a) + (1 - \theta)a$.

(a) Let ν be a prior distribution. What is the associated Bayesian estimator in the absence of experiment? Verify that its Bayesian risk is majorized by 1/2; what is this risk for $\nu = \delta_{1/2}$?

(b) An n-sample from $b(\theta)$ is observed. What are the Bayesian estimators associated with the following prior distributions: $\delta_{1/2}$, $\mathcal{U}_{[0,1]}$, $\beta(a,b)$ for $a > 0$ and $b > 0$? Show that the estimator 1/2 is minimax. Show that the sample mean is also minimax.

E.5. Consider two independent r.v.'s with distributions $b(p_1)$ and $b(p_2)$. The prior distribution of (p_1,p_2) is uniform on $]0,1[^2$. What is the Bayes estimator of $p_1 - p_2$?

E.6. *Conjugate distributions for an exponential family.* Consider a family of exponential distributions on $\mathbb{R}$: Θ is an interval of $\mathbb{R}$, and for $\theta \in \Theta$, F_θ has, with respect to a σ-finite measure μ of $\mathbb{R}$, the density $x \longmapsto G(\theta)h(x)\exp[\phi(\theta)x]$, where G, h, and ϕ are given r.v.'s on $\mathbb{R}$. Assume that for all θ, F_θ has a finite second moment.

(a) A prior distribution $\nu(a,b)$ of θ concentrated on Θ is given, defined for $a \in \mathbb{R}$ and $b > 0$, the density of which is proportional on Θ, to $\theta \longmapsto (G(\theta))^b\exp(a\phi(\theta))$ (assuming $\int_\Theta (G(\theta))^b\exp(a\phi(\theta))d\theta$ is finite). This family of distributions is the "conjugate" of the exponential family. Show that the posterior distribution of θ is $\nu_x = \nu(a + x, b + 1)$. What is the posterior distribution of θ after observing an n-sample $(X_1, ..., X_n) = (x_1, ..., x_n)$ from the distribution F_θ? Use this process to find families of Bayesian estimators, and check that they are admissible in order to estimate θ with the help of an n-sample from the following distributions: $(b(\theta))_{\theta\in]0,1[}$; $(N(\theta,1))_{\theta\in\mathbb{R}}$; $(p(\theta))_{\theta>0}$; $(\mathcal{E}(\theta))_{\theta>0}$. Amongst these estimators look for one which has a constant risk and is minimax.

E.7. *Unbiased non-Bayesian estimators.* (a) In Section 8.2.6 show that for every P_ν-integrable r.v. Y defined on $(\Omega,\mathcal{A})$, we have (denoting the expectation associated with P_ν by E_ν) $E_\nu(Y|\theta) = E_\theta(Y)$.

(b) Show that an unbiased Bayesian estimator has a zero Bayesian risk. Deduce from this that the sample mean cannot be a Bayesian estimator of θ for an n-sample from $b(\theta)$ (or from $N(\theta,1)$, or from $p(\theta)$).

E.8. *Admissible non-Bayesian estimators.* (a) An n-sample from $N(\theta,1)$ is observed. Show that the risk $R(\theta,T)$ of every estimator T is a continuous function of θ. Let $\hat{\theta}_\sigma$ be the Bayes estimator associated with the prior distribution $N(0,\sigma^2)$. What is its risk? Prove that it is admissible. If T is an estimator, denote its Bayesian risk for $N(0,\sigma^2)$ by $R_\sigma(T)$. For T better than $\overline{X}$, prove

$$\overline{\lim_{\sigma\to\infty}}\ \sigma(R_\sigma(T) - R_\sigma(\overline{X})) < 0.$$

Deduce from this that $\overline{X}$ is admissible.

(b) Copy this proof, to prove that if an r.v. X from distribution $p(\theta)$ is observed, X is an admissible estimator of θ (use the conjugate prior distributions of Exercise 8.2.6).

E.9. An n-sample $(X_1, ..., X_n)$ from $N(\theta,1)$ is observed in order to estimate θ. Show (by using Exercise 8.2.8) that the sample mean is a minimax estimator with constant risk. Show that it is the unique minimax estimator.

E.10. *Estimation for a hypergeometric distribution.* Let $1 \leqslant n \leqslant N$ be two fixed integers, and θ a parameter, $\theta \in \{0,1, ..., N\}$. In order to estimate θ, an r.v. X from distribution F_θ on $\{0,1, ..., n\}$ is observed (the hypergeometric distribution) defined by

$$F_\theta(k) = \frac{\binom{\theta}{k}\binom{N-\theta}{n-k}}{\binom{N}{n}}.$$

For $(\alpha,\beta) \in \mathbb{R}^2$, what is the risk of $\alpha X + \beta$? What are the Bayesian estimators associated with a beta binomial prior distribution (Exercise 6.1.8). Give an estimator with constant risk of the form $\alpha X + \beta$ and verify that it is minimax.

E.11. *Bayesian tests.* For $0 < \theta_1 \leqslant \theta_2 < \infty$, an n-sample $(X_1, ..., X_n)$ from the $U_{[0,\theta]}$ distribution and an Erlang prior distribution of order $n + 1$ (Exercise 4.3.6) with parameter λ, determine a Bayesian test of "$\theta \in [\theta_1,\theta_2]$" against "$\theta \notin [\theta_1,\theta]$" for given costs.

E.12. An engineer must build a dyke against the flooding due to the rising levels of a river. If he decides to build a dyke of height h, it will cost ch; if there is a rise in the level, there will not be any damage if the height H is less than h, and damage valued at $\Gamma(H - h)$ if the height H is greater than h. H is an exponential random variable with parameter $1/\theta$, and $\Gamma > c$. The levels H and h are measured above the mean level.

(a) Explain why, by stopping the observation at the time of the first flood, a natural loss function is

$$(\theta,h) \longmapsto ch + \Gamma\theta \exp\left(-\frac{h}{\theta}\right).$$

If θ is known, which height h is to be chosen?

(b) n independent rises, of heights $H_1, \ldots, H_n$, are observed. Let $\bar{H} = (H_1 + \ldots + H_n)/n$. The engineer, who does not know θ, decides to fix the height of the dyke at $d(H_1, \ldots, H_n) = k\bar{H}$. What is the risk of this decision?

E.13. *The Kalman-Bucy Filter.* Consider the sequence (ξ_n) (resp. (η_n)) of independent r.v.'s from an $N(0,1)$ (resp. $N(0,\sigma^2)$) distribution. These two sequences are independent of each other and independent of an r.v. X with distribution $N(\hat{x}_0,\hat{\sigma}_0^2)$, where $\hat{x}_0$ and $\hat{\sigma}_0$ are known. We are given a real sequence (θ_n) and study a signal (X_n) defined by $X_{n+1} = \theta_n X_n + \xi_n$ ($n \geq 0$). Only the noisy signal $Z_n = X_n + \eta_n$ can be observed. Show that the distribution of X_n conditional on $(Z_0, \ldots, Z_n) = (z_0, \ldots, z_n)$ is a Gaussian distribution $N(\hat{x}_n,\hat{\sigma}_n^2)$, where $\hat{\sigma}_n^2$ depends only on n (and not on the observations), and where $\hat{x}_n$ is a linear function of the observations, "the Kalman–Bucy filter." Give the recurrence relations which allow $\hat{x}_n$ and $\hat{\sigma}_n^2$ to be calculated. What is the best approximation of X_n by a function of the past observations $Z_0, \ldots, Z_n$ in the least squares sense? What is the quadratic error?

E.14. *Bayesian estimators for a uniform distribution.* The notations are those of Exercise 3.3.17.

(a) An *n-sample* $(X_1, \ldots, X_n)$ from a uniform distribution on $[0,\theta]$ is observed. The prior distribution of θ is a one-sided Pareto distribution with density $f_{r,\alpha}$ ($r > 0$, $\alpha > 0$). What is the posterior distribution of θ? What is the Bayesian estimator of θ associated with this distribution? Is it admissible?

(b) An n-sample $(X_1, ..., X_n)$ from the uniform distribution on the interval $[\theta_1,\theta_2]$ of $\mathbb{R}$ is observed. Assume that the prior distribution of (θ_1,θ_2) is the two-sided Pareto distribution with density $f_{r,s,\alpha}$ $(r < s;\ \alpha > 0)$. What is the posterior distribution? Give the Bayesian estimators of θ_1, of θ_2, and of $(\theta_2 - \theta_1)$. Are they admissible?

E.15. *The Dirichlet distribution* (following from Exercises 3.3.13, 3.3.14, and 3.3.3). Let $\alpha_1, ..., \alpha_k$ be k strictly positive numbers and $Z_1, ..., Z_k$ independent r.v.'s with respective distributions $\gamma(\alpha_1,1), ..., \gamma(\alpha_k,1)$, and let $Z = Z_1 + ... + Z_k$. The distribution of $(Y_1, ..., Y_k)$ with $Y_i = Z_i/Z$ concentrated on

$$S = \{(y_1, ..., y_k);\ 0 < y_i < 1 \quad \text{for } i = 1, ..., k,\ \sum_{i=1}^{k} y_i = 1\}$$

is called the Dirichlet distribution $\mathcal{D}(\alpha_1, ..., \alpha_k)$. Let $\alpha = \alpha_1 + ... + \alpha_k$.

(a) Show that the density of $(Y_1, ..., Y_{k-1})$ is

$$(y_1, ..., y_{k-1}) \longmapsto \frac{\Gamma(\alpha)}{\Gamma(\alpha_1) \dots \Gamma(\alpha_k)} \left(\prod_{j=1}^{k-1} y_j^{\alpha_j - 1}\right) \left(1 - \sum_{j=1}^{k-1} y_j\right)^{\alpha_k - 1}.$$

(b) What is the distribution of Y_i, its mean and its variance? Calculate $E(Y_iY_j)$ for $i \neq j$.

(c) Let X be an r.v. taking values in $(1, ..., k)$ with distribution $p = (p_1, ..., p_k)$, where $P_p(X = i) = p_i$. Assume that the prior distribution of p is $\mathcal{D}(\alpha_1, ..., \alpha_k)$. Prove that posterior distribution of p, when $X = i$ has been observed, is $\mathcal{D}(\alpha_1, ..., \alpha_{i-1}, \alpha_i + 1, \alpha_{i+1}, ..., \alpha_k)$.

(d) An n-sample $(X_1, ..., X_n)$ from p is observed. Determine the posterior distribution of p as a function of the sufficient statistic

$$\left(N_i = \sum_{p=1}^{n} 1_{(X_p = i)}\right)_{1 \leq i \leq k}.$$

What is the Bayesian estimator of p? Is it admissible?

8.3. Optimality Properties of Some Likelihood Ratio Tests

8.3.1. A Test of Two Simple Hypotheses

Theorem 8.3.10. (Neyman–Pearson Lemma). *Let P_0 and P_1 be two distinct probability distributions on the experiment space (Ω,A). Let p_0 and p_1 be two versions (likelihood) of their densities with respect to a dominant μ, for example $(P_0 + P_1)$. P_0 is tested against P_1. For every $\alpha \in]0,1[$, constants γ and C can be determined such that the following function ϕ:*

$$\phi = 1_{[p_1>Cp_0]} + \gamma 1_{[p_1=Cp_0]}$$

is a test of size α, optimal amongst tests of level α. This test is unbiased ($E_1(\phi) \geqslant \alpha$). Every test ϕ of size α equal to 1 on $[p_1 > Cp_0]$ and to 0 on $[p_1 < Cp_0]$ has these properties. We say that it is a Neyman–Pearson test of size α.

How Can We Determine C and γ? The function $x \longmapsto P_0(p_1 > xp_0)$ decreases from 1 to 0; we will take $C = \inf\{x; P_0(p_1 > xp_0) < \alpha\}$. Then

$$P_0[p_1 > Cp_0] \leqslant \alpha \leqslant P_0[p_1 \geqslant Cp_0].$$

When $P_0[p_1 = Cp_0]$ is zero, γ can be taken as zero; ϕ is a non-randomized test, with rejection region $\{p_1 > Cp_0\}$. In general, where $x \longmapsto P_0(p_1 > xp_0)$ can jump at the point C, we take $\gamma P_0[p_1 = Cp_0] = \alpha - P_0[p_1 > Cp_0]$.

Proof of the Theorem. Let ϕ^* be a test of P_0 against P_1 of level α. Let us calculate the difference between the powers, $\int(\phi^* - \phi)p_1 d\mu$. Ω is split into three parts: $A_1 = \{p_1 > Cp_0\}$, where ϕ equals 1, hence $\phi^* - \phi$ is negative; $A_2 = \{p_1 = Cp_0\}$; $A_3 = \{p_1 < Cp_0\}$ where ϕ is zero, hence $\phi^* - \phi$ is positive. We deduce from this, for $i = 1,2,3$,

$$\int_{A_i}(\phi^* - \phi)p_1 d\mu \leqslant C\int_{A_i}(\phi^* - \phi)p_0 d\mu.$$

Summing over i,

$$\int(\phi^* - \phi)p_1 d\mu \leqslant C\left(\int\phi^* p_0 d\mu - \int\phi p_0 d\mu\right)$$

$$\leqslant C(\alpha - \alpha) = 0.$$

Taking the test of level α associated with the test function ϕ^* identical to α, we obtain $\alpha = E_1(\phi^*) \leq E_1(\phi)$; ϕ is unbiased.

Example. Consider an exponential model, where Θ is included in $\mathbb{R}^k$, T is measurable from (Ω,A) into $\mathbb{R}^k$, and p_θ has density $L(\theta)$: $x \longmapsto \exp(-\psi(\theta) + \langle\theta,T(x)\rangle)$ with respect to a σ-finite measure μ. Here θ can take two values, θ_0 or θ_1:

$$\text{Log}\,\frac{L(\theta_1)}{L(\theta_0)} = \langle\theta_1 - \theta_0,T\rangle + \Psi(\theta_0) - \Psi(\theta_1).$$

The Neyman–Pearson test is defined by

$$\phi = \mathbf{1}_{(\langle\theta_1-\theta_0,T\rangle>C)} + \gamma\mathbf{1}_{(\langle\theta_1-\theta_0,T\rangle=C)};$$

$$C = \inf\{x;\ P_{\theta_0}(\langle\theta_1 - \theta_0,T\rangle > x) < \alpha\}.$$

For $k = 1$ and, for example $\theta_0 < \theta_1$, we obtain

$$\phi = \mathbf{1}_{(T>C)} + \gamma\mathbf{1}_{(T=C)},$$

with

$$C = \inf\{x;\ P_{\theta_0}(T > x) < \alpha\}.$$

If the distribution of T is continuous ϕ can be taken to be nonrandomized.

Particular Case. (a) *An n-sample* $(X_1, \ldots, X_n)$ *from a Bernoulli distribution* $b(p)$. We test "$p = p_0$" against "$p = p_1$" with $p_0 < p_1$ by

$$\phi = \mathbf{1}_{(X_1+\ldots+X_n>C)} + \gamma\mathbf{1}_{(X_1+\ldots+X_n=C)}$$

$$C = \sup(x;\ b(n,p)\{x, x+1, \ldots, n\} > \alpha),$$

$$\gamma b(n,p)\{C\} + b(n,p)[C + 1, n] = \alpha.$$

(b) *An n-sample from a Gaussian distribution* $N(m,\sigma^2)$, where σ^2 is given and we test "$m = m_0$" against "$m = m_1$" with $m_0 < m_1$. Here, $\phi = \mathbf{1}_{(\overline{X}>C)}$, with $N(m_0,\sigma^2/n)([C,\infty[) = \alpha$, where $C = m_0 + (\sigma/\sqrt{n})\phi_\alpha$

Note. *An optimal test of level* $\alpha \in]0,1[$ *is* $P_0 + P_1$ a.s. *equal to a Neyman–Pearson test* (this follows from the above proof).

8.3.2. Families with Monotone Likelihood Ratio

In the preceding examples, if the statistic $\overline{X}$ is denoted by T, we see that the test is of the form $[T > C(\alpha,\theta_0)]$, where θ_0 is either m_0 or p_0. However, the threshold of the test C does not depend on the alternative hypothesis θ_1 as long as θ_1 is greater than θ_0. (For $\theta_1 < \theta_0$, the sense of the inequality in the test region must be changed.) This leads us to a class of models, indexed by an ordered set in $\mathbb{R}$, for which this situation again arises.

Definition 8.3.11. We are given $(\Omega,A,P_\theta)_{\theta\in\Theta}$, a statistical model dominated by Q, where Θ is a subset of $\mathbb{R}$, and $(L(\theta))_{\theta\in\Theta}$ a likelihood. Assume that there exists a real r.v. T and for every pair $\theta < \theta'$ an increasing function $\phi_{\theta,\theta'}$ from $\mathbb{R}$ into $\mathbb{R}_+$, for which (Q a.s.):

$$\frac{L(\theta')}{L(\theta)} = \phi_{\theta,\theta'}(T).$$

The family $\{L(\theta)\}_{\theta\in\Theta}$ is then said to have **increasing likelihood ratio** (increasing may be replaced by decreasing or monotone).

Examples. Let us reconsider the exponential model of Section 8.3.1, assuming here that Θ is an interval of $\mathbb{R}$, $\{\theta; \int\exp(\theta,T(x))d\mu(x) < \infty\}$. For θ and θ' in Θ,

$$\text{Log}\,\frac{L(\theta')}{L(\theta)} = (\theta' - \theta)T - \Psi(\theta') + \Psi(\theta).$$

There are other examples; this is the case for an n-sample $(X_1, ..., X_n)$ from a uniform distribution on $[0,\theta]$ with $\theta > 0$, since we then obtain

$$\frac{L(\theta')}{L(\theta)} = \frac{\theta}{\theta'} \quad \text{on } \{0 < \sup X_i = T < \theta\}$$

$$= \infty \quad \text{on } \{T > \theta\}.$$

Given a family with increasing likelihood ratio, for each pair $\theta < \theta'$, a test of "θ" against "θ'" of the form $\phi = \mathbf{1}_{(T>C)} + \gamma\mathbf{1}_{(T=C)}$ is a Neyman–Pearson test. Consequently $E_\theta(\phi) \leq E_{\theta'}(\phi)$ and the function $\theta \longmapsto E_\theta[\mathbf{1}_{(T>C)} + \gamma\mathbf{1}_{(T=C)}]$ is increasing.

Let $\theta_0 \leqslant \theta_1$ and let α be given. Let ϕ be the Neyman–Pearson test of θ_0 against θ_1 and ϕ^* a test of level α of $\theta \leqslant \theta_0$ against $\theta > \theta_0$. ϕ is also a test of level α of $\theta \leqslant \theta_0$ against $\theta > \theta_0$, since $\theta \longmapsto E_\theta(\phi)$ is increasing. However, ϕ being the Neyman–Pearson test of θ_0 against θ_1, it is, for $\theta > \theta_0$, at least as powerful as ϕ^* for testing "θ_0" against "θ":

$$E_\theta(\phi^*) \leqslant E_\theta(\phi).$$

Theorem 8.3.12. *For the family with monotone likelihood ratio of Definition* 8.3.11 *and for* $\theta_0 \leqslant \theta_1$, *the Neyman–Pearson test*

$$\phi = 1_{(T>C)} + \gamma 1_{(T=C)}$$

$$P_{\theta_0}(T > C) + \gamma P_{\theta_0}(T = C) = \alpha$$

is a test of size α, *uniformly most powerful amongst all tests of* "$\theta \leqslant \theta_0$" *against* "$\theta > \theta_1$" *of level* α. *Its power function* $\theta \longmapsto E_\theta(\phi)$ *is increasing on* Θ. *In order to test* "$\theta \geqslant \theta_0$" *against* "$\theta < \theta_2$," $\theta_2 \leqslant \theta_0$, $1 - \phi$ *is a test of size* $1 - \alpha$, *uniformly most powerful amongst tests of this level.*

Note. The smallest power corresponds to $\theta = \theta_1$. If the model is exponential, the set of parameters being an open interval of $\mathbb{R}$, the Kullback information (Section 7.2.2) has been calculated

$$K(\theta,\theta') = \psi(\theta') - \psi(\theta) - (\theta' - \theta)\psi'(\theta).$$

Thanks to the convexity of ψ, $\theta' \longmapsto K(\theta,\theta')$ increases for $\theta' > \theta$, and $\theta \longmapsto K(\theta,\theta')$ decreases for $\theta < \theta'$. Hence for $\theta_0 < \theta_1$, we have

$$K(\theta_0,\theta_1) = \inf\{K(\theta,\theta'); \theta \leqslant \theta_0, \theta' \geqslant \theta_1\}.$$

The uniformly most powerful test of "$\theta \leqslant \theta_0$" against "$\theta \geqslant \theta_1$" is the most powerful test of the same level for testing the least dissimilar hypotheses "$\theta = \theta_0$" against "$\theta = \theta_1$."

In more complex cases, this idea reappears, which gives the Neyman-Pearson lemma its importance. It is rare that a test problem of two simple hypotheses "$\theta = \theta_0$" against "$\theta = \theta_1$" arises. However, *it often happens that, testing* "$\theta \in \Theta_0$" *against* "$\theta \in \Theta_1$," *the most unfavorable pair* $\theta_0 \in \Theta_0$ *and* $\theta_1 \in \Theta_1$ *can be found. The Neyman–Pearson test of* "$\theta = \theta_0$" *against* "$\theta = \theta_1$" *can then be used.*

8.3.3. Two-Sided Tests

Let us remain within the framework of a family with increasing likelihood ratio described in Section 8.3.2. We often try to test "$\theta = \theta_0$" against "$\theta \neq \theta_0$", or "$\theta \in [\theta_1,\theta_2]$" against "$\theta \notin [\theta_1,\theta_2]$": this is the case for goodnesss of fit tests such as those of Section 2.4.4 or 5.1.5. Here we can no longer hope to find a test which is optimal amongst those of a given level. Assume that the distributions P_θ are equivalent. Let ϕ be optimal of level α of "$\theta = \theta_0$" against "$\theta \neq \theta_0$"; it is then necessarily optimal of level α of "$\theta = \theta_0$" against "$\theta = \theta'$," for arbitrary $\theta' \neq \theta_0$. Taking $\theta' > \theta_0$, it is a.s. of the form $\mathbf{1}_{(T>C)} + \gamma \mathbf{1}_{(T=C)}$, $C \in \mathbb{R}$, and γ an r.v.. Taking $\theta' < \theta_0$ it is a.s. of the form $\mathbf{1}_{(T<C)} + \gamma \mathbf{1}_{(T=C)}$. These conditions are incompatible, except in the case of a test identical with α which is not, in general, optimal. We shall state without proof the principal result (see Exercise 8.3.5).

Theorem 8.3.13. *Let $(\Omega,\mathcal{A},P_\theta)_{\theta\in\Theta}$ be an exponential model, where Θ is an interval of $\mathbb{R}$ and where, for an* r.v. *T the distribution of which is continuous, a likelihood is* $\exp[-\psi(\theta) + \theta T]$. *Then there exists a test ϕ of $\theta \in [\theta_1,\theta_2]$ against $\theta \notin [\theta_1,\theta_2]$, uniformly most powerful amongst unbiased tests of level α. This test has a rejection region $D = \{T \notin [C_1,C_2]\}$, where C_1 and C_2 are solutions of the following equations:*

(a)
$$\begin{cases} P_{\theta_0}(D) = \alpha \\ P'_{\theta_0}(D) = 0 \end{cases}, \quad \textit{for } \theta_1 = \theta_2 = \theta_0 \quad \left[P'_\theta(D) = \frac{d}{d\theta} P_\theta(D)\right].$$

(b)
$$P_{\theta_1}(D) = P_{\theta_2}(D) = \alpha, \quad \text{for } \theta_1 \neq \theta_2$$

As a shorthand we denote UMP (resp. UMPU) for "uniformly most powerful" amongst tests of that level (resp. "and unbiased").

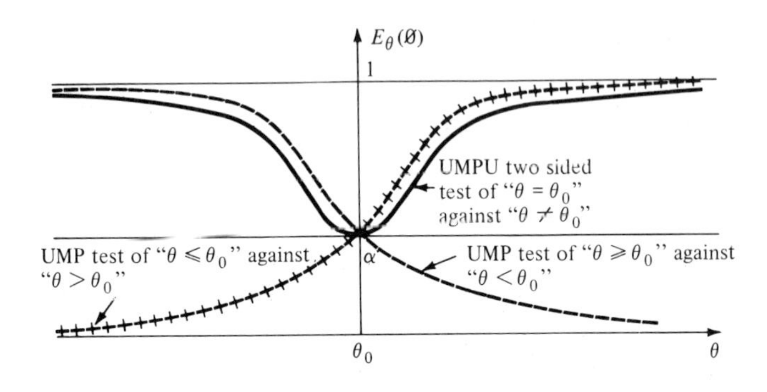

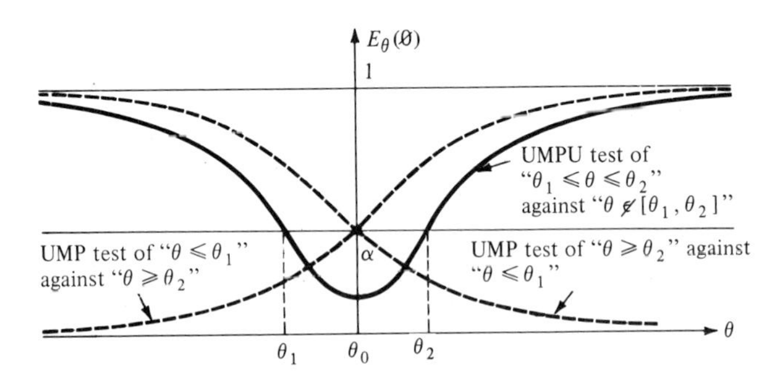

Exercises 8.3

E.1. We consider an n-sample from $(\mathcal{N}(\theta,1))_{\theta\in\mathbb{R}}$ with sample mean $\overline{X}$, and test "$\theta = \theta_0$" against "$\theta \neq \theta_0$." Show that the UMPU tests of Theorem 8.4.13 have in this case, rejection regions of the form $\{|\overline{X} - \theta_0| \geq A\}$. Give an UMPU test of level 0.05 of "$\theta = \theta_0$" against "$\theta \neq \theta_0$" or of "$\theta \in [\theta_1,\theta_2]$" against "$\theta \notin [\theta_1,\theta_2]$."

E.2. *Tests on breakdown times.* Recall the experiment described in Exercise 7.3.2. We observe $(T_{(1)}, \ldots, T_{(r)})$. Prove that the ratio of the likelihoods at θ_0 and at θ_1 may be written as

$$\left(\frac{\theta_1}{\theta_0}\right)^r \exp\left\{-T_{r,n}\left(\frac{1}{\theta_0} - \frac{1}{\theta_1}\right)\right\}$$

where $2T_{r,n}/\theta$ is distributed as $\chi^2(2r)$ if the parameter equals θ. With the help of quantiles of the χ^2 distribution, express the most powerful test of "$\theta = \theta_0$" against "$\theta = \theta_1$" of level α

and its power. What are its properties if it is used for testing "$\theta \geqslant \theta_0$" against "$\theta \leqslant \theta_1$" for $\theta_1 < \theta_0$?

E.3. *Tests on a Poisson process.* Recall the experiments of Exercise 7.3.1.

(a) The first r breakdown times are observed. Let $\theta_0 > \theta_1$. Express with the help of quantiles from the χ^2 distribution, the most powerful test of "$\theta = \theta_0$" against "$\theta = \theta_1$" and its power. What properties does it have for testing "$\theta \geqslant \theta_0$" against "$\theta \leqslant \theta_1$"?

(b) Answer similar questions when the observations are stopped at time t.

E.4. Let f be a strictly positive density on $\mathbb{R}$. Show that a necessary and sufficient condition that the family of distributions, the densities of which are the translates of f, $\{f(\cdot-\theta)\}_{\theta\in\mathbb{R}}$, has increasing likelihood ratio, is that Log f be concave.

E.5. *Families of Polya distributions and two-sided tests.* Consider a continuus function: $(\theta,x) \longmapsto p_\theta(x)$ from $\mathbb{R}^2$ into $\mathbb{R}_+$, p_θ being the density with respect to a σ-finite measure μ of a distribution on $\mathbb{R}$. The family of distributions is Polya of type $n \geqslant 1$, if, for $\theta_1 < \dots < \theta_n$, $x_1 < \dots < x_n$, the determinant $|p_{\theta_i}(x_j)|_{1\leqslant i,j\leqslant n}$ is strictly positive.

(a) What does the condition imply for $n = 2$? Show that an exponential family $\{p_\theta(x) = C(\theta)e^{\theta x}\}_{\theta\in\mathbb{R}}$ is Polya for all n ($n = 3$ is sufficient for what follows).

(b) Assume, in what follows, that *the family is Polya of type 3 and that for* $\theta < \theta'$, $x \longmapsto p_{\theta'}(x)/p_\theta(x)$ *is strictly increasing.* Let $\theta_1 < \theta < \theta_2$ and k_1, k, and $k_2 > 0$. Show that the function $k_1p_{\theta_1} + k_2p_{\theta_2} - kp_\theta$ has at most two zeros x_1 and x_2, and that in this case, it is negative on $]x_1,x_2[$.

(c) Consider on $\mathbb{R}$, $c_1 < c_2$, $\theta_1 < \theta_2$, and θ different from θ_1,θ_2. Resolve the system:

$$k_1p_{\theta_1}(c_i) + k_2p_{\theta_2}(c_i) = p_\theta(c_i) \quad (i = 1,2).$$

Show

$$]c_1,c_2[\; = \{x;\; p_\theta(x) > k_1 p_{\theta_1}(x) + k_2 p_{\theta_2}(x)\},$$

for $\theta_1 < \theta < \theta_2$ and

$$]c_1,c_2[\; = \{x;\; p_\theta(x) < k_1 p_{\theta_1}(x) + k_2 p_{\theta_2}(x)\},$$

for $\theta \notin [\theta_1,\theta_2]$.

(d) Generalization of the Neyman–Pearson Theorem. Let $(f_i)_{0\leqslant i\leqslant k}$ be a sequence of functions in $L^1(\Omega,\mathcal{A},\mu)$ for a σ-finite measure μ, $(a_i)_{1\leqslant i\leqslant k}$ a positive sequence and ϕ an integrable r.v. which equals 1 on $\{f_0 > \Sigma_{i=1}^k a_i f_i\}$ and 0 on $\{f_0 < \Sigma_{i=1}^k a_i f_i\}$. Show that if ϕ^* is an r.v. taking values in [0,1], the inequality $\int \phi^* f_i d\mu \leqslant \int \phi\, f_i d\mu$ for all $1 \leqslant i \leqslant k$, implies

$$\int \phi^* f_0 d\mu \leqslant \int \phi f_0 d\mu.$$

(e) Let $\alpha \in \,]0,1[$; ψ_u being a UMP test of size α of $\theta \leqslant \theta_1$ against $\theta > \theta_1$. Study the power of the tests $\psi_u + 1 - \psi_{1-\alpha+u}$. Show that real numbers c_1 and c_2, and γ_1 and γ_2 in [0,1] can be found such that for

$$\phi = 1_{[c_1,c_2]^c} + \gamma_1 1_{\{c_1\}} + \gamma_2 1_{\{c_2\}}$$

we have $E_{\theta_1}(\phi) = E_{\theta_2}(\phi) = \alpha$. Prove that this test is UMP amongst tests ϕ^* of $\theta \in [\theta_1,\theta_2]$ against $\theta \notin [\theta_1,\theta_2]$, such that $E_{\theta_1}(\phi^*)$ and $E_{\theta_2}(\phi^*)$ are α. Show that ϕ is UMP of size α.

(f) Assume that, for every test ϕ of the type defined in (d), $\theta \longmapsto E_\theta(\phi)$ is differentiable and that a test of this type can be obtained such that

$$E_{\theta_0}(\phi) = \alpha \quad \text{and} \quad \frac{d}{d\theta_0} E_{\theta_0}(\phi) = 0.$$

Prove that this test is UMPU of "$\theta = \theta_0$" against "$\theta \neq \theta_0$."

E.6. *A test in the presence of nuisance parameters and Neyman structure.* Let $(\Omega,\mathcal{A},P_\theta)_{\theta\in\Theta}$ be a statistical model for which Θ is given a topology such that $\theta \longmapsto \int \phi dP_\theta$ is continuous for every test function ϕ. Let H_0 and H_1 be two disjoint sets of Θ and $\Delta = \overline{H}_0 \cap \overline{H}_1$ the communal boundary of the hypothesis to be tested.

Let ϕ be a test of H_0 against H_1. It is said to be *similar* of size α if $E_\theta(\phi)$ equals α for all $\theta \in \Delta$. Let T be a sufficient statistic for the model $(\Omega,A,P_\theta)_{\theta\in\Theta}$. We say that (ϕ,T) defines a *Neyman structure* for the test problem if $E_\theta(\phi|T)$ equals α, P_θ-a.s. for $\theta \in \Delta$.

(a) Show that under the above hypotheses every UMP test in the class of similar tests of size $\leqslant \alpha$ is also UMPU. Show that a necessary and sufficient condition, in order that every similar test ϕ defines a Neyman structure (ϕ,T), is that T is b-complete in the model $(\Omega,A,P_\theta)_{\theta\in\Delta}$.

(b) Let Θ be an open set in $\mathbb{R}^k$ and $T = (T_1, ..., T_k)$ a random vector of dimension k which has for a probability P a distribution $f. \lambda_1 \otimes ... \otimes \lambda_k$ ($\lambda_1, ..., \lambda_k$ σ-finite measures on $\mathbb{R}$). Take P_θ, a probability having a density with respect to P, proportional to $\exp(<\theta,T>)$ (c.f. Exercise 6.1.12). Let θ_1 be the first coordinate of θ and let $a \in \mathbb{R}$ we want to test H_0 "$\theta_1 \leqslant a$" against H_1 "$\theta_1 > a$." Show that $U = (T_2, ..., T_k)$ allows a Neyman structure to be defined for every similar test ϕ. Give a UMP test of level α of H_0 against H_1 conditional on $\{U = u\}$; deduce from this a UMPU test of level α of H_0 against H_1. Let $a < b$; give a UMPU test of "$\theta_1 \in [a,b]$" against "$\theta_1 \notin [a,b]$."

(c) Hence, recover Student's test introduced in Chapter 5.

(d) Given an n-sample from $U_{(\theta_1,\theta_2)}$, construct a UMPU test of "$\theta_1 \leqslant 0$" against "$\theta_1 > 0$."

8.4. Invariance

8.4.1. The Translation Parameter

Let f be a density on $\mathbb{R}^k$. If a random vector X has density f, its translation $X + c$ has, for every $c \in \mathbb{R}^k$, the density $x \longmapsto f(x - c)$. Let $(X_1, ..., X_n)$ then be an n-sample from F_θ, the distribution with density $x \longmapsto f(x - \theta)$, with the aid of which statistics on θ will be carried out, θ a parameter in $\mathbb{R}^k$; for all $c \in \mathbb{R}^k$, $(X_1 + c, ..., X_n + c)$ has distribution $F_{\theta+c}$. Hence it seems natural to choose an estimator $\hat{\theta}$, a function from $\mathbb{R}^{kn}$ into $\mathbb{R}^k$, which respects this translation law:

$$\hat{\theta}(x_1 + c, ..., x_n + c) = \hat{\theta}(x_1, ..., x_n) + c.$$

We have used a similar idea in Section 5.1.7, in order to choose tests relating to Gaussian samples, where the envisaged class of tests has been constrained by invariance properties. We are going to make these ideas more precise.

8.4.2. Preliminaries

Consider G, a group of measurable transformations of a measure space $(E,\mathcal{E})$: G is a group under composition of transformations. The **orbit** of $x \in E$ is $0(x) = \{g(x); g \in G\}$. Let T be a function from E into a set F; T is **invariant** under G if it is constant on each orbit; it is a **maximal invariant** if, moreover, T takes different values in each orbit; if U is a maximal invariant, every invariant T factorizes over U, i.e., there exists a function T' for which $T = T' o U$. The following groups of transformations of $\mathbb{R}^n$ are often used.

(a) Let $\mathbf{1}$ be the vector $(1, 1, ..., 1)$ in $\mathbb{R}^n$. Consider G, the set of translations of the vector $c\mathbf{1}$, for $c \in \mathbb{R}$, and $0(x) = \{x + c\mathbf{1}; c \in \mathbb{R}\}$. Each of the following mappings is a maximal invariant, denoting

$$x = (x_1, ..., x_n) \quad \text{and} \quad \overline{x} = \frac{1}{n}\sum_{i=1}^{n} x_i;$$

$$U_1(x) = x - \overline{x}\mathbf{1}; \quad U_2(x) = x - x_n\mathbf{1};$$

$$U_3(x) = (x_1 - x_2, x_2 - x_3, ..., x_{n-1} - x_n).$$

(b) Let G be the group of orthogonal transformations of $\mathbb{R}^n$: the Euclidean norm is a maximal invariant.

(c) Let $G = \{h_c; c > 0\}$, where h_c is the dilatation of ratio c. We can take

$$U_1(x_1, ..., x_n) = \frac{1}{\|x\|}(x_1, ..., x_n) \quad \text{on } \{x \neq 0\};$$

or

$$U_2(x_1, ..., x_n) = \left(\frac{x_1}{|x_n|}, ..., \frac{x_{n-1}}{|x_n|}, \frac{x_n}{|x_n|}\right) \quad \text{on } \{x_n \neq 0\}.$$

Let $(F_\theta)_{\theta\in\Theta}$ be a family of distributions on $(E,\mathcal{E})$, which we will assume to be identifiable. To two distinct θ's correspond two distinct distributions F_θ. This family is said to be invariant under G if, to every g in G and every θ, we can associate $\bar{g}(\theta) \in \Theta$ such that if X has distributions F_θ, $g(X)$ has distribution $F_{\bar{g}(\theta)}$. The identifiability of the family $(F_\theta)_{\theta\in\Theta}$ implies that $\bar{g}(\theta)$ has a unique value, and $\bar{g}$ is a function. Let $\bar{G}$ be the set of transformations of Θ thus defined; $\bar{G}$ is a group under the composition of transformations ($\bar{g} \circ \bar{g}' = \overline{g \circ g'}$). In particular, $\bar{g}$ is bijective (its inverse is $\overline{g^{-1}}$). We can thus define orbits on Θ, $\{\bar{g}(\theta); \bar{g} \in \bar{G}\}$ for each $\theta \in \Theta$; the notion of an invariant and maximal invariant under $\bar{G}$ is the same as that under G.

Examples. (1) Let $(X_1, ..., X_n)$ be an n-sample from (m,σ^2), and let $\Theta = \{(m,\sigma^2); m \in \mathbb{R}, \sigma > 0\}$.

(a) For G, the set of translations t_c of the vector $c1$ with $c \in \mathbb{R}$, it is easily shown that $\bar{t}_c(m,\sigma) = (m + c,\sigma)$. A maximal invariant under $\bar{G}$ is $\lambda: (m,\sigma) \longmapsto \sigma$.

(b) For G a set of dilatations of ratio c, $c > 0$, $\bar{h}_c(m,\sigma) = (cm,c\sigma)$; a maximal invariant under $\bar{G}$ is $\lambda: (m,\sigma) \longmapsto m/\sigma$.

(2) If we return to the framework of Section 8.4.1, where $(X_1, ..., X_n)$ is a sample from a distribution with density $f(\cdot - \theta)$ with $\theta \in \mathbb{R}^k$, then to translations $t_c: (x_1, ..., x_n) \longmapsto (x_1 + c, ..., x_n + c)$ defined for all $c \in \mathbb{R}^k$, corresponds the translations $\bar{t}_c: \theta \longmapsto \theta + c$. Here there is only one orbit and the identity is a maximal invariant of Θ under $\bar{G}$.

Proposition 8.4.14. *Within the framework defined above, if T is an invariant of G the image measure $T(F_\theta)$ is invariant on the orbits of Θ under $\bar{G}$ (it only depends on $\lambda(\theta)$, if λ is a maximal invariant under $\bar{G}$).*

Proof. Let $\phi \circ T$ be a bounded r.v., and let $g \in G$:

$$\int \phi \circ T \, dF_\theta = \int \phi \circ T \circ g \, dF_\theta = \int \phi \circ T \, dF_{\bar{g}(\theta)},$$

$$\int \phi \, dT(F_\theta) = \int \phi \, dT(F_{\bar{g}(\theta)}).$$

8.4.3. Invariant Statistic for a Translation Parameter

Returning to the framework of Section 8.4.1 we look for an estimator $\hat{\theta}(X_1, ..., X_n)$ equal, for all c, to $\hat{\theta}(X_1 + c, ..., X_n + c) - c$. In other words, $\hat{\theta}(X_1, ..., X_n) - X_1$, is an invariant statistic under the translations t_c; it is a function of the maximal invariant

$$Y = (X_2 - X_1, ..., X_n - X_1) = (Y_2, ..., Y_n);$$

$$\hat{\theta}(X_1, ..., X_n) = X_1 - \phi(Y).$$

We shall minimize its quadratic risk

$$E_\theta[X_1 - \theta - \phi(Y)]^2 = E_0[X_1 - \phi(Y)]^2.$$

The estimator of this type having a minimum risk (constant) is obtained for $\phi(Y) = E_0(X_1|Y)$. For $\theta = 0$, the density of $(X_1, Y_2, ..., Y_n)$ is $(x_1, y_2, ..., y_n) \longmapsto f(x_1, x_1 + y_2, ..., x_1 + y_n)$, and a version of ϕ is given by

$$\phi(y_2, ..., y_n) = \frac{\int u f(u, u+y_2, ..., u+y_n)du}{\int f(u,u+y_2, ..., u+y_n)du}.$$

Taking $\theta = x_1 - u$ and $y_i = x_i - x_1$ $(2 \leqslant i \leqslant n)$:

$$\phi(y_2, ..., y_n) = \frac{\int (x_1-\theta)f(x_1-\theta,x_2-\theta, ..., x_n-\theta)d\theta}{\int f(x_1-\theta,x_2-\theta, ..., x_n-\theta)d\theta}$$

$$\hat{\theta}(X_1, ..., X_n) = \frac{\int \theta f(X_1-\theta, ..., X_n-\theta)d\theta}{\int f(X_1-\theta, ..., X_n-\theta)d\theta}.$$

This is the **Pitman estimator**: this is the analogue of a Bayesian estimator where the set of parameters $\mathbb{R}$ would have for prior distribution Lebesgue measure, the unique measure invariant under translation.

In order to test H_0 "$(X_1, ..., X_n)$ is an n-sample from $f(\cdot - \theta)$ for a θ in $\mathbb{R}$" against H_1, this is "an n-sample from $g(\cdot - \theta)$ for a θ in $\mathbb{R}$," the invariance of the hypotheses to be tested can be used if an observation is replaced by its transform t_c. Hence we can restrict ourselves to invariant tests, which only depend on the experiment Y. The distribution of Y has, under H_0, the density f_Y: $(y_2, ..., y_n) \longmapsto \int f(t, t+y_2, ..., t+y_n)dt$ and, under H_1, the analogous density g_Y with f replaced by g. Thus, if we restrict ourselves to invariant tests of H_0 against H_1, we are led to a test of two simple hypotheses H_0' "Y has density f_Y" against H_1' "Y has density g_Y."

Setting

$$\Lambda(X_1, \ldots, X_n) = \frac{\int d\theta\ g(X_1-\theta, \ldots, X_n-\theta)}{\int d\theta\ f(X_1-\theta, \ldots, X_n-\theta)},$$

we obtain the *Neyman–Pearson test*

$$1_{[\Lambda>c]} + \gamma 1_{[\Lambda=c]} .$$

This is the most powerful *invariant test* of its level.

8.4.4. Invariant Tests

Let us generalize the above. Let Θ_0 and Θ_1 be two disjoint subsets of Θ, which we assume to be globally invariant under $\underline{G}$. An observation X is made on $(\Omega, A, P_\theta)_{\theta\in\Theta}$ in order to carry out a test of Θ_0 against Θ_1. It is then natural to consider that the observation of $g(X)$ would have led to the same decision as that of X. In other words, we limit ourselves to **invariant tests** $\phi(X)$, for which ϕ is a constant on the orbits of G; this is the **invariance principle.** Let U be a maximal invariant of G: the function ϕ may be written as $\phi = \phi' \circ U$. An invariant test will thus use only the observation $U(X)$, the distribution of which depends only on the orbit of θ. Thus if λ is a maximal invariant under G, the problem can be rephrased in terms of a simpler problem dealing with λ and using the experiment $U(X)$. To test "λ_0" against "λ_1", the Neyman–Pearson theorem gives, for example, the most powerful test at its level amongst the invariant tests.

Examples. *Tests dealing with a Gaussian sample.* Let $(X_1, \ldots, X_n)$ be an n-sample from $N(m,\sigma^2)$. We know that the statistic $(\bar{X}, \Sigma_{i=1}^n(X_i - \bar{X})^2)$ is sufficient, which leads to replacing the experiment by observation of this statistic.

(a) *A variance test for a Gaussian sample.* To the translations of the observations $(X_1, \ldots, X_n) \longmapsto (c + X_1, \ldots, c + X_n)$, $c \in \mathbb{R}$, corresponds the transformation

$$\left[\bar{X}, \sum_{i=1}^{n}(X_i - \bar{X})^2\right] \longmapsto \left[c + \bar{X}, \sum_{i=1}^{n}(X_i - \bar{X})^2\right]$$

of the sufficient statistic, thus the maximal invariant $\Sigma_{i=1}^n(X_i - \bar{X})^2$. The maximal invariant of the parameter space is σ.

Hence every invariant test dealing with σ should only use the statistic $\Sigma_{i=1}^{n}(X_i - \bar{X})^2$. For $\sigma_0 \leqslant \sigma_1$, the rejection region of the most powerful invariant test of level α of "$\sigma \leqslant \sigma_0$" against "$\sigma > \sigma_1$" is

$$\left\{\sum_{i=1}^{n} (X_i - \bar{X})^2 \geqslant \sigma_0^2 \chi_\alpha^2\right\}.$$

(b) *Student's test.* To the dilatations $(X_1, ..., X_n) \longmapsto (cX_1, ..., cX_n)$ of ratio $c > 0$, correspond the transformations

$$\left(\bar{X}, \sum_{i=1}^{n} (X_i - \bar{X})^2\right) \longmapsto \left(c\bar{X}, c^2 \sum_{i=1}^{n} (X_i - \bar{X})^2\right)$$

hence a maximal invariant is

$$\frac{\bar{X}\sqrt{n}}{\left[\sum_{i=1}^{n} (X_i - \bar{X})^2\right]^{1/2}} = T.$$

From Section 5.1.5 it has a Student's T distribution on $n - 1$ degrees of freedom with noncentrality parameter $\lambda\sqrt{n} = m\sqrt{n}/\sigma$. Let $t \longmapsto h(\lambda\sqrt{n},t)$ be the density of such a distribution. The family $(h(\lambda,\cdot))$ has increasing likelihood ratio. Hence, in order to test "$m/\sigma \leqslant \lambda_0$" against "$m/\sigma > \lambda_1$" *a* UMP *test amongst invariant tests at a given level is obtained by taking a rejection region of the form* $\{T \geqslant K\}$; for example $\{T \geqslant t_{\alpha,n-1}\}$ *gives a* UMP *test amongst invariant tests of* "$m \leqslant 0$" *against* "$m > 0$" *of level* α. If now we want to carry out a two-sided test "$m = 0$" against "$m \neq 0$" we note that the problem is invariant under dilatations with ratio $c \neq 0$ (and not only with ratio $c > 0$). A maximal invariant will then be $|T|$. The density of $|T|$ is, for $T > 0$, $t \longmapsto h(\lambda,t) + h(\lambda,-t) = k(\lambda,t) = k(-\lambda,t)$: it can be shown that for all $\lambda \neq 0$, $t \longmapsto k(\lambda,t)/k(0,t)$ is increasing.

The two-sided test of "$m = 0$" *against* "$m \neq 0$" *can hence be replaced by the one-sided test* "$\lambda = 0$" *against* "$|\lambda| > 0$." *By taking the rejection region* $\{|T| \geqslant t_{n-1,\alpha/2}\}$, *we thus have a test of level* α *which is unbiased and uniformly most powerful amongst invariant tests.*

We have thus justified the one-sided Student tests "$m \leqslant 0$" against "$m > 0$" and the two-sided "$m = 0$" against "$m \neq 0$" introduced in Section 5.1.5. They are optimal as invariant tests.

Exercises 8.4

We shall denote UMP and UMPU as in Exercises 8.3, and UMPI for "uniformly most powerful amongst invariant tests of its level."

E.1. Calculate the Pitman estimator for the following translation models (Section 8.4.3): An n-sample from a uniform distribution on $[\theta - 1/2, \theta + 1/2]$ or from $N(\theta,1)$.

E.2. $X = (X_1, ..., X_n)$ and $Y = (Y_1, ..., Y_n)$ are two independent samples, the distributions of which admit, respectively, the following densities with respect to Lebesgue measure:

$$f_{a,b}(x) = ae^{-a(x-b)}1_{[b,\infty]}(x)$$

and $f_{\alpha,\beta}$ where a and α are positive, b and β are real numbers. We are concerned with testing H_0 "$\alpha/a \leq \tau$" against H_1 "$\alpha/a > \tau$" with a test invariant under $X \longmapsto cX + d$, $Y \longmapsto cY + e$, $c > 0$, d and e real numbers. Find a test of H_0 against H_1 uniformly most powerful amongst invariant tests. What is the distribution of the statistic on which it depends?

E.3. (a) Show that every UMPU test of "$\theta = \theta_0$" against "$\theta \neq \theta_0$" is admissible. (b) On the other hand, a UMPI test is not necessarily admissible. *Counter-example*: Let $X = (X_1,X_2)$ and $Y = (Y_1,Y_2)$ be two independent random vectors taking values in $\mathbb{R}^2$, with respective distributions $N_2(0,\Sigma)$ and $N_2(0,\Delta\Sigma)$, with $\Delta \in \mathbb{R}^+$ and Σ an invertible covariance matrix. We are concerned with testing $(\Delta = 1)$ againgst $(\Delta > 1)$ and the experiment consists of a single observation of a pair (X,Y). Let $[X,Y]$ be the matrix, the column vectors of which are X and Y. Show that it takes a.s. its values in the set E of 2×2 matrices of rank 2.

(a) Let G be the set of transformations $e \longmapsto ge$ of E, for $g \in E$. Show that G is a group under which the problem is invariant. Show that the only functions invariant under G are the constants. What is the UMPI test of level α?

(b) By considering only the first coordinates of X and of Y, give a test ϕ_1 of $(\Delta = 1)$ against $(\Delta > 1)$, better than ϕ; deduce from this that ϕ is not admissible.

E.4. *Rank tests.* (a) Let $\mathbb{R}_*^n$ be the set of vectors of $\mathbb{R}^n$, all of the coordinates of which are distinct, let G be the set of strictly increasing transformations of $\mathbb{R}$, and let G_n be the set of functions $f^{\otimes n}$ defined on $\mathbb{R}^n$ with $f \in G$. Verify that G_n is a group and that the rank statistic R is a maximal invariant. Let $\mathcal{F}_0$ be the set of continuous distributions on $\mathbb{R}$, and let $\mathcal{F}_0^{(n)}$ be the set of distributions of n-samples from the distributions of $\mathcal{F}_0$. Show that $\mathcal{F}_0^{(n)}$ is invariant under G_n.

(b) Two independent samples are observed, an n-sample X from $F \in \mathcal{F}_0$ and an n-sample Y from $F' \in \mathcal{F}_0$. We test H_0 "$F = F'$" against H_1 "$F < F'$", which means to say that "the distribution function F is larger than that of F'." Show that the problem is invariant under G_{n+m}. Give a maximal invariant of G_{n+m} and hence justify the use of rank tests made in Section 4.5.

(c) In (b) H_1 is replaced by the *Lehmann alternative* H_1' "$F' = F^{\Delta}$" for $\Delta > 0$. What is the distribution of $F(X_1)$? Under H_1', what is the density of $F(Y_1)$? and the probability of $\{Y_1 \geq X_1\}$? The sequence of $n + m$ observations is ordered and we denote $Z_i = 1$ or 0 according to whether the observation of rank i is a Y or an X. Determine the distribution of Z under H_0 and under H_1' and deduce from this UMPI tests.

E.5. *Invariance and sufficiency.* Let $(X_1, ..., X_n)$ be an n-sample from $N(m,\sigma^2)$. A maximal invariant statistic associated with translations is $U = (X_1 - X_n, ..., X_{n-1} - X_n)$. Calculate the density of U and show that $\sum_{i=1}^n (X_i - \bar{X})^2$ is a sufficient statistic for U. Similarly, calculate the density of

$$V = \left(\frac{X_1}{|X_n|}, ..., \frac{X_{n-1}}{|X_n|}, \frac{X_n}{|X_n|}\right)$$

with respect to the product of Lebesgue measure on $\mathbb{R}^{n-1}$ and of $\{\delta_{+1} - \delta_{-1}\}$; show that the statistic

$$T = \frac{\sqrt{n}\ \bar{X}}{\left[\sum_{i=1}^{n} (X_i - \bar{X})^2\right]^{1/2}}$$

is sufficient for V.

Conclusion

In Section 8.4.4 for invariant tests on Gaussian samples, the n-sample $(X_1, \ldots, X_n)$ could have been used instead of its associated sufficient statistic. The sufficient statistics and the maximal invariants are the same (thus also the invariant tests). This is a particular case of a general theorem (due to Stein).

Bibliographic Notes

The theory of tests is a quite difficult part of statistics. The decision theory point of view is introduced particularly well in Ferguson, which provided the inspiration for this chapter.

The Bayesian viewpoint is developed in DeGroot. Blackwell and Girshick can also be consulted for the links between decision theory and the theory of games.

A more statistical outlook can be found in Raoult; Lehmann is a work with very rich content at a more difficult mathematical level.

BIBLIOGRAPHY

ANDERSON T. W. *An introduction to multivariate analysis*, 1974, Wiley.

ASH R. *Information theory*, 1967, Wiley.

BARLOW R. and PROSCHAN F. *Statistical theory of reliability and life testing*, 1975, Holt Rinehart Winston.

BARNDORFF-NIELSEN O. *Information and exponential families in statistical theory*, 1978, Wiley.

BARRA J. R. *Mathematical basis of statistics*, 1980, Academic Press.

BERGE C. and GOUILA HOURI A. *Programmes, Jeux et Réseaux de transport*, 1962, Dunod.

BENZECRI J. P. *L'analyse des données*. 1. La taxinomie. 2. L'analyse des correspondances, 1976, Dunod.

BICKEL P. and DOKSUM K. *Mathematical statistics*, 1977, Holden Day.

BILLINGSLEY P. [1]*Convergence of probability measures*, 1968, Wiley

BILLINGSLEY P. [2] *Statistical inference for Markov processes*, 1961, Univ. of Chicago.

BILLINGSLEY P. [3] *Probability and measure*, 1968, Wiley.

BLACKWELL D. and GIRSHICK M. A. *Theory of games and statistical decisions*, 1954, Wiley.

BOROVKOV A. A. *Statistical process in queuing theory*, 1976, Springer-Verlag.

BREIMAN L. [1] *Probability and stochastic processes with a view towards applications*, 1969, Houghton Mifflin.

BREIMAN L. [2] *Statistics with a view towards applications*, 1973, Houghton Mifflin.
BREIMAN L. [3] *Probability*, 1968, Addison Wesley.
CHOW V. S. and TEICHER H. *Probability theory*, 1978, Springer-Verlag.
CHUNG K. L. *Elementary probability theory with stochastic processes*, 1974, Springer-Verlag.
COCHRAN W. G. *Sampling techniques*, 1963, Wiley.
COURSOL J. *Techniques statistiques des modeles lineaires*, 1981, cours du CIMPA.
COX D. R. and HINKLEY D. [1] *Theoretical statistics*, 1974, Chapman-Hall.
COX D. R. and HINKLEY D. [2] *Problems and solutions in theoretical statistics*, 1979, Chapman-Hall.
DAGNELIE P. *Theorie et methodes statistiques* (2 volumes), 1976, Presses Agronomiques de Gembloux.
DEGROOT M. H. *Optimal statistical decisions*, 1970, McGraw Hill.
EKELAND I. *Theorie des jeux*, 1974, P.U.F.
EWENS W. J. *Mathematical population genetics*, 1979, Springer-Verlag.
FALCONER *Introduction to quantitative genetics*, 1967, Olivier Boyd.
FELLER W. *An introduction to probability theory and its applications*, (3rd ed.), 1968, Wiley, (2 volumes).
FERGUSON T. *Mathematical statistics, a decision theoretic approach*, 1967, Academic Press.
FREIBERGER W. and GRENANDER U. *A short course in computational probability and statistics*, 1971, Springer-Verlag.
GNEDENKO B. V. [1] *The theory of probability*, 1967, Chelsea.
GNEDENKO B. V. [2] *Mathematical methods of reliability theory*, 1969, Academic Press.
GNEDENKO B. V. [3] *Introduction a la theorie statistique*, 1964, Ed. de Moscou.
GRANT E. L. and LEAVENWORTH R. S. *Statistical quality control*, (5th ed.), 1980, McGraw-Hill.
HABERMAN, S. *Analysis of qualitative data*, Vol.1: *Introductory topics*, 1978. Vol. 2: *New developments*, 1979, Academic Press.
HAJEK J. and SIDAK. *Theory of rank tests*, 1967, Academic Press.
HALMOS P. R. *Measure theory*, 1969, Van Nostrand.
HAMMERSLEY, J. M. and HANDSCOMB D. C. *Monte Carlo methods*, 1979, Chapman-Hall.

HODGES, J. L., KRECH D. and CRUTCHFIELD R. *Stat. Lab.*, 1979, Economica.

HODGES J. L. and LEHMANN E. L. [1] *Basic concepts of probability and statistics*, 1970, Holden Day.

HODGES J. L. and LEHMANN E. L. [2] *Elements of finite probability*, 1965, Holden Day.

HOEL P., PORT S. and STONE C. [1] *Introduction to probability theory*, [2] *Introduction to statistical theory*, [3] *Introduction to stochastic processes*, 1972, Houghton Mifflin.

HOLLANDER M. and WOLFE, D. *Non-parametric statistical methods*, 1973, Wiley.

JACQUARD A. *The genetic structure of populations*, 1974, Springer-Verlag.

KARLIN S. and TAYLOR H. M. [1] *A first course in stochastic processes*, 1966, Academic Press.

KEMENY J. G. and SNELL J. L. *Finite Markov chains*, 1976, Springer-Verlag.

KENDALL M. G. and STUART, A. *The advanced theory of statistics* (vol. 1-3), 1961, Griffin.

KRICKEBERG, K. and ZIEZOLD H. *Methodes statistiques*, 1980, D.I.A.

KULLBACK S. *Information theory and statistics*, 1968, Dover.

LEBART L. and FENELON J. P. *Statistique et informatique appliquees* (3rd ed.), 1975, Dunod.

LeCAM. *Notes on asymptotic methods in statistical decision theory*, University de Montreal.

LEHMANN E. L. [1] *Testing statistical hypothesis*, (4th ed.), 1966, Wiley.

LEHMANN E. L. [2] *Non parametrics. Statistical methods based on ranks*, 1975, Holden Day.

LEVY P. *Calcul des probabilites*, 1925.

LEVY P. *Theorie de l'addition des variables aleatoires*, 1937, Gauthier-Villars.

LIPSCHUTZ. *Probabilites* (exercises), 1977.

LOEVE M. *Probability theory*, (4th ed.), 1977, Springer-Verlag.

MALINVAUD E. *Statistical methods of econometrics*, 1980, North-Holland.

METIVIER M. *Notions fondamentales de la theorie des probabilites*, 1968, Dunod.

MOSTELLER M. and ROURKE R. *Sturdy statistics*, 1973, Addison Wesley.

NEVEU J. *Mathematical foundations of the calculus of probabilities*, 1968, Holden Day.

PARTHASARATHY K. R. *Probability measures on metric spaces*, 1967, Academic Press.
PEARCE, S. C. *Biological Statistics*, 1965, McGraw-Hill.
RAO C. R. *Linear statistical inference and its applications*, 1973, Wiley.
RAOULT J. P. *Structures statistiques*, 1975, P.U.F.
RENYI A. *Probability theory*, 1970, North-Holland.
RIESZ F. and NAGY B. *Functional analysis*, 1955, Ungar.
ROMANOVSKI V. I. *Discrete Markov chains*, 1970, Walters, Noordhoff publishing.
ROSS S. M. [1] *Applied probability models with optimization applications*, 1970, Holden Day.
ROSS S. M. [2] *An introduction to probability models*, 1972, Academic Press.
ROZANOV Y. *Processus aleatoires*, 1975, Ed. de Moscou.
RUDIN W. *Real and complex analysis*, 1975, McGraw-Hill.
SAVAGE. *Statistics uncertainty and behavior*, 1968, Houghton Mifflin.
SCHMETTIERER. *Introduction to mathematical statistics*, 1974, Springer-Verlag.
SNEDECOR G. and COCHRAN W. *Statistical methods*, 1971, Iowa State.
SPIEGEL. *Statistics (Schaum Outline Series)*, 1977, McGraw-Hill.
THEIL H. *Principles of econometrics*, 1971, Wiley.
TOPSOE F. *Information theorie*, 1974, Teubner.
TUKEY J. W. *Exploratory data analysis*, 1977, Addison Wesley.
VAJDA S. *The theory of games and linear programming*, 1956, Masson.
VAN DER WAERDEN B. L. *Mathematical statistics*, 1969, Allen and Unwin.
VENTSEL H. [1] *Theorie des probabilites*, 1977, ed. de Moscou.
VENTSEL H. [2] *Elements de la theorie des jeux dans Yaglom, Initiations aux mathematiques.*
WOLFOWITZ J. *Coding theorems of information theory*, 1964, Springer-Verlag.
YAGLOM A. M. and YAGLOM I. M. *Probability and information*, 1983, Reidel.
YAGLOM I. M., TRAKHTENBROT B., VENTSEL H. and SOLOVNIKOV A. *Initiations aux mathematiques*, 1975, Mir-Moscou.
ZACKS S. *The theory of statistical inference*, 1971, Wiley.

Statistical Tables

PEARSON E. S. and HARTLEY H. O. *Biometrika tables for statisticians* (3rd ed.), 1979, Cambridge Univ. Press.

NOTATION AND CONVENTIONS

Mathematical Notations

$\mathbb{R}$ real numbers

$\overline{\mathbb{R}}$ extended real line

$\mathbb{C}$ complex numbers

$\mathbb{Q}$ rational numbers

$\mathbb{N}$ integers $\geqslant 0$

$\mathbb{Z}$ integers

$\mathfrak{S}$ or $\mathfrak{S}_n$ set of permutations of $\{1, ..., n\}$

$n!$ n factorial

$\binom{n}{p}$ number of combinations of n objects, taken p at a time

$\geqslant 0$ means positive; > 0 means strictly positive;

$\leqslant 0$ means negative; < 0 means strictly negative

For a and b real, and f and g real-valued functions:

$\wedge$ minimum, $a \wedge b = \inf(a,b)$, $f \wedge g = \inf(f,g)$;

$\vee$ maximum, $a \vee b = \sup(a,b)$, $f \vee g = \sup(f,g)$;

$a_+ = a \vee 0$ or $f_+ = f \vee 0$, positive part;

$a_- = -a \wedge 0$ or $f_- = -f \wedge 0$, negative part;

$|a|$ or $|f|$, modulus of a or f.

For $A \subset \mathbb{R}$ or F a set of real-valued functions:

$A_+ = \{a;\ a \in A,\ a \geqslant 0\}$, $F_+ = \{f;\ f \in F,\ f \geqslant 0\}$, positive parts of A or F.

For (a_n) a sequence of real numbers of (f_n) or a sequence of real functions.

$\overline{\lim}\, a_n$ or $\overline{\lim}\, f_n$ is the lim upper limit

$\underline{\lim}\, a_n$ or $\underline{\lim}\, f_n$ is the lim lower limit

$o(x)$, of order less than x, for $x \to 0$

$O(x)$, of order x, for $x \to 0$

C_0, C_b, C_k 120

x a vector of $\mathbb{R}^n$ is also the $n \times 1$ matrix of which it is the column vector

$\|x\|$ Euclidean norm of x

$\bar{x} = \frac{1}{n}(x_1 + \dots + x_n)$ for $x = (x_1, \dots, x_n)$

$<x,y>$ scalar product of x and y

1_n or 1 is the vector $(1, \dots, 1)$ of $\mathbb{R}^n$

I_n or I is the $n \times n$ identity matrix

${}^tM$ transposed matrix of M; t transposition

$|M|$ determinant of the square matrix M

J_ϕ Jacobian of ϕ

For $(x_{ij})_{1 \leqslant j \leqslant n_i \cdot 1 \leqslant i \leqslant k}$ denote

$x_{i\cdot} = \sum_{i=1}^{n_i} x_{ij}$ in Chapter 1 and 5.3

$x_{i\cdot} = \frac{1}{n_i} \sum_{j=1}^{n_i} x_{ij}$ and $x_{\cdot\cdot} = \frac{1}{k} \sum_{j=1}^{k} x_{i\cdot}$ in Sections 5.1 and 5.5.

Measure Theory

$A \cup B$ 104

$A \cap B$ 104

$A \supset B$ 104

$A \subset B$ 104

$A \backslash B$ 104

A^c 104

1_A 104

$\overline{\lim}\, A_n$ 107

$\underline{\lim}\, A_n$ 112

$\bar{A}$ closure of A

ess sup A_i 123

$f^{-1}(A) = (f \in A) = \{\omega; f(\omega) \in A\}$ 105

$f^{-1}(\mathcal{C})$ 113

$\sigma(\mathcal{C})$, $\sigma(X)$ (generated σ-algebra) 104, 105

Probabilities

Distributions

$\mathcal{M}(n,p)$ 226

$\mathcal{N}(m,\sigma^2)$ 130

$\mathcal{N}(m,\Gamma')$ 219

$p(\lambda)$ 56

$t(n)$ 171, 202

$t'(n,a)$ 204

$\mathcal{U}(a,b)$ 129

Convergence for Sequences of r.v.'s

$\xrightarrow{a.s.}$ 107

$\xrightarrow{P}$ 108, 119

$\xrightarrow{\mu}$ 107, 119

$\xrightarrow{L^p}$ 118

$\xrightarrow{\mathcal{D}}$ 147

$\xrightarrow{v}$ 140

$\xrightarrow{w}$ 140

$\xrightarrow{n}$ 140

Samples

$\bar{X}$ or $\bar{X}_n$ 173

$\bar{F}_n$ 177

S^2 or S^2_n 179 $\bar{\sigma}^2_n$ 179 $\bar{\Gamma}_n$ 180 $\bar{\rho}_n$ 180

$(X_{(1)}, \ldots, X_{(n)})$ 181 $(R_1, \ldots, R_n)$ 181

Abbreviations

P.C.A.	17
a.s.	107
UMP	331
UMPU	331
UMPI	341
MVUE	287
r.v.	105

INDEX